Polymer Surfaces

Polymer Surfaces

Edited by

D. T. Clark and W. J. Feast

Department of Chemistry
Durham University

A Wiley–Interscience Publication

JOHN WILEY & SONS
Chichester · New York · Brisbane · Toronto

Library of Congress Cataloging in Publication Data:
Main entry under title:

Polymer surfaces.

   'A Wiley-Interscience publication.'
   Includes bibliographical references and index.
   1. Polymers and polymerization – Surfaces – Congresses.
I. Clark, David Thomas. II. Feast, William James
QD380.P65   547'84   77-17426

ISBN 0 471 99614 9

Typeset in IBM Press Roman by
Preface Ltd., Salisbury, Wilts.
Printed and bound in Great Britain by
Pitman Press Ltd., Bath.

Contributors

ANDREWS, E. H. *Department of Materials, Queen Mary College, London, E1 4NS.*

BRISCOE, B. J. *Cavendish Laboratory, University of Cambridge.*

CLARK, D. T. *Department of Chemistry, University of Durham, South Road, Durham City.*

CLARK, P. J. *Dow Corning Ltd., Barry, Glamorgan.*

CLEGG, P. L. *I.C.I. Plastics Division, Welwyn Garden City.*

CURSON, A. D. *I.C.I. Plastics Division, Welwyn Garden City.*

DAVIS, G. T. *Institute for Materials Research, National Bureau of Standards, Washington, DC 20234, U.S.A.*

DILKS, A. *Department of Chemistry, University of Durham, South Road, Durham City.*

DOWSON, D. *Institute of Tribology, Department of Mechanical Engineering, The University of Leeds, Leeds LS2 9JT.*

DUKE, C. B. *Xerox Corporation, Webster Research Center, 800 Phillips Road, Webster, New York 14580, U.S.A.*

FABISH, T. J. *Xerox Corporation, Webster Research Center, 800 Phillips Road, Webster, New York 14580, U.S.A.*

GAMLEN, G. A. *I.C.I. Fibres Ltd. and Department of Chemistry and Applied Chemistry, University of Salford.*

HOFFMAN, J. D. *Institute for Materials Research, National Bureau of Standards, Washington, D.C. 20234, U.S.A.*

KING, N. E. *Department of Materials, Queen Mary College, London, E1 4NS.*

LEWIS, T. J. *School of Electronic Engineering Science, University College of North Wales, Bangor, Gwynedd, LL57 1UT.*

RÅNBY, B. *Department of Polymer Technology, The Royal Institute of Technology, Stockholm, Sweden.*

SCHONHORN, H. *Bell Telephone Laboratories, 600 Mountain Avenue, Murray Hill, New Jersey 07974, U.S.A.*

SHUTTLEWORTH, D. *Department of Chemistry, University of Durham, South Road, Durham City.*

TABOR, D., F.R.S. *Cavendish Laboratory, University of Cambridge.*

TIGHE, B. J. *Department of Chemistry, University of Aston in Birmingham, Birmingham B4 7ET.*

WATSON, P. K.   *Xerox Corporation, Rochester, New York, U.S.A.*
WILLIS, H. A.   *I.C.I. Plastics Division, Welwyn Garden City.*
WILSON, N.   *Shirley Institute, Didsbury, Manchester, M20 8RX.*
WRIGHT, A. N.   *General Electric Company, Corporate Research and Development, Schenectady, New York 12301, U.S.A.*
ZICHY, V. J. I.   *I.C.I. Plastics Division, Welwyn Garden City.*

Contents

9. The application of plasmas to the synthesis and surface modification of polymers   185

D. T. CLARK, A. DILKS, and D. SHUTTLEWORTH

10. Surface modification of polymers for adhesive bonding   213

H. SCHONHORN

17. Defects of the surface zone — their origin and inhibition with particular reference to PVC
P. L. CLEGG and A. D. CURSON

18. Photodegradation and photo-oxidation of polymer surfaces
B. RÅNBY

19. Tribological characteristics of polymers with particular reference to polyethylene
D. DOWSON

Preface

The present volume is an attempt to provide a review of the current state of knowledge of polymer surfaces from the standpoint of their chemical, physical, electrical, and mechanical properties and modifications thereof; their characterization; and their uses. In planning the meeting on which the text is based we attempted to identify the most important aspects of the subject from both an academic and technological viewpoint with the underlying theme that an interdisciplinary approach was mandatory. The contributors to this volume are drawn from academia and industry in roughly equal proportion, and the great variety of approaches and disciplines involved should be evident from the chapter headings.

In planning the meeting and subsequently the book, the difficulties in a multidisciplinary approach readily become apparent. Thus it is clear that the appellation 'surface' implies something different to an electron spectroscopist compared, for example, with a mechanical engineer. As a starting point, however, we may take the O.E.D. definition as : 'The outermost part of a material body, considered with respect to its form, texture or extent'. In practice, the subject matter of the volume is largely concerned with the surface regions extending to a depth of 1 μm (although in some instances a greater depth of material is considered) and may be classified in terms of a three-dimensional grid representing preparation, characterization, and properties.

The first six chapters cover fundamental aspects of the mechanical, physical, and electrical properties of surfaces and provide an introduction to most of the terminology used in succeeding chapters. Polymerization at surfaces by photopolymerization and plasma techniques are covered in Chapters 8 and 9, and although some consideration was given to the inclusion of pyrolitic routes and solution phase polymerization, it was decided that these topics are adequately described elsewhere. In addition the main emphasis in the part of the book devoted to the preparation of surfaces is on the *in-situ* controlled polymerization at surfaces rather than solution- or powder-coating techniques using preformed polymers. Surface characterization is covered in Chapters 14–16, whilst Chapter 13 presents a synopsis of some aspects of the surface morphology of polymers.

Chapters 10 and 11 deal with modification of surface properties for technological applications, whilst the in-service deterioration of polymers consequent upon processes initiated at the surface are covered in the last three chapters on respectively, defects of the surface zone, photooxidation and photodegradation, and tribological characteristics of polymer surfaces. Two chapters might be singled out as presenting novel aspects of polymer surfaces; namely, Chapter 7 dealing with static charges on textile surfaces and Chapter 12 which describes the development of epitropic fibres.

As editors we started out with the intention of ensuring that units and symbols were presented in a consistent manner. In the event, however, it is clear that workers in a given area are very conservative about changing time-honoured units which adequately express the typical accuracy of their measurements. Polymer surfaces would seem to be yet another area of science where SI units are not uniformly employed. In the case of symbols used for various quantities we have decided in the time-honoured 'damned if you do, damned if you don't' philosophy, to leave the authors as best judges of the commonly accepted symbols for their particular fields. This results in the same symbol representing different things in different chapters; however, perusal of the manuscript will reveal that this does not cause any real confusion and in any case readers who are interested to pursue background material in the references will at least start from the basis of knowing the peculiarities in terms of symbols for the given field.

The book as a whole provides a balanced state of the art review of most of the academically and technologically important areas relating to polymer surfaces, and as such should provide a broadly based introduction to the literature. References pertinent to a given article are collected together at the end of each article and the book should provide a useful source to the original literature. In this connection we have attempted to construct a fairly detailed index which should allow ready access to information from different fields.

We were fortunate indeed to be able to persuade the authors, all of whom are acknowledged leaders in their fields, to present lectures at the Durham Polymer Surfaces Symposium and to contribute to this book. We would like to thank the Officers and the Committee of the Chemical Society Macromolecular Group, particularly the chairman, Professor C. H. Bamford, F.R.S. (Liverpool University), for the advice and encouragement proffered throughout this endeavour. Thanks are also due to numerous friends in both industry and academia who provided advice on our choice of topics, and in this connection our thanks are especially due to Professor Ian Ward (Leeds University) for many helpful suggestions and for agreeing to open the meeting. We trust that some of the excitement and activity in this important field, which was evident from the lectures and discussion sessions of the meeting is embodied in the text.

D. T. CLARK
W. J. FEAST

1

Friction and Wear of Polymers

B. J. Briscoe and D. Tabor, F.R.S.

Cavendish Laboratory, University of Cambridge

1. INTRODUCTION

This review article deals with certain aspects of the friction and wear of polymers and is mainly concerned with unlubricated surfaces. During the last few years several rather specialist surveys on analogous themes have appeared,[1-9] and it is natural that some of the basic ideas described in these reviews should be repeated. However, we shall bear in mind that this volume is not exclusively devoted to friction and wear and we will not therefore deal in detail with some of the contentious issues which concern professional tribologists. We shall attempt to emphasize those concepts which have some bearing on other parts of the material covered in this volume, and where possible introduce themes specifically relevant to the role of the surface in the friction and wear processes. Finally, although the bulk of polymers dealt with in this volume are thermoplastics we shall include some references to elastomers since their frictional behaviour provides some insight into the behaviour of the more conventional polymers.

2. THE BASIC MECHANISMS OF FRICTION

Although many mechanisms are involved in the friction of solids there are two processes which are generally of major importance.[8] The first is the ploughing of asperities of the harder solid through the surface of the softer; the second is the shearing of adhesive bonds formed at the interface at the regions of real contact. In general, the ploughing mechanism deals with relatively large volume-deformations and small strains (see Figure 1(a)), whilst the shearing mechanism involves very thin interfacial regions and large strains. However, in the sliding of a polymer over a relatively smooth, hard surface the ploughing or deformation process may be almost non-existent and may involve very small volumes (Figure 1(b)). For convenience we shall separate the deformation and shearing processes and treat them as two independent terms so that the forces due to each process can be simply added together to give the total frictional force. Naturally, this is an over-simplification and in some cases can lead to misleading conclusions.

We shall treat the wear of polymers in a similar way. We shall first discuss the

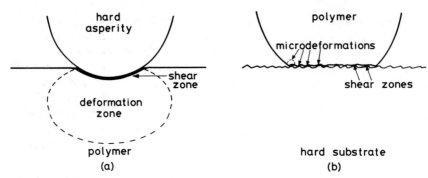

Figure 1. (a) Sliding of a hard sphere (representing a portion of an asperity) over a polymer: shearing takes place in a thin interfacial layer where the shear strains are large. On the other hand the grooving of the polymer involves gentle deformation of a relatively large volume. (b) Sliding of a polymer over a fairly smooth hard surface. The grooving of the polymer by the surface roughness involves volume deformations comparable in size with the interfacial layer in which shearing occurs.

type of wear which arises from a ploughing or deformation mechanism. Then wear which is largely due to interfacial adhesion, and finally certain types of wear in which both mechanisms interact in a significant way. We shall also deal specifically with the friction and wear properties of PTFE and high-density polyethylene (HDPE).

3. THE DEFORMATION OR PLOUGHING TERM IN FRICTION

If a hard slider (representing a single asperity) traverses the surface of a polymer, energy is fed into the polymer ahead of the slider and some of this is restored at the

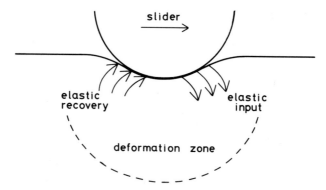

Figure 2. When a hard slider traverses the surface of an elastomer in the absence of interfacial adhesion, work is expended on the front of the contact region and energy is restored at the rear. The overall energy loss is related to the input energy and the hysteretic loss properties of the elastomer.

rear of the slider because of elastic recovery and urges it forward (Figure 2). The energy lost in this process accounts for the force (or work) required to maintain sliding.[10] With rubber-like materials where an evanescent groove is formed the net energy loss is related to the input energy and the loss properties of the material at the particular temperature, contact pressure, and rate of deformation involved[11-13] (see below). Similar conclusions also apply to polymers. However, with such materials a certain amount of plastic grooving (i.e. permanent set) may occur, and in this case it may be more convenient to treat the polymer as though it were a plastic solid (like a metal) with an equivalent yield-pressure. In most situations, however, it is the loss property of the polymer that is relevant.

If the elastomer or polymer is subjected to simple tension or simple shear at a deformation rate comparable to that involved in the ploughing process, we obtain hysteresis loops of the type shown in Figure 3 where the energy lost in the loading–unloading cycle is a fraction α of the total input energy of the cycle. (In a sinusoidally driven system α is equivalent to $\pi \tan \delta$, where $\tan \delta$ is the loss tangent, and this depends on temperature and deformation rate in a well-defined way.) If now the elastic work done in deforming the polymer by the slider per unit distance of sliding is ϕ it seems reasonable to expect the energy loss in the grooving process to be $\alpha\phi$. This is equivalent to the force F_d needed to move the slider unit distance.[10] Consequently,

$$F_d = \alpha\phi \tag{1}$$

For a sphere of radius R traversing a surface under a load W

$$\phi = 0.17 W^{4/3} R^{-2/3} (1 - \nu^2)^{1/3} (E)^{-1/3} \tag{2}$$

where E is the real part of Young's modulus of the solid and ν is Poisson's ratio

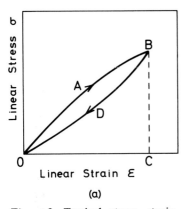 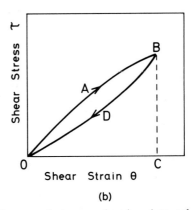

Figure 3. Typical stress–strain curves for an elastomer showing internal friction or hysteretic losses: (a) tensile experiment, (b) shear experiment. The input energy is the area under the curve OABCO: the energy loss is the area under the curve OABDO. The fractional energy loss α is the ratio of these two areas. For a given material and fixed rate of deformation α is the same for both types of deformation.

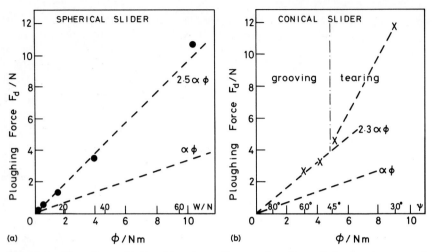

Figure 4. The observed ploughing force F_d as a function of the input energy ϕ per unit distance of sliding for a hard slider traversing a well-lubricated rubber surface of high loss factor $\alpha = 0.35$. (a) Spherical slider of fixed radius $R = 3.2$ mm. The input energy ϕ is varied by varying the normal load W. It is seen that the observed values of F_d are more nearly equal to $2.5\alpha\phi$ than to $\alpha\phi$. (b) Conical sliders of various semiapical angles ψ at a constant load $W = 16$ N. It is seen that the observed values of F_d are more nearly equal to $2.3\alpha\phi$ than $\alpha\phi$ when smooth grooving of the rubber occurs. When the rubber is torn with the more sharply pointed cones the sliding force is very much higher.

(Greenwood and Tabor).[13] For a conical indenter of semi-apical angle ψ, provided it does not tear the rubber,

$$\phi = (W/\pi) \cot \psi \tag{3}$$

Combining these equations with equation (1) it should thus be possible to calculate the grooving force F_d. Such an analysis was first applied to rubber-like materials and was found to be qualitatively very satisfactory.[13] However, it was always found to be two or three times smaller than the observed value (see Figure 4). The reason is simple if we consider the path of any element in the rubber as it passes under the slider,[14] and for simplicity we consider here a spherical slider (see Figure 5). In the undisturbed part of the rubber is a square element A. As it reaches region B it undergoes deformation, the greater part of which corresponds to a shear parallel to the surface. At region C the element is now under almost pure compression. It turns out that between B and C there is little change in elastic energy, but a $45°$ rotation in the direction of shear. The sequence is reversed as the element passes to position D and finally to the undeformed condition at position E. Thus the element A has been virtually subjected to a horizontal shear between A and B, a removal of the horizontal shear between B and C, a $45°$ shear angle between B and D, and a further parallel shear cycle between D and E. Indeed, both theory and experiment show that the effective loss is about $3\alpha\phi$ instead of $\alpha\phi$ per unit distance of traverse.

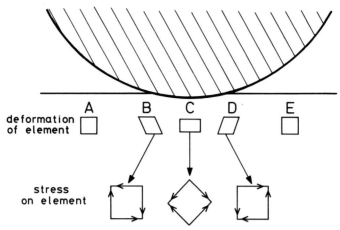

Figure 5. The rolling of a hard sphere or cylinder over a rubber surface. The square element A is first sheared parallel to the surface (B), then compressed (C), then sheared back parallel to the surface (D), and finally restored to its original shape (E). The main stresses on the element are shown for positions B, C, and D.

This concept is of fundamental importance in any attempt to calculate deformation losses for systems in which elements of the material are subjected to a complex strain path. It has been used very successfully by Wannop and Archard[15] in their study of the frictional behaviour of two discs of perspex rolling freely against one another. The energy dissipation calculated by the model indicated in Figure 5 gave good agreement with observation. Furthermore, the study emphasized a point of considerable importance inherent in the analysis, namely that the region of maximum energy dissipation is not in the interface itself but at a small distance below the surface. In this region there is cumulative heating and ultimate failure of the polymer. Similar effects lead to the degradation of lignin in the fibrilization of woodpulp[16,17] and to the blistering of automobile tyres.

When polymers possess loss characteristics which are markedly dependent on deformation rate or temperature the ploughing or deformation force shows a similar dependence. For example, Figure 6 shows the ploughing force F_d as a function of temperature for a sphere moving over the surface of a PTFE specimen.[18] (In order to reduce any adhesion component to a minimum the deformation was produced by rolling a sphere over the surface of the polymer rather than sliding it.) In a separate experiment the loss factor α and the real part of the elastic modulus E of the polymer were determined over the same range of temperature at a deformation rate comparable with that applying to the friction experiment. Since ϕ is proportional to $E^{-1/3}$ (see equation (2)) we should expect $\alpha\phi$ to vary in a way that resembles the variation of F_d. This is seen to be roughly true. Of course, quantitative agreement demands a more sophisticated treatment. As already mentioned the complex stress path will increase the effective loss. Again

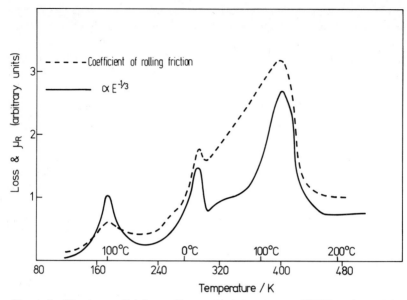

Figure 6. The tangential force F_d required to groove a PTFE surface with a hard sphere as a function of bulk temperature. (To minimize the adhesion component the motion of the sphere was achieved by rolling rather than sliding.) Load 900 g, speed 0.1 mm s^{-1}. On the same curve using arbitrary units the quantity $\alpha E^{-1/3}$ is drawn. This is a measure of the hysteretic losses in the PTFE, both α and E being determined at the appropriate temperatures.

the local pressures in the contact region may exceed 1000 atmospheres; these pressures reduce the free volume in the polymer and may produce significant changes in the modulus and the loss factor.

4. THE ROLE OF THE PLOUGHING TERM IN THE WEAR OF POLYMERS: FATIGUE AND ABRASION

In the complete absence of adhesion the ploughing process can contribute to the wear of polymers in several ways. If the hard asperity approximates to a portion of a smooth sphere the deformations are relatively small, involving strains of the order of only a few per cent. As we saw above the energy dissipation is a maximum a short distance below the polymer surface and this can lead to thermal softening, chemical degradation, or some other form of weakening of the material which ultimately leads to the production of wear fragments. This process may be described as fatigue. Sometimes the wear fragment is chunky, sometimes it is flaky.[19] With polymers containing local inclusions such as fibre reinforcements the subsurface stresses may readily produce detachment of the polymer from the fibre. Indeed, with fibre-reinforced polymers it is very difficult to prevent some form of delamination from occurring under almost any sort of stress. The only way fatigue can be avoided is by operating under conditions such that even the highest local stresses around the asperities are below some critical value. Lancaster's work[1] suggests that this can be achieved by using surfaces for which the average surface roughness is less than about 1 μm.

If the hard asperity has sharp corners or edges it may cut or tear the surface. The ploughing force is increased (see Figure 4(b)) and the surface seriously damaged. With rubbers, as Schallamach has shown[20] a sharp asperity stretches the material around it until local *tensile* failure occurs: a lip is left protruding for later removal by subsequent asperity interactions.[21] With polymers a sharp asperity may act as a miniature cutting tool and remove material: wear produced in this way may be described as abrasion.[22]

It is often difficult to distinguish between abrasion and fatigue and in many practical situations the process is further complicated by the presence of interfacial adhesion (see below). With metals rubbing on polymers the wear increases with the roughness of the metal in a manner shown in Figure 7, taken from Lancaster's work.[1] In a discussion of these results Lancaster suggests that if S is the breaking stress of the polymer and ϵ is the elongation to break, the work required to complete the process of producing a wear fragment is proportional to the product $S\epsilon$. If we ignore the differences in hardness and friction the wear rate of various polymers might be expected to be proportional to some function of $(1/S\epsilon)$: in fact Figure 8 suggests a linear relation. Whether this is to be regarded as abrasion or fatigue (plus some adhesion) remains unresolved.

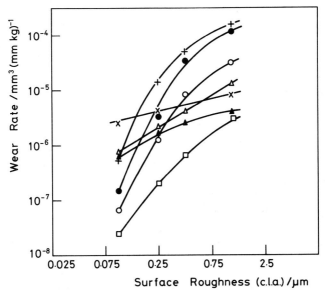

Figure 7. Variation of wear rate with surface roughness for polymers during single traversals over mild steel rings (from Lancaster[1]): + polystyrene, ● PMMA, ○ acetal hompolymer, △ polypropylene, × PTFE, ▲ polyethylene, □ nylon 6.6.

8

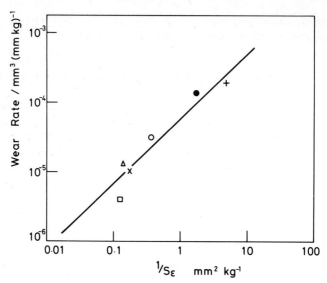

Figure 8. Correlation between wear rates of polymers and the reciprocal of the energy parameter $S\epsilon$. The symbols are the same as in Figure 7. The wear rates refer to single traversals over a rough steel surface (c.l.a. = 1.2 μm) and the results show that all the data of Figure 7 lie close to a single curve.

Fatigue and abrasion play an important part in the wear of polymer fibres. Indeed, the wear resistance of fabrics is often assessed by some type of abrasion tester. However, fatigue can occur simply from the continuous flexure of fibres and yarns. In this case tensile stresses (which are a maximum at the surface of the fibres) can initiate cracks at the surface: these spread through the fibre and cause it to break. In such a situation the failure of the fibre will be greatly accelerated if it is present in a 'stress-corrosion' environment.

Finally, we may ask what role the polymer surface itself plays in the ploughing, abrasion, and fatigue processes. Since (in the absence of adhesion) the deformation process occurs in a finite volume it is evident that it is the bulk rather than the surface properties which are involved. However, if the asperities on the counterface are extremely fine the grooving process may be restricted to a very thin layer (see Figure 1(b)). In that case changes in the structure of the polymer near the surface may have a significant effect on the microdeformations occurring during sliding. For example, the oxidation or chlorination of a rubber surface may produce a thin surface layer with deformation properties very different from those of the bulk. In the same way the crystallinity and morphology and hence the deformation properties of the outermost layers of a polymer may be very different from those of the bulk. For example Eiss and Warren[23] have recently shown that crystalline PCTFE has a higher wear rate than the amorphous polymer because the energy of rupture of the crystalline polymer is appreciably smaller. Clearly, if the surface layers are comparable in thickness with the depth of deformation produced by the asperities they may have a marked effect on the processes in this section.

5. THE ADHESION TERM IN THE FRICTION OF ELASTOMERS

All solids when brought into contact will experience some type of adhesion and the force (or work) involved in overcoming the adhesive forces may be regarded as the adhesive component of friction. With elastomers and polymers the adhesion arises from two basic sources. One is electrostatic, and this has been the subject of a great deal of dispute.[24-26] The other source of adhesion arises from van der Waals forces[27] and, if there are certain polar atoms present, from dipole interactions and hydrogen bonding.

The adhesion of elastomers will be discussed in a later chapter: in this section we discuss its role in the friction of rubber-like solids. The most striking experimental results are those of Grosch[28] who studied the friction of rubber sliding over a glass surface (possessing a slightly irregular topography) over a wide range of speeds and

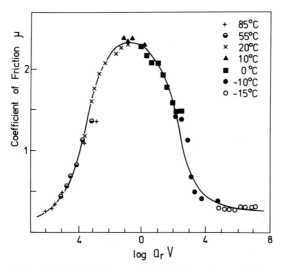

Figure 9. Coefficient of friction for rubber sliding over a glass surface possessing a slightly irregular topography. The experiments were carried out at various temperatures from -15°C to $+85^\circ$C and at sliding speeds ranging from 10^{-3} to 10 mm sec (over this speed range frictional heating may be neglected). The whole of the data may be plotted against a parameter $a_T V$ where a_T is a function of the temperature so that $a_T V$ is a temperature-compensated velocity. This is known as the Williams–Landell–Ferry (WLF) transform and it is seen that all the data lie on a single master curve.

temperatures. He did not, however, exceed a sliding speed of about 10 mm s^{-1} so as to avoid complications arising from frictional heating. His results showed that each friction–velocity curve could be displaced on the velocity axis by an amount determined by the temperature and the known viscoelastic properties of the rubber using the well-known Williams, Landel, and Ferry (WLF) transform to give a single master curve. A typical result is shown in Figure 9. Similar results have been obtained by many other workers and it is therefore natural that attempts should have been made to explain the behaviour in terms of the viscoelastic properties of the rubber. As we have already seen in a previous section of this paper such a correlation has been established for the ploughing or deformation part of the friction process.

The attempt to explain the adhesion part has, however, been far less satisfactory. Molecular theories have been proposed in terms of the formation and breaking of atomic bonds at the interface, using Eyring's rate-process theory in one form or another (Schallamach,[29] Bartenev[4]). Other attempts have been used to apply macroscopic concepts of adhesion and shearing or tearing at the interface (Bulgin,[30] Ludema and Tabor,[18] Savkoor,[31] Kummer[32]). These theories have been described by Schallamach, Bartenev, and Moore and discussed critically by Tabor.[33] There is little point in repeating this since new developments in the last five years have suggested that, however ingenious these theories may be they do not correspond to what actually happens at the sliding interface. In 1971 Schallamach[34] described experiments in which a smooth hemispherical rubber slider passed over a clean, smooth glass surface: he observed that in the presence of adhesion between the surfaces the application of a tangential force causes the surface of the rubber to buckle and generate 'waves of detachment' which traverse the contact area from front to rear at a very high speed. The motion of the rubber over the glass surface does *not* involve interfacial sliding but resembles the passage of a 'ruck' through a carpet, or the motion of a caterpillar. There is continuous dehesion on one side of the ruck and readhesion on the other side of the ruck as it passes through the contact zone (see Figure 10). Several workers have now

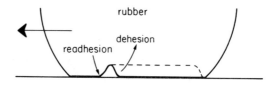

Figure 10. Schematic diagram showing how a wave of detachment travels through the contact zone when a soft rubber slider moves over a clean, smooth glass surface. The adhesion and readhesion of the rubber–glass interface around the fold in the rubber enables relative motion between the rubber and the glass to occur without interfacial sliding actually taking place.

confirmed this observation[35-37] and it is evident that the frictional work is associated with the energy lost during the continuous dehesion—readhesion process. Since as will be shown in Chapter 2 this is a rate- and temperature- dependent process the friction similarly depends on temperature and sliding speed.

These studies thus provide a macroscopic theory of 'sliding friction' in terms of interfacial adhesion; the interfacial adhesion itself involves both molecular forces and viscoelastic losses. As the hardness of the rubber increases the rucks become finer and their speed higher. Whether this process always applies to the 'sliding' of rubber over another surface is not known. It is possible that when the ratio of the modulus of the rubber to the surface adhesion exceeds a certain critical value true sliding occurs. On the other hand it is possible that in this limiting case the rucks are of molecular dimensions and the waves of detachment continue to operate, but on a molecular scale.

6. THE ROLE OF ADHESION IN THE WEAR OF ELASTOMERS

When rubbers of high strength slide over clean glass surfaces at moderate speeds, very little wear of the rubber occurs. On the other hand, with weak rubbers, lumps of rubber may sometimes be torn out of the surface particularly at higher speeds.[38] In some cases, especially if the other surface is covered with gentle undulations, there is tensile failure of the rubber behind each undulation.[3] This will lead to a ribbed pattern on the rubber surface and the ultimate detachment of fine rubber filaments (as in the use of rubber as an eraser). At higher sliding speeds the frictional energy which is generated very close to the interface leads to excessive heating, softening, and chemical degradation. If the interfacial adhesion can be decreased these processes are reduced in scale. This can be achieved by the use of liquid lubricants or by incorporating certain types of fillers in the rubber which provide solid-film lubrication. Naturally, the friction is also reduced.

The role of hard fillers which maintain a fair level of friction but reduce the wear is part of the technology of rubber-tyre production and will not be discussed further here.

7. THE ADHESION TERM IN THE FRICTION OF POLYMERS

As with elastomers, the adhesion of polymers is responsible for an adhesive component of friction. If a polymer sphere is slid over a clean, smooth, hard surface (so that there is virtually no ploughing of the polymer) the whole of the frictional force F_a must be due to this factor.[39-42] However, careful studies have not shown any evidence for the occurrence of waves of detachment. They may occur at temperatures very much above the glass transition temperature T_g: or they may occur on a molecular scale at lower temperatures. On the whole, however, it would seem that true sliding occurs. If the attachment of the polymer to the counterface is weaker than the polymer, sliding occurs truly at the interface: if it is stronger than the polymer shear will occur a short distance from the interface within the polymer itself. This distance may range from a few nanometres to a few micrometres. If τ is the relevant shear strength per unit area and A is the true area of contact we may write

$$F_a = A\tau$$

It is possible to study the factors influencing the shear strength τ in a relatively simple way.[43] A thin film of the polymer is deposited on to a smooth, hard glass surface. A very smooth glass hemisphere is then loaded on to the film and the force to produce sliding is measured. Since the film is extremely thin (ranging from a few nanometres to a fraction of a micrometre) the area of contact is determined by the elastic deformation of the glass surfaces and may be easily calculated. If the force to produce sliding is divided by this area, and if during sliding the sphere does not penetrate the film to produce glass–glass contact, the resultant value is the relevant shear strength τ. By changing the radius of curvature of the sphere and the load the contact pressure can be raised over a large range. The sliding speed may also be raised over a wide range, but if it exceeds more than, say, 10 mm s^{-1} the temperature rise due to frictional heating may no longer be neglected. Finally, the bulk temperature of the whole system may be varied. In this way it is possible to study τ as a function of contact pressure p, sliding speed v, and temperature T.

The most striking observation is that for a wide range of polymers (and other solids too) the shear strength at constant v and T is related to the contact pressure p by an expression of the form

$$\tau = \tau_0 + \alpha p \tag{4}$$

as is seen in Figure 11.[44-48] This relation resembles qualitatively the bulk behaviour of the polymer although the value of τ_0 is considerably smaller than the bulk value. This may be because of the high degree of orientation produced in the film during sliding or to the fact that in the thin-film experiments the shear is forced to follow a very well-defined plane. On the other hand the value of α is very close to that of the bulk polymer.

The effect on τ of speed of sliding in the low-velocity range is rather small. On the other hand the effect of temperature is, as we should expect, rather striking. Below T_g, temperature has little influence on τ, but above T_g the value of τ falls in a well-defined way with increasing T. Pressure affects the temperature dependence but we shall not discuss this further here. We may, however, remark that in many practical situations involving the sliding of unlubricated polymers, the motion is intermittent. 'Stick-slip' motion appears to arise from an interaction between the time dependence of τ and the mechanical properties of the sliding system.

Assuming that τ has a well-defined value we now show how it may be used to calculate that part of the coefficient of friction which is due to adhesion.[48] If a polymer specimen is placed on a smooth, hard surface under a normal load W, elastic, plastic, or viscoelastic flow will occur at the interface until the area of true contact A is sufficient to support the applied load. Under these conditions the average contact pressure p is simply $p = W/A$ or

$$A = W/p \tag{5}$$

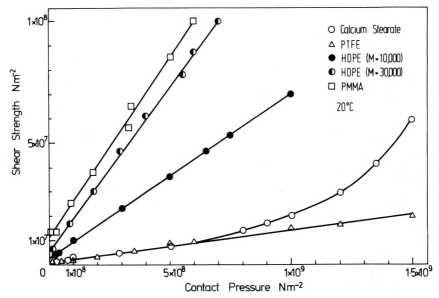

Figure 11. Shear strength τ of a number of organic materials as a function of mean contact pressure p. (Sliding speed $ca.$ 0.1 mm s^{-1}.) Over a wide range of materials and pressures τ increases linearly with p so that one can write $\tau = \tau_0 + \alpha p$.

If sliding involves the shearing process described above the frictional force F_a arising from the adhesion component is

$$F_a = A\tau = A\,(\tau_0 + \alpha p) = (W/p)(\tau_0 + \alpha p) \tag{6}$$

Hence the coefficient of friction μ_a is

$$\mu_a = F_a/W = \tau_0/p + \alpha \tag{7}$$

It turns out that, in general, the mean pressure p increases with load so that at high loads the first term becomes small compared with the second. Thus, at high loads

$$\mu_a \approx \alpha \tag{8}$$

Some typical results given in Table 1 show that for a wide range of thermoplastics relations (7) and (8) are reasonably well obeyed.

There is one point that should be mentioned at this stage. The equations used in these derivations say nothing about the nature of the shear process. They do not distinguish between shear at the interface itself or shear within the polymer. Whichever process occurs in this thin-film shear experiment also occurs in the bulk friction experiment so that the correlation finally expressed in equations (7) and (8) must be true. For reasons which are not yet clear, even when shearing appears to occur truly at the interface the shear strength τ varies with pressure according to equation (4) and agrees qualitatively with the shear properties of the polymer in bulk and with its dependence on pressure.

On the other hand the location of the shear zone has a profound effect on the adhesive wear of the polymer.

Table 1 The shear properties of thin polymer films and the coefficient of friction μ of the same polymers sliding on clean glass and on themselves. Sliding speed ~ 0.2 mm s^{-1}. Load sufficient to produce yielding of the polymer at contact pressure p_0

Polymer	Thin film experiments			Friction experiments	
	τ_0/p_0	α	$\tau_0/p_0 + \alpha$	μ:Polymer on glass	μ:Polymer on polymer
Low density polythene (LDPE)	0.39[a]	0.14	0.53	0.52	0.42
High density polythene (HDPE)	0.06	0.10	0.16	0.15	0.08
PTFE	0.04	0.08	0.12	0.16	0.13
Polypropylene (PP)	0.10	0.17	0.27	0.26	0.26
PMMA (Perspex)	0.03	0.36	0.39	0.36	0.41
Polyvinyl chloride (PVC)	−0.09[b]	0.57	0.46	0.55	0.54
Polystyrene (PS)	0.03	0.45	0.48	0.42	0.45

[a]With LDPE the first term of friction equation (7) is abnormally high. With all the other polymers the major contribution to the coefficient of friction arises from the term α.
[b]The negative value has no physical meaning: it is the result of extrapolating the $\tau:p$ straight line back to zero pressure.

8. THE ROLE OF ADHESION IN THE WEAR OF POLYMERS

With amorphous polymers such as polystyrene, PMMA (Perspex) and PVC and with semicrystalline polymers such as polypropylene sliding at low speeds on clean, smooth hard surfaces such as stainless steel or glass, the interfacial adhesion is strong but does not appear to be as strong as the polymer itself. Consequently, although the friction may be relatively high there is no gross transfer of the polymer to the counterface. Minute fragments may appear at the interface, but these are more likely the result of local fatigue. If, however, the temperature is raised the bulk strength of the polymer decreases and may reach a point where it is less than the interfacial shear strength. At that stage lumps of polymer are transferred to the counterface and the polymer wear becomes very heavy. In general this occurs around the glass transition temperature (see Figure 12). Consequently, high sliding speeds which generate heating may encourage the onset of appreciable wear.

Below their glass transition temperature many polymers are relatively brittle. Even under conditions of simple normal loading against another surface, tensile stresses may be set up around each contact zone and these can initiate cracks which spread into the polymer. This type of behaviour is well known in fracture mechanics. If, for example, a sphere (representing an individual asperity) is loaded against a flat surface the maximum tensile stress occurs around the edge of the circle of contact, and if it exceeds a critical value a cone-shaped crack will form in one or both of the solids. If sliding takes place and there is appreciable adhesion between the surfaces the augmented tensile stress in the surfaces greatly facilitates the onset of cracking.[49-51] For example, in the macroscopic loading of a sphere on to a glass surface it was found that a normal load of 10 kg was just sufficient to

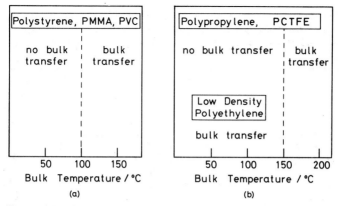

Figure 12. Slow-speed sliding of polymer over a clean, smooth glass transition temperature; (a) amorphous polymers, (b) negligible frictional transfer to marked bulk transfer when the bulk temperature exceeds a critical value of the order of the glass transition temperature; (a) amorphous polymers (b) semicrystalline polymers.

produce the cone-crack. In sliding, where the coefficient of friction μ was only 0.3, the normal load required to initiate cracking was reduced to less than 2 kg.[52] It might be thought, therefore, that reducing the friction by the use of a lubricant would reduce the danger of surface cracking. This is true in principle. But many liquids that would function as effective lubricants are also active as stress-corrosion agents.[53] In that case the 'cure would be worse than the disease'.

9. THE ROLE OF ADHESION IN ENERGY DISSIPATION

We have already seen that the grooving process involves energy dissipation due to hysteresis and that it is concentrated below the surface. By contrast the adhesion process involves energy dissipation at, or very close to the surface. When both processes are involved the resultant behaviour is complex and will depend on the ratio of the adhesion to the grooving component of friction: the higher the adhesion component the nearer the maximum energy dissipation is to the surface. However, this ignores the important role of temperature on the flow properties of thermoplastics. If energy dissipation produces heating of the polymer in a given region its resistance to flow decreases: consequently it becomes easier to maintain flow in that region and a run away situation may arise. Even in the presence of some adhesion, the maximum temperature rise may occur below the surface. This resembles the process of adiabatic shear observed in the high rate deformation of metals.

At sufficiently high sliding speeds[54] surface melting of the polymer may occur and if the temperature rise is not allowed to diffuse into the bulk of the polymer, the friction and wear may be small as is the case, for example, in the sliding of ice.[55,56] On the other hand the generation of high temperatures during sliding may lead to oxidation or other types of degradation of the polymer. In that case the surrounding environment may be of considerable importance.

10. THE REDUCTION OF THE ADHESION COMPONENT OF FRICTION AND WEAR

In later chapters in this volume it will be shown that, if we ignore possible electrostatic effects, the *normal* adhesion between polymers or elastomers and another surface is determined by two basic properties. The first is the free surface energy γ of the polymer since this is a measure of the atomic forces at the surface. The second is the viscoelastic or hysteretic property of the polymer. When the surfaces are pulled apart energy is expended, not only in overcoming the surface forces but as viscoelastic losses in the stressed material lying just behind the interface. These losses are often much greater than the surface energy but appear to be proportional to it.

If similar considerations apply to the adhesion component in shear (i.e. in friction experiments) we may ask whether it may be possible, starting with a polymer of given viscoelastic properties, to modify the surface so as to reduce the surface energy. If the polymer is non-polar its surface energy, as judged by wetting experiments[57] and other studies is of order $< 20–30$ mJ m^{-2}. If it is polar the surface energy may have values of order $40–50$ mJ m^{-2}. It may thus be possible to reduce the surface energy by suitable chemical treatments.[58] However, the lowest surface energy obtainable is probably not less than *ca.* 20 mJ m^{-2}. Furthermore, in a sliding system the modified surface will be easily worn away.

A different approach is to use lubricants. We do not intend to deal with lubricated sliding in any detail but mention one interesting development, namely the possibility of incorporating the lubricant within the polymer itself. For example, it is common practice now to add extremely fine flakes of MoS$_2$ to nylon and similar polymers.[59] The MoS$_2$ is exposed at the rubbing surface and provides effective solid-film lubrication. Again, with low-density polyethylene, specimens may be prepared containing a few hundred parts per million of oleamide or stearamide. These amides diffuse to the surface and provide a thin film which is readily sheared and which prevents intimate polymer—counterface contact.[61-62] Incidentally, the diffusion of long-chain molecules through the bulk and over the surface of polymers presents a number of fascinating and challenging problems which have not yet been resolved.[63] Another approach is to dispense extremely small pockets of silicone throughout the polymer:[64] these may burst under the stresses induced during sliding and release small but adequate quantities of liquid lubricant.

11. THE FRICTION AND WEAR OF PTFE AND HDPE

These two polymers have exceptional frictional properties.[65-68] If a slider of the polymer is slid over a smooth clean surface at low speeds the initial coefficient of friction is high ($\mu = 0.2–0.3$) and a lump of polymer is transferred to the counterface: the thickness of the lump is of the order $0.1-10 \mu m$. Once this initial movement has taken place the friction falls to a value of about $\mu = 0.08$ and the transfer is now in the form of a very thin film of polymer (Figure 13(b)) the thickness of which is of order $50-100$ Å. This film is fairly firmly attached to the counterface and consists of molecules highly oriented in the direction of sliding.[42] Apparently once the initial stick has been overcome the material around the contact region of the slider is pulled into a convenient orientation which provides easy drawing of the molecular chains out of the crystalline portions of the polymer. The behaviour does not appear to depend on the degree of crystallinity or on the molecular weight of the polymer. It seems to be associated with the smooth molecular profile of the polymer chain. For example, if bulky side groups are introduced at frequent intervals into the polymer chain the friction and transfer are high and the behaviour resembles that of typical thermoplastics above their glass transition temperature (Figure 13). The smooth profile also implies that the sliding of the polymer over the transferred film is an easy process and does not necessarily involve the drawing out of additional molecular chains. Consequently, not only is the friction low: there is little tendency for the transferred film to grow in thickness.[42] This implies that once the transferred film has been laid down, further motion involves almost true interfacial sliding with an extremely small wear rate of the polymer. Under some conditions, however, particularly at higher temperature, the transferred film may grow in thickness and Tanaka attributes the behaviour to the drawing out of lamellae rather than molecular chains.[69]

It is natural to ask whether other molecules with similar molecular architecture behave in a similar way. Two possible candidates are polyoxymethylene and isostatic polystyrene, but no detailed studies of these materials have been described.

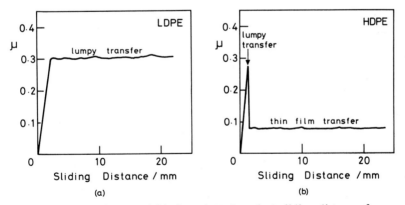

Figure 13. Coefficient of friction plotted against sliding distance for a polyethylene (PE) hemisphere sliding over a clean smooth glass surface at $20°C$. Load 1 kg, sliding speed 1 mm s^{-1}; (a) low-density PE (LDPE), (b) high-density PE (HDPE). For HDPE the initial (static) coefficient of friction is large, but as soon as sliding commences the friction falls to a low value and transfer is in the form of a very thin film. LDPE shows s high friction and lumpy transfer throughout (from Pooley and Tabor[42]).

The limited evidence so far available suggests that both a smooth molecular profile and low intermolecular forces are necessary in order to obtain the behaviour observed with PTFE and HDPE.

There are two conditions under which PTFE and HDPE lose their low friction and wear properties. First, if the roughness of the countersurface exceeds a few micrometres. Presumably this interferes with the easy drawing of the molecular chains out of the polymer. Secondly if the sliding speed is increases to say 0.1 m s^{-1} or the bulk temperature reduced to say $-20°C$ the force to draw out molecular chains or to slide them over one another increases to such an extent that it becomes easier for the polymer to yield by the tearing out of relatively large lumps. (This is because the movement of the molecular chains is a stress-aided, thermally activated process.) Under these conditions the friction and wear become high. At higher sliding speeds, say a few metres per second, frictional heating appears to compensate for the increased shear rate and the sliding of the chains over one another again becomes easy. The basic problem now is how to secure the firm attachment of the transferred film to the countersurface.

12. THE EFFECT OF FILLERS ON THE WEAR OF HDPE AT HIGHER SLIDING SPEEDS

The main reason that HDPE and PTFE cease to be low-wear materials at typical engineering speeds (several metres per second) is that the transferred film does not survive repeated traversals. It is either wiped away or removed by fatigue. Whatever the process it leads to the continuous laying down and removal of a film of the order of 50 Å thick. If certain fillers are added to the polymer the transferred film may become firmly attached to the countersurface and the wear of the polymer may be reduced a hundredfold.[70] The fillers fall into two classes, chemical and mechanical. In the first class a mixture of oxides (typically 5% CuO and 30% Pb_3O_4) is incorporated in the HDPE. The filler is ineffective against glass or aluminium, but highly effective against a steel counterface provided it is not too rough. With the unfilled polymer (see Figure 14) the wear is associated with the continuous formation and removal of the transferred film: with repeated traversals the wear rate actually increases probably because of thermal softening of the polymer. By contrast the filled polymer gradually forms a transferred film which remains firmly attached to the steel countersurface and further sliding over this film now leads to a very low polymer wear rate.[71] The detailed processes are undoubtedly very complicated and probably involve chemical reaction between the polymer and the steel counterface, forward and backward transfer of polymer and chemical changes in the polymer itself. It is interesting to note, however, that the basic frictional properties of the filled polymer are not substantially different from those of the unfilled polymer. For example, once the firmly attached film has been formed on the steel surface the slider may be replaced by an unfilled specimen of HDPE: this will give very low polymer wear for a protracted period.[71]

The other type of filler consists of small hard particles, typically glass fibres or various types of carbon black. They also lead to the formation of a strongly

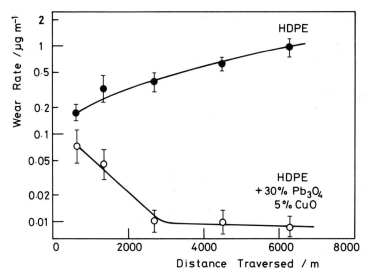

Figure 14. Wear rate as a function of distance travelled over a rotating disc of mild steel (diameter 10 cm, surface finish 0.05 μm) in air. With repeated traversals over the same track the wear rate of unfilled HDPE slowly increases with number of traversals, probably as a result of some thermal softening. With filled HDPE (containing 30% Pb_3O_4 and 5% CuO) after a short running-in period during which a strongly attached film is formed on the steel surface the wear rate of the polymer becomes extremely small. Load 5 kg, sliding speed 2 m s^{-1} (from Briscoe et al.[71]).

attached polymer film and a low wear rate. There is a marked change in the roughness of the counterface and appreciable oxidation of the metal. It would seem that although the fillers are relatively inert the hot spots at the regions of rubbing contact cause crucial chemical changes at the interface.[72]

Finally, we may note that in a recent study of the wear of unfilled HDPE against stainless steel (as part of a general investigation of materials for prosthetic joints) Dowson et al.[73] found that low wear rates of the polymer could be achieved at an optimum surface roughness of the steel of order 0.1 μm. Evidently wear is a complex process and a single mechanism is unlikely to explain all wear phenomena.

13. CONCLUSIONS

It may be useful in summary to consider some of the aspects of friction and wear discussed in this chapter which have relevance to other chapters to be presented in this volume.

The first general observation is that the friction of polymers or elastomers is bound to involve their bulk properties as well as those of the surface. In situations where polymers slide on other, harder surfaces, the work of sliding, if there is little surface adhesion, is largely expended in dragging asperities of the harder solid through the polymer. The major part of this energy is dissipated as viscoelastic or hysteretic losses in the subsurface material. Consequently, the surface itself plays little part in the process. However, if the surface roughnesses are small the subsurface material may be at a very small depth below the free surface. In that case changes in the surface layers of the polymer – e.g. oxidation, crystallinity, morphology – may have an appreciable influence on the amount of energy dissipated. In this process, as in all processes involving the deformation of polymers, any attempt to calculate the energy loss must take into account the complex strain path through which each element of the polymer passes.

The second general observation is that the adhesion part of the friction and wear process involves possible electrostatic effects, van der Waals forces, as well as viscoelastic losses. The electrostatic effects and the surface forces may be modified to some extent by chemical treatments, but the modified surface layer is likely to be worn away by the frictional process itself. Adhesive forces between the sliding surfaces introduce relatively large surface tractions. With polymers below their glass transition temperature the associated tensile forces may lead to the formation of cracks in the surface with the subsequent detachment of relatively large pieces. This may be greatly accelerated in the presence of stress–corrosive environments. Friction and wear may be reduced by the incorporation of additives which reduce the adhesion at the interface. These include MoS_2, long-chain amides, or minutely dispersed pockets of silicones which can burst and release small amounts of lubricant. Thirdly, we note that with high-density polythene and PTFE the wear can be reduced by the use of fillers which ensure firm attachment of the transferred film to the counterface. But, as is true with the whole field of wear, no single mechanism can explain all the observed results.

ACKNOWLEDGEMENTS

The authors wish to thank Dow Corning (Barry, Glamorgan), General Electric (Schenectady) and the Science Research Council for providing financial support for much of the work described.

REFERENCES

1. J. K. Lancaster, 'Friction and wear', Ch. 14 in A. D. Jenkins, (ed.) *Polymer Science, A Materials Science Handbook*, North-Holland Publ. Co., 1972.
2. S. S. Voyutski, *Autoadhesion and Adhesion of High Polymers*, Wiley, New York, 1963.
3. A. Schallamach, 'Recent advances in knowledge of rubber friction and tyre wear', *Rubber Chem. Tech.*, **41**, 209–223, 1968.
4. G. M. Bartenev and V. V. Lavrentev, *Trenye i Iznos Polimerov*, Khimiya, Leningrad, 1972.
5. D. F. Moore, *The Friction and Lubrication of Elastomers*, Pergamon, Oxford, 1972.
6. Lieng-Huang Lee (ed.) *Advances in Polymer Friction and Wear*, Plenum Press, New York and London, 1975.
7. D. Tabor, 'The wear of non-metallic materials', Paper 1, 3rd Leeds–Lyons Symposium on Tribology, Leeds, Sept., 1976.

8. F. P. Bowden and D. Tabor, *Friction and Lubrication of Solids*, Clarendon Press, Oxford, 1950.

9. V. A. Belii, A. I. Sviridenok, M. I. Petrokovetz, and V. G. Savkin, *Friction and Wear of Polymer-Based Materials* (in Russian), Nauka i Tekhnika, Minsk, 1976.

10. D. Tabor, 'The mechanism of rolling friction. II. The elastic range', *Proc. Roy. Soc.*, **A229**, 198–220, 1955.

11. D. G. Flom and A. M. Bueche, 'Theory of rolling friction for spheres', *J. Appl. Phys.*, **30**, 1725–1730, 1959.

12. W. D. May, E. L. Morris, and D. Atack, 'Rolling friction of a hard cylinder over a viscoelastic material', *J. Appl. Phys.*, **30**, 1713–1724, 1959.

13. J. A. Greenwood and D. Tabor, 'The friction of hard sliders on lubricated rubber: the importance of deformation losses', *Proc. Phys. Soc.*, **71**, 989–1001, 1958.

14. J. A. Greenwood, H. Minshall, and D. Tabor, 'Hysteresis losses in rolling and sliding friction', *Proc. Roy. Soc.*, **A259**, 480–507, 1961.

15. G. L. Wannop and J. F. Archard, 'Elastic hysteresis and a catastrophic wear mechanism for polymers', *Proc. Instn. Mech. Engrs.*, **187**, 615–623, 1973.

16. D. Atack and D. Tabor, 'The friction of wood', *Proc. Roy Soc.*, **A246**, 539–555, 1958.

17. D. Atack and W. D. May, 'Frictional mechanisms in the grinding press', *Pulp and Paper Magazine of Canada*, 1958.

18. K. C. Ludema and D. Tabor, 'Friction and viscoelastic properties of polymeric solids', *Wear*, **9**, 329–348, 1966.

19. N. P. Suh, 'The delamination theory of wear', *Wear*, **25**, 111–124, 1973.

20. A. Schallamach, 'Abrasion of rubber by a needle', *J. Polymer Sci.*, **9**, 385–396, 1952.

21. E. Southern and A. G. Thomas, 'Some recent studies in rubber abrasion', Paper 22, 3rd Leeds–Lyons Symposium on Tribology, Leeds, Sept. 1976.

22. A. J. Sedricks and T. O. Mulhearn, 'Mechanics of cutting and rubbing in simulated abrasive processes', *Wear*, **6**, 457–466, 1963.

23. N. S. Eiss and J. H. Warren, 'On the influence of the degree of crystallinity of PCTFE on its transfer to steel surfaces of different roughnesses', Paper 3, 3rd Leeds–Lyon Symposium on Tribology, Leeds, Sept., 1976.

24. B. V. Derjaguin, 'Problems of adhesion', *Research* (London), **8**, 70–74, 1955.

25. S. M. Skinner, R. L. Savage, and J. E. Rutzler, 'Electrical phenomena in adhesion: I. Electron atmospheres in dielectrics', *J. Appl. Phys.*, **24**, 438–450, 1953.

26. H. G. Von Harrach and B. N. Chapman, 'Charge effects in thin-film adhesion', *Proc. Int. Conf. Thin Films*, Venice, **II**, 157–161, 1972.

27. J. N. Israelachvili and D. Tabor, 'Van der Waals forces: theory and experiment', in *Progress and Membrane Science*, J. F. Danielli, M. D. Rosenberg, and D. A. Cadenhead, eds., Academic Press, New York, pp. 1–55, 1973.

28. K. A. Grosch, 'The relation between the friction and viscoelastic properties of rubber', *Proc. Roy. Soc.*, **A274**, 21–39, 1963.

29. A. Schallamach, 'A theory of dynamic rubber friction', *Wear*, **6**, 375–382, 1963.

30. D. Bulgin, G. D. Hubbard, and M. H. Walters, 'Road and laboratory study of friction of polymers', *Proc. 4th Rubber Technology Conf., London*, pp. 173–188, 1962.

31. A. R. Savkoor, 'On the friction of rubber', *Wear*, **8**, 222–237, 1965.

32. H. W. Kummer and W. E. Meyer, 'Skid or slip resistance', *J. Mater. Sci.*, **1**, 667, 1966.

33. D. Tabor, 'Friction, adhesion and boundary lubrication of polymers', in

Lieng-Huang Lee (ed.) *Advances in Polymer Friction and Wear*, Plenum Press, New York and London, pp. 5–30, 1975.

34. A. Schallamach, 'How does rubber slide?', *Wear*, **17**, 301–312, 1971.
35. A. D. Roberts and A. G. Thomas, 'The adhesion and friction of smooth rubber surfaces', *Wear*, **33**, 45–46, 1975.
36. G. A. D. Briggs and B. J. Briscoe, 'The dissipation of energy in the friction of rubber', *Wear*, **35**, 357–364, 1975.
37. M. Barquins and R. Courtel, 'Rubber friction and the rheology of viscoelastic contact', *Wear*, **32**, 133–150, 1975.
38. K. C. Ludema, 'The wear of rubber', Paper 19, 3rd Leeds–Lyon Symposium on Tribology, Leeds, Sept., 1976.
39. K. V. Shooter and D. Tabor, 'The frictional properties of plastics', *Proc. Phys. Soc.*, **B65**, 661–671, 1952.
40. R. F. King and D. Tabor, 'The effect of temperature on the mechanical properties and the friction of plastics', *Proc. Phys. Soc.*, **B66**, 728–736, 1953.
41. M. W. Pascoe and D. Tabor, 'The friction and deformation of polymers', *Proc. Roy. Soc.*, **A235**, 210–224, 1956.
42. C. M. Pooley and D. Tabor, 'Friction and molecular structure: behaviour of some thermoplastics', *Proc. Roy. Soc.*, **A329**, 251–274, 1972.
43. B. J. Briscoe, B. Scruton, and F. R. Willis, 'The shear strength of thin lubricant films', *Proc. Roy. Soc.*, **A333**, 99–114, 1973.
44. I. V. Kraghelsky and V. P. Sabelnikov, 'Experimental check of elementary laws of friction', Paper 7, *Proc. Inst. Mech. Engrs. Conference on Lubrication and Wear*, London, pp. 247–251, 1957.
45. B. J. Briscoe and D. Tabor, 'Rheology of thin organic films', *A.S.L.E. Trans.*, **17**, 158–165, 1974.
46. B. J. Briscoe and D. Tabor, 'Shear properties of thin organic films', Symposium on lubricant properties, American Chem. Soc., *Preprints*, **21**, 10–25, 1976.
47. N. Adams, 'Friction and deformation of nylon 1', *J. Appl. Polymer. Sci.*, **7**, 2075–2103, 1963.
48. J. K. A. Amuzu, B. J. Briscoe, and D. Tabor, 'Friction and shear strength of polymers', *Preprint No. 76-AM-IA-3*. 31st Annual Meeting A.S.L.E., Philadelphia, 1976.
49. G. M. Hamilton and L. E. Goodman, 'The stress-field created by a circular sliding contact', *Trans. A.S.M.E., Series E: J. Appl. Mech.*, **33**, 371–390, 1966.
50. F. C. Frank and B. R. Lawn, 'On the theory of Hertzian fracture', *Proc. Roy. Soc.*, **A299**, 291–306, 1967.
51. B. R. Lawn, 'Partial cone crack formation in a brittle material loaded with sliding spherical indentor', *Proc. Roy. Soc.*, **A209**, 307–316, 1976.
52. P. R. Billinghurst, C. A. Brookes, and D. Tabor, 'The sliding process as a fracture inducing mechanism', in *Conf. Proc. 'Physical Basis of Yield and Fracture'*, 253–268, Oxford, 1966.
53. A. R. C. Westwood, 'Environment sensitive fracture processes', *J. Mater. Sci.*, **9**, 1871–1895, 1974.
54. F. P. Bowden and E. H. Freitag, 'The friction of solids at very high speeds', *Proc. Roy. Soc.*, **A248**, 350–367, 1958.
55. F. P. Bowden, 'Friction on snow and ice', *Proc. Roy. Soc.*, **A217**, 462–478, 1953.
56. D. C. B. Evans, J. F. Nye, and K. J. Cheeseman, 'The kinetic friction of ice', *Proc. Roy. Soc.*, **A347**, 493–512, 1976.
57. W. A. Zisman, 'Contact angle, wettability and adhesion', *Advances in Chemistry Series*, No. 43, 1, American Chem. Soc. 1964.
58. J. M. Senior and G. H. West, 'Interaction between lubricants and plastic bearing surfaces', *Wear*, **18**, 311–323, 1971.

59. T. E. Powers, 'Molybdenum disulfide in nylon for wear resistance', *Modern Plastics*, **37**, 148−154, 1960.
60. S. C. Cohen and D. Tabor, 'The friction and lubrication of polymers', *Proc. Roy. Soc.*, **A291**, 186−207, 1966.
61. B. J. Briscoe, V. Mustafaev, and D. Tabor, 'Lubrication of polythene by oleamide and stearamide', *Wear*, **19**, 399−414, 1972.
62. D. Allan, B. J. Briscoe, and D. Tabor, 'Lubrication of polythene by oleamide and stearamide. II', *Wear*, **25**, 1973.
63. J. Klein and B. J. Briscoe, 'Diffusion of large molecules in polymers: a measuring technique based on microdensitometry in the infra-red', *Polymer*, **17**, 481−484, 1976.
64. M. P. L. Hill, P. L. Millard, and M. J. Owen, 'Migration phenomena in silicon modified polystyrene', in Lieng-Huang Lee (ed.) *Advances in Polymer Friction and Wear*, Plenum Press, New York and London, pp. 469−478, 1975.
65. R. C. Bowers, W. C. Clinton, and W. A. Zisman, 'Frictional behaviour of polyethylene, poltetrafluoroethylene and halogenated derivatives', *Lub. Engng.*, **9**, 204−208, 1953.
66. K. R. Makinson and D. Tabor, 'The friction and transfer of polytetra-fluoroethylene', *Proc. Roy. Soc.*, **A281**, 49−61, 1964.
67. R. P. Steijn, 'The sliding surface of polytetrafluoroethylene: an investigation with the electron microscope', *Wear*, **12**, 193−212, 1968.
68. H. U. Mittman and H. Czichos, 'Reibungsmessungen und Oberflächen-untersuchungen an Kunstoff-Metall Gleitpaarungen', *Materialprüf.*, **17**, 366−372, 1975.
69. K. Tanaka, Y. Uchiyama, and S. Toyooka, 'The mechanism of wear of polytetrafluoroethylene', *Wear*, **23**, 153−172, 1973. (a) R. C. Bowers and W. A. Zisman, 'Frictional properties of tetrafluoroethylene-perfluoro (propyl-vinyl ether) copolymers', *2nd. Eng. Chem. Prod. Res. Develop.*, **13**, 115−118, 1974. (b) D. L. Hunston, J. R. Griffith, and R. C. Bowers, 'Surface properties of fluoro-epoxies', A.C.S. reprints, *Plastics and Plastic Coatings*, March, 1977.
70. G. C. Pratt, 'Recent developments in polytetrafluoroethylene-based dry bearing materials and treatments', *Trans. J. Plastics Inst.*, **32**, 255−271, 1964.
71. B. J. Briscoe, A. K. Pogosian, and D. Tabor, 'The friction and wear of high density polythene: the action of lead-oxide and copper-oxide fillers', *Wear*, **27**, 19−34, 1974.
72. B. J. Briscoe and M. Steward, 'The wear of carbon-filled PTFE in controlled environments', Paper 30, 3rd Leeds−Lyon Symposium on Tribology, Leeds, Sept., 1976.
73. D. Dowson, J. M. Challen, K. Holmes, and J. R. Atkinson, 'The influence of counterface roughness on the wear rate of polyethylene', Paper 14, 3rd Leeds−Lyon Symposium on Tribology, Leeds, Sept., 1976.

2

Some Aspects of the Autoadhesion of Elastomers

B. J. Briscoe

Cavendish Laboratory, University of Cambridge

1. INTRODUCTION

If two solid bodies are brought into contact a finite force or energy is generally required to separate them again. This is the phenomenon of adhesion. If the solids have molecularly smooth and clean surfaces then the strength of the interface will be comparable with the cohesive strengths of the two bodies. Most solid surfaces are neither molecularly smooth nor clean, and as a consequence we do not observe strong adhesive junctions in our common experiences. For example, two iron nails do not spontaneously adhere when brought into contact in air. If, however, the nails are carefully polished and cleaned in ultra-high vacuum (*ca.* 10^{-10} Torr) and the same experiment is carried out in vacuum the measured adhesion is extremely large and the junction has a strength comparable with the tensile strength of iron.[1] Close examination of the contact after the experiment often reveals that failure has occurred in the bulk iron. The adhesion of elastomers to themselves and other solids is controlled by similar, but not identical factors to those which governed the adhesion of iron, but there are three important differences. First, elastomers or rubbers are van der Waals solids and they have low surface free energies (*ca.* 30–40 mJ m^{-2}; the value for iron is about 1500 mJ m^{-2}) and are not readily contaminated by the environment. Thus, simple and meaningful experiments can be carried out in the open laboratory. Second, as rubbers are quite compliant they can conform to gentle irregularities on adjacent solid surfaces. It is thus possible to achieve molecular areas of contact which are close to the apparent area of contact. Finally, since rubber can accommodate large reversible strains, it is possible to conduct apparently wholly reversible adhesion experiments where the locus of failure is precisely defined close to or at the original interface. Cohesive failure within the elastomer is not normally observed. For these reasons elastomers are particularly suitable for the study of certain general aspects of adhesion, particularly those factors which influence the adhesion of other elastic solids such as diamond, titanium carbide, and organic glasses.

The adhesion of ductile materials such as iron presents other problems which are not relevant here. In addition the adhesion of elastomers is an interesting problem in its own right and has practical importance in the science and technology of structural adhesives, although this aspect is not considered here. This paper reviews some of the factors which influence the autoadhesion or 'tack' of elastic and viscoelastic solids. The strength of an adhesive junction is described by one of two means. The most fundamental is the work required to destroy or create unit apparent area of adhesive junction under specific conditions. This is termed the 'energy release rate' by Andrews. Alternatively, we may use the force per unit crack length required to maintain a constant peeling velocity. The two definitions are equivalent. Although adhesion between solid bodies arises because of the interaction of surface forces, the magnitude of the measured adhesion often depends critically on the other factors. These are the bulk viscoelastic properties of the elastomer, the environment, and the surface topography of the solids. Surface forces interact with or are attenuated by these factors to provide a wide range of adhesion strengths. In the absence of these factors the strength of elastomer adhesive junctions would all be in the range 70–100 mJ m^{-2}. In practice the strengths range from near zero to several joules per square metre. Where specific chemical forces are created at the interface the strengths may be many kilojoules per square metre.

This paper will concentrate on the empirical and theoretical aspects which can account for this range of strength. Three sections deal with the phenomenology of static contacts, theoretical aspects of the adhesion of static contacts, and the origin of surface forces and surface free energy. This is followed by three sections which deal with dynamic and viscoelastic nature of adhesive strengths. Two final sections cover the influence of fluids and surface topography on the strength of adhesion. We begin with a description of perhaps one of the simplest adhesion experiments.

2. THE SIMPLE EXPERIMENT

If an optically smooth, clean, spherically shaped rubber specimen is brought into contact with an optical glass flat, Figure 1(a), then the stress–displacement curve obtained is shown by Figure 1(b).[2] At A, the surfaces are separated. They make contact at B and the surfaces are pulled together along BC by the action of surface forces. At C the loading stress becomes tensile and the further compression follows CD. If the surfaces are now pulled apart the line DCE is followed until the surfaces separate at E with a critical tensile force P_c. If the rubber surface is dusted with talc to reduce the adhesion the simple Hertzian stress–displacement curve BG is reproduced. We also note that P_c is:

(a) a strong function (generally increasing) of the rate of change of displacement;
(b) influenced by the length of time (dwell time) that the solids are held at position D; longer contact times usually increase P_c;
(c) reduced by the presence of fluids (if we neglect peripheral capillary effects);
(d) dependent on the roughness of the counterface;
(e) sometimes a function of the mechanical properties of the polymer;
(f) related to the surface free energy of the solids;

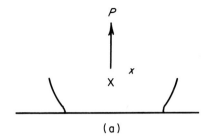

(a)

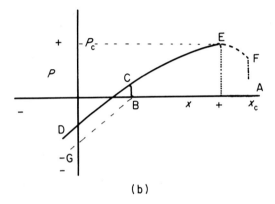

(b)

Figure 1. (a) The contact of an elastic sphere against a rigid flat. The normal load is P, and x is the displacement of a point X in the bulk of the elastomer. (b) Load against displacement curves for Figure 1(a). The curve is somewhat schematic, but closely resembles that obtained by experiment. Johnson[1,2] has computed theoretical curves which are similar in shape. In the presence of surface forces, when the surfaces are brought together (decreasing x) along AB they make contact at B. The load then becomes tensile (positive). Further loading turns the load into a compression along CD. Unloading (increasing x again) first gives a tensile load and at E the contact between the elastomer and the glass breaks at a maximum load, P_c. Under some circumstances, for example with rough surfaces, the contact may fail at a critical value of displacement at F, although the portion of the curve EF has not been observed directly.

In the absence of surface forces the load—displacement curve follows the Hertzian analysis along BG in a reversible manner.

(g) directly proportional to the mutual radius of curvature of the surfaces if the surfaces are smooth;

(h) sometimes a function of the loading history (the level of stress at D, Figure 1(b)) and the presence of electrostatic charges.

These points will be discussed at greater length shortly, but before doing so we will mention another aspect of this simple experiment. Johnson et al.[2] have carried out a comprehensive series of measurements similar to those described above. The experiments were carried out on optically smooth surfaces and the radius of the contact area was measured as a function of the normal load under equilibrium conditions. They observed that a finite force was required to separate the surfaces, but in addition the area of contact was significantly greater than that predicted by the Hertz theory[3] which neglects the effect of adhesive forces. The surface forces were indeed sufficient to produce a finite contact area under zero applied load. Their theoretical treatment of this problem is outlined in section 3 along with alternative approaches presented by other authors.

3. THEORETICAL ANALYSIS OF THE CONTACT OF SMOOTH ELASTIC SOLIDS IN THE PRESENCE OF ADHESIVE FORCES

There are two approaches to this problem: those based on considering the action of surface forces, and those which treat the problem in terms of surface energetics. Tabor[4] has recently reviewed those theories based on the action of surface forces. These theories include those due to Tomlinson,[5] Bradley,[6] and Derjaguin et al.[7] Those due to Tomlinson and Bradley simply sum the surface forces in the contact region, while Derjaguin et al. consider the balance of compressive elastic forces in the contact and the attractive forces in the region of close approach of the surface just outside the contact. They all predict that the pull-off force, P_c, is proportional to the mutual radius of curvature of the surfaces and independent of the moduli of the solids. This is confirmed by experiment. The Derjaguin analysis also predicts that the area of contact should exceed the Hertzian value, but unfortunately this analysis does not accurately fit the published experimental data.[2,8] The analysis developed by Johnson et al.[2] based on surface energy and contact mechanics is much more successful in this respect. In contrast to the Derjaguin approach, their method assumes that surface forces only operate *within* the contact region: in the region just outside the contact the interactions are negligible. Tabor[4] has likened the adhesive process (of a rubber sphere on a rigid flat) to stitching the surfaces together with a thread of limited extensibility. If within the contact the separation of the surfaces has a uniform value (perhaps a few tenths of a nanometre), then as one moves out from the centre of the contact the stress on the thread increases until it breaks at the edge of the contact. Contact mechanics provides a unique solution for this problem[2] and the stress distribution for zero contact load of an elastic sphere on a rigid flat is shown in Figure 2. The analysis in fact minimizes the total energy of the contact which arises from three contributions: mechanical potential energy, stored elastic energy, and surface free energy. Three results of this analysis are of interest.

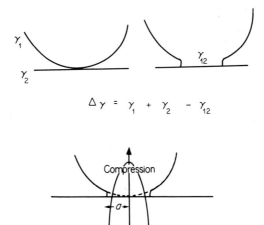

$$\Delta \gamma = \gamma_1 + \gamma_2 - \gamma_{12}$$

Figure 2. (a) Schematic diagrams of a section through a contact of a rubber sphere on a rigid flat under zero load. In the presence of surface forces (solid line) the contact area is finite and the contact angle at the rubber–glass–air triphase line is close to $\pi/2$. The stress distribution (after Johnson et al.[2]) is shown: at the centre is a modified Hertzian compressive stress while at the edge of the contact tensile stresses are operating. Under these conditions of zero load the tensile and compressive stress are equal. In the absence of surface forces the area of contact is zero and the contact angle measured through the rubber is 2π, broken line (Hertz analysis). These are theoretical calculations which have been substantially confirmed by experiment. (b) Definition of the thermodynamic work of adhesion $\Delta\gamma$ after Dupré. The γ's are surface free energies which are also surface tensions for liquids and perhaps elastomers.

(i) The predicted profile of the contact differs markedly from the Hertzian case without surface forces (Figure 2) in that the contact angle at the rubber–glass–air junction is $\pi/2$, not zero. A contact angle $\pi/2$ would lead to infinite stresses at the triphase line and one anticipates values rather less than $\pi/2$ (see later).

The height, h', of the characteristic 'neck' is, to a good approximation, given by[4]

$$h' \approx (R\Delta\gamma^2/E^2)^{1/3} \tag{1}$$

where R is the mutual radius of curvature of the surfaces $[R_1 R_2/(R_1 + R_2)]$, $\Delta\gamma$ the change in surface free energy on forming the contact (see below) and E is Young's modulus. This expression is useful for testing the assumption that the interaction of surface forces is not important just outside the contact.[4]

(ii) The radius of contact, a, in the presence of surface forces is given by

$$a = \left\{ \frac{R}{K} [P + 3\Delta\gamma\pi R + (6\Delta\gamma\pi RP + \langle 3\Delta\gamma\pi R \rangle^2)^{1/2}] \right\}^{1/3} \tag{2}$$

where P is the normal load and K is

$$(4/3) \left[\frac{1 - \nu_1^2}{E_1} + \frac{1 - \nu_2^2}{E_2} \right]$$

where ν is Poisson's ratio. Equation (2) has been found accurately to describe experimental data for a chosen value of $\Delta\gamma$ in several studies.[2,8] When $\Delta\gamma = 0$ (no adhesive forces) equation (2) reverts to the classical Hertz equation.

(iii) The analysis predicts that the critical force, P_c, required to separate the bodies is

$$P_c = \tfrac{3}{2}\pi R\Delta\gamma \tag{3}$$

The rigorous testing of this analysis requires an accurate and independent value of $\Delta\gamma$. These values are not available. In Figure 2, $\Delta\gamma$ is

$$\Delta\gamma = \gamma_1 + \gamma_2 - \gamma_{12} \tag{4}$$

which is the familiar Dupré equation. For many rubbers in contacts with themselves P_c is consistent with a value of $\Delta\gamma$ between 70 and 90 m Jm^{-2} (neglecting for the moment those other factors listed in section 2). In this case $\gamma_1 = \gamma_2$, and one can by definition put $\gamma_{12} = 0$ (the work of cohesion) and thus γ_1 is in the range 35–45 m Jm^{-2}. As this value is close to the maximum value of the surface tension of liquids which wet the rubber (the critical surface tension of wetting)[9] it is thought to be reasonable.

Savkoor[10] treated this problem in terms of a stress intensity factor and his work leads to similar results to those quoted in sections 2 and 3 above. In addition, he has considered the influence of tangential stresses on the area of contact. The application of tangential stress increases the stored elastic energy at the expense of surface energy and the area of contact decreases.[11]

Johnson[12] and Fuller and Tabor[39] have extended the original analysis to describe the form of Figure 1. Part of the analysis is shown in Figure 1. Here P_c is the maximum adhesive load and x_c is the maximum adhesive strain. When two smooth adhesive bodies are pulled apart by increasing x, the load, P follows DC until at E, P is a maximum P_c, and the contact fails. Under certain conditions the contact may also fail at F at a maximum strain x_c. We will return to this curve when we discuss the adhesion of rough surfaces.

Before we leave this brief description of the theoretical aspect it is worth remembering that we have been considering the special and perhaps hypothetical case of the adhesion of smooth ideal elastic bodies in a reversible adhesion process. In addition the equations (2) and (3) apply where $R \gg a$. Tabor[4] has considered this problem by making use of equation (1). The contact mechanical approach fails when h', the height of the 'neck', is such that surfaces can act strongly over this distance in the regions just outside the contact. For van der Waals forces this value is say 0.5 nm (for electrostatic forces which have longer range h' will be larger), and this gives a minimum value of R for a typical elastomer as 20 nm. In the event this analysis is limited by a restriction which requires a_0 (the contact radius under zero applied load) to be less than $R/2$, and this sets the limit of R at *ca.* 100 nm. Although the analysis has not been developed it is possible that equation (3) will have the form

$$P_c = \frac{3}{2}\pi R \Delta\gamma\phi\left(\frac{R\Delta\gamma^2}{E^2 h^2}\right) \qquad : \frac{R}{2} > a_0 \qquad (5)$$

where ϕ $(R\Delta\gamma^2)/(E^2 h^2)$ is a function which tends to unity for large values of $(R\Delta\gamma^2)/(E^2 h^2)$. Although these effects do not influence the large-scale experiments described here they are important in the study of the adhesion of small particles.

4. SURFACE FORCES AND SURFACE ENERGY

The treatment of Johnson *et al.* does not specify the nature of the surface forces responsible for adhesion. The success of the analysis has much to do with the use of the non-specific concept of surface free energy. Surface forces for elastomers can arise from (i) van der Waals forces, (ii) electrostatic forces, and (iii) hydrogen bonds. For rubber contacts which we hold together for long periods there is also the possibility of diffusion of chain segments across the interface. Which forces provide the major source of bonding is not clear, but since van der Waals dispersion forces, which are certainly present, can account for most of the observed values of $\Delta\gamma$ it seems that the contributions from other effects are not significant. Roberts[13] argues, for example, that electrostatic forces provide only up to 15% of $\Delta\gamma$ for many elastomeric contacts. Davis[14] and Derjaguin[15] maintain that electrostatic effects can often predominate.

A good deal of literature[16,17] exists on the influence of contact time and molecular mobility on autoadhesion (or tack). It seems certain that these processes are diffusion controlled and P_c (or something like it) is proportional to the square root of the contact time.

There are also many examples where adhesive forces may be significantly increased by chemical bonding.[18] This mechanism will not be considered here; however, further details are available in Chapter 3.

5. THE SIMPLE DYNAMIC EXPERIMENT

So far we have considered a quasi-static experiment and the force, P_c, required to completely separate a rubbers sphere from a rigid flat. This value of P_c gives a value of $\Delta\gamma$ via equation (3). In a typical experiment $\Delta\gamma$ would be found *ca.* $500-1000$ mJ m^{-2} and not the anticipated value of around $80-100$ mJ m^{-2}. The reason is that $\Delta\gamma$ is a strong function of the rate at which the surfaces peel apart. Normally, $\Delta\gamma$ increases with increasing crack-tip velocity, v. Unfortunately, there is not a satisfactory theoretical analysis for this phenomenon.

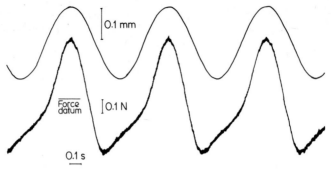

Figure 3. Force (P)–displacement (x) traces for the dynamic analogue of Figure 1(a), save that the surfaces are not completely separated. The displacement, x, is sinusoidal. The work of adhesion per unit area W_A is given by $(1/A)\int P \, dz$ which is a function of the peeling velocity, v, of the glass–rubber–air triphase line. The asymmetry of the force trace is due to the difference in magnitude of the works to make, W_M, and break, W_B, the rubber–glass contact, $W_A = W_B - W_M$. Trace from Briggs.[20]

If, in Figure 1, the rubber sphere is subjected to a sinusoidally varying displacement[19] normal to the counterface the contact area will change by ΔA during each cycle. The experiment is shown in Figure 3; the strain (or displacement) is sinusoidal but the force on the counterface shows severe distortion.[20] This is because the work to peel is much greater than the work to adhere (see later). The work dissipated per cycle, W_A, is

$$W_A(v) = \int F \, dx.$$

where v is the crack velocity (in this case the r.m.s. velocity). Figure 4 shows $W_A(v)$ as a function of v. The net work of adhesion, W_A, increases strongly with v. We shall return to this effect shortly, but first it is useful to review other types of adhesion experiments as they provide similar experimental data.

6. EQUIVALENT ADHESION EXPERIMENTS

Figure 5 shows eight adhesion experiments. Figure 5(a) is the peel test where F is the force required to peel the surfaces apart,[18,21-29] or θ the angle at which the surfaces spontaneously re-adhere.[23] Figures 5(b) and 5(c) are classical tests for measuring the strength of adhesive joints.[26-28] Figure 5(d) shows the bubble test[31-33] which we will use later in an amended form. Figures 5(e) and 5(f) depict two types of rolling experiment.[16,23,34,35] The free-rolling experiment (Figure 5(e)) is a very convenient method of studying adhesion at low peeling velocities. A simple dynamic adhesion experiment can be carried out at high peel rates by measuring the rebound height of a rigid sphere on an elastic solid.[34] In this experiment, as in Figure 5(e) bulk deformation sample losses can be estimated by coating the elastomer with talc to reduce the adhesion to negligible proportions.

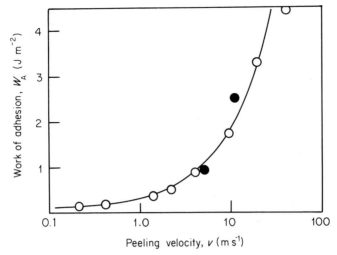

Figure 4. The net work of adhesion, W_A, as a function of the mean velocity, v, of the rubber–glass–air triphase line. Open circles are for the experiment described in Figure 3; the close circles are obtained from friction experiments and the study of Schallamach waves. The rubber is Dow Corning dielectric gell. Here, W_A often shows an increase as the peeling velocity, v, is increased. Data from Briggo and Briscoe.[19] (Reproduced by permission of Elsevier Sequoia S.A.)

Finally, Figures 5(h) and 5(i) show two experiments which involve the motion of macroscopic dislocations at elastomeric interfaces. Figures 5(h) illustrates the phenomenon of Schallamach waves[36] which will be referred to in section 10. Figure 5(i) shows a simple experiment due to Kendall[37] where a lubricated roller forces a 'ruck' along an interface in a manner which is not unlike the motion of a caterpillar when it moves along a leaf.

With the exception of (a) to (c) in Figure 5, all these tests correspond to the same process; the adhering and de-adhering of a contact. In each process W_A, the net work of adhesion per unit area, is

$$W_A = W_B - W_M \tag{7}$$

where W_M and W_B are, respectively, the works to make and break the unit of the contact. In experiment (a), Figure 5(a), W_B and W_M have been studied independently.

Before moving on to describe the factors which influence W_A it is worth describing the chemistry of the elastomers that have been studied and how they are fabricated. Several procedures are available, including those which use rubber cured at high temperature (*ca.* 160°C) and those which are cross-linked at room temperature. Roberts and Tabor[38] have used high-temperature curing of natural and synthetic rubbers in conjunction with optically smooth moulds. The other alternative is a small number of room-temperature-curing, silicone-based rubbers

34

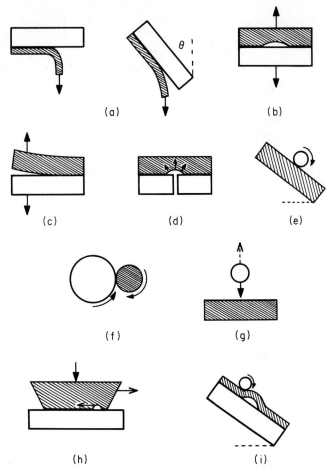

Figure 5. Various adhesion experiments which involve making and breaking elastomeric contacts. (a) Simple peel test for measuring either the work to break, W_B, or to make, W_M, unit area of contact.[23-30] (b) The growth of a crack in tension at an elastomeric interface.[27,28] (c) The propagation of a crack in an elastic solid or at an elastic interface.[27,28] (d) The 'bubble test', similar to (b) where a lens of gas or liquid is trapped at an interface.[31-33,46] (e) Free-rolling experiment where a rigid cylinder rolls down an inclined elastomeric plane.[23,35] (f) High-speed rolling machine or 'Tackometer'.[16] (g) Simple rebound.[34] (h) Schallamach waves or curves of detachment moving through an elastic interface.[19,34] (i) A special case of a macro dislocation being 'pushed along' an interface by a lubricated roller.[37] Experiments (e), (f), (g), (h), and (i) give W_A, while experiments (a), (b), (c), and (d) may give either W_M or W_B according to the sign of the imposed stress.

which have been used by Fuller and Tabor[39] and Briggs and Briscoe.[19] Roberts[40] has pointed out that certain rubbers are particularly prone to the formation of surface 'bloom' or surface impurity layers, presumably by diffusion from the bulk. Otherwise, the choice at present seems to depend upon the range of modulus and loss property required. Rubbers are interesting in that the usual correlation between mechanical strength and surface free energy does not hold. The elasticity is entropic in origin and the surface energy is a consequence of molecular interactions. The two are not directly related and this accounts for the lack of correlation.

7. THE INFLUENCE OF THE RATE OF PEELING ON THE WORK OF ADHESION

This aspect of adhesion was introduced in section 6 and is discussed in detail in Chapter 3. It has been the subject of many studies particularly by Kendal,[23] Andrews and Kinloch,[27] Gent and Schultz,[25] Roberts and Thomas[34] and Gent and Petrich[24] as well as many others. Figure 6 shows data for peeling (experiment of Figure 5(a)) which illustrates this effect again. Perhaps of more interest here is the data of Gent and Petrich[24] (again for peeling) shown in Figure 7. The reduced peeling force P (plotted as $296P/T$ after Ferry) is plotted against the reduced rate of

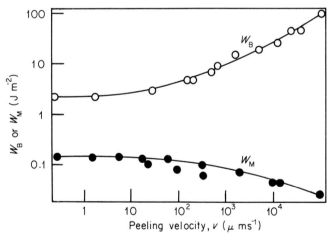

Figure 6. Works of breaking, W_B, and making, W_M, unit contact area of a rubber–glass interface in simple peeling and re-adhering (Figure 5(a)). The difference between W_B and W_M is W_A, the net work of adhesion. Here, W_B was measured after a dwell or contact time of 300 s. As the crack velocity, v, increases W_B increases but W_M decreases. The effect on W_B, is, however, much greater than that on W_M, and W_A shows a marked increase (see Figure 4). The rubber is cross-linked ethylene propylene. Data from Kendall.[23] (Reproduced by permission of Elsevier Sequoia S.A.)

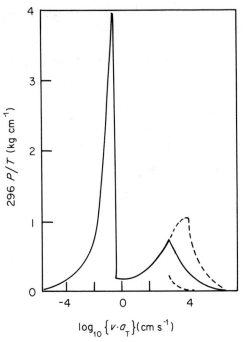

Figure 7. Master curve for peel force, P, reduced peeling velocity, $v a_T$. Here, a_T is the WLF relation given by $\log_{10} a_T = -17.4(T - T_g)/(52 + T - T_g)$ for a standard temperature of $23°C$. Using this relationship data for P against v for various temperatures, T, can be superimposed on the same master curve. The ordinate is plotted as $(296/T)P$. The fact that the data reduces to a single master curve is very good evidence for the viscoelastic nature of adhesion. The peak at low velocities is associated with a change in the nature of adhesive failure (from cohesive to interfacial failure as the velocity is increased). The peak at higher velocities has been attributed to a true bulk viscoelastic effect. Where P is a decreasing function of v, stick—slip or irregular peeling takes place. Data from Gent and Petrich.[24] (Reproduced by permission of the Royal Society.)

peeling Va_T to produce the familiar 'master curve' relationship found with many visocelastic measurements.

Clearly, the work of adhesion is a strong function of the viscoelastic properties of the polymer. Andrews and Kinloch have suggested on empirical grounds that

$$W_B = \Delta\gamma \cdot F(v) \tag{8}$$

where $F(v)$ is a temperature-rate parameter which follows the Williams, Landel, and Ferry (WLF) transform. If W_B is made up of two terms, the loss in surface energy, $\Delta\gamma$, and the viscoelastic losses, S, in the regions close to the edge of the contact, S, then we have

$$W_B = \Delta\gamma + S \tag{9}$$

so

$$F(v) = \frac{S}{\Delta\gamma} + 1 \tag{10}$$

Equations (8)–(10) are useful in that they specify the observed behaviour and rationalize the interaction of the surface and bulk properties. Other workers including Gent and Schultz,[25] and Kendall[41] have arrived at similar conclusions. Whilst this approach is useful it does not, however, allow the *a priori* calculation of W_A. Such a comprehensive solution of this problem, essentially that of the viscoelastic crack, is not available. Johnson[42] has suggested that since very high rates of deformation are involved at the crack tip it would be sensible to use a value of the high-frequency modulus in a Griffith-type analysis. This leads to $W_A \gg \Delta\gamma$, but not the detailed rate dependences observed in Figures 6 or 7. This problem remains unresolved for the time being.

There is one other important aspect of the dynamics of adhesion – the energy recovered on making unit area of contact, W_M. It is found that $W_M \ll W_B$ and Figure 6 shows data for W_M as a function of contact time (or adhering velocity, using method (a) Figure 5). It is the difference between W_B and W_M (equation (7)) which governs the response of the experiments depicted in Figures 5(e), (f), (d), (g), (h), and (i). In this review we call this the net work of adhesion, W_A. Why $W_B > W_M$ is not clear, and it is also uncertain whether W_B tends to W_M at low peeling velocities. Some workers maintain that at very low velocities $W_B = W_M$. Others feel that this is not so and that W_B is always rather greater than W_M. For a perfectly elastic system one's intuition favours $W_A = 0$ and thus $W_B = W_M$, but this may never occur in practice. The same uncertainty seems to exist in the similar situation where liquid drops have differing advancing and receding contact angles on solid surfaces. Indeed, the difference between W_B and W_M at low peeling rates may be in part a special case of contact-angle hysteresis. In a rather different context Kendall[29] has also noted the similarity in geometry of liquid–solid and elastomer–solid contacts. We also note that W_M increases with contact time in Figure 6. Whether this is a general result is not known, but it is associated with creep or permanent 'set' increasing the contact area in some experiments. In other cases the effect may be due to diffusion of polymer segments across the interface.

The dynamics of the adhesion process are also found to depend upon the previous loading history of the contact. Roberts and Thomas[34] have studied this effect in detail.

In those experiments where W_A or W_B increase with peeling velocity we have a stable experiment. This is why we measure a constant rolling velocity in experiment (e) Figure 5. If W_A or W_B decreases with increasing peeling velocity we will see irregular or 'stick–slip' motion. This effect was clearly observed by Gent and Petrich (Figure 7). The treatment of this problem, the interaction of a viscoelastic machine and a velocity-dependent variable, has been treated by many authors for the case of sliding friction and the occurrence of stick–slip motion.[43,44]

8. ADHESION IN THE PRESENCE OF FLUIDS

It is clear that many liquids can modify W_A and W_B. Gent and Schultz[25] have measured W_B, using a peel test in the presence of various liquids. Johnson et al.[2] also estimated $\Delta\gamma$ using equation (3) by measuring P_c in the presence of water and aqueous sodium dodecyl sulphate. In the case of water which had a finite contact angle of ca. 60° on their rubber, they were able to do the following experiment. For two rubber spheres in air $\Delta\gamma_S$, was found to be 71 ± 4 m Jm^{-2}; in water, $\Delta\gamma_L$ was 3.4 ± 0.2 m Jm^{-2}. These experiments give (Figure 2)

$$\Delta\gamma_L = \gamma_{SL} + \gamma_{SL} - \gamma_{SS}$$

$$\Delta\gamma_S = \gamma_S + \gamma_S - \gamma_{SS}$$

where γ_{SL}, γ_S, and γ_{SS} are the free surface energies of the solid–liquid, solid–air, and solid–solid interfaces, respectively. If we assume that surface free energy is the same as surface tension we can test the Young equation (Figure 8(a)), where

$$\gamma_S = \gamma_{SL} + \gamma_L \cos \theta_1 \tag{11}$$

where γ_L is the surface tension of the liquid (in this case water, 72 m Jm^{-2}) and θ_1 the contact angle in Figure 8(a). Here θ_1 is calculated as $64 \pm 2°$ which is a unique confirmation of the Young equation. The experiment also suggests the equivalence of surface free energy and surface tension for elastomers. Surface tension and surface free energy are equivalent for liquids, but this is not generally the case for solids.[45] For dilute sodium dodecyl sulphate solution equation (11) predicts wetting which is indeed observed. Briscoe and McClune[46] have extended these ideas and also included the case where a liquid drop on a lens is trapped at an elastomeric interface. Figure 8(b) shows the case where a liquid lens is trapped at an elastomer glass interface. If the lens is circular in plan with a radius of a and height of ω, then the two dimensions are approximately related by[46]

$$\omega^2 \simeq 1.9a \frac{\Delta\gamma}{E} \tag{12}$$

This relationship was developed for the simple treatment of the bubble test for

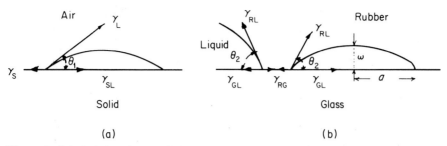

Figure 8. (a) A drop of partially wetting liquid at a solid surface. (b) The contact of a rubber sphere at a glass surface in the presence of a fluid. A liquid lens of radius a and height ω is shown.

adhesive strength (Figure 5(d)) where surface work is balanced against the elastic work. A plot of ω against $a^{1/2}$ is shown in Figure 9 for perfluorodecalin, at a butyl rubber–glass interface. From the slope of the line in Figure 9 we may calculate $\Delta\gamma$ and compare the value with that obtained in simple pull-off tests in the presence of fluids, using equation (3). The agreement between the two methods is only fair,

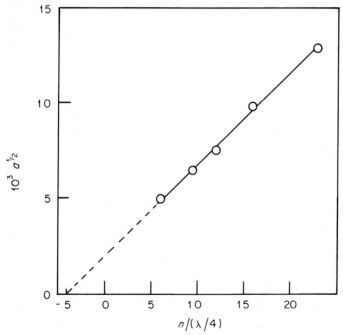

Figure 9. The height, ω, of liquid lenses trapped at a rubber–glass interface in the presence of various fluids plotted as a function of the square root of the mean radius, $a^{1/2}$ (Figure 8(b)). The height, ω, measured by optical interference is plotted in units of $\lambda/4$, where $\lambda = 400$ nm. This experiment clearly shows the difficulty in obtaining θ_2 (Figure 8(b)) by interference methods. Data from Briscoe and McClune.[46]

reflecting the rather inexact form of equation (12). However, the data do show that in the presence of fluids $\Delta\gamma$ is reduced; the values for $\Delta\gamma$ in air, perfluorodecalin, and silicone fluids are 70, 45, and 3 m Jm^{-2}, respectively. Since it is difficult to change $\Delta\gamma$, for dry rubber contacts by more than 30%, the use of fluid affords a useful method of changing $\Delta\gamma$. Gent and Schultz[25,47,48] have also shown that W_B for constant temperature and peeling rate is proportional to a 'theoretical' value of $\Delta\gamma$.

The estimation of 'theoretical' values of $\Delta\gamma$ is interesting and requires the use of the Dupré and Young equations (equations (4) and (11), respectively.) In Figure 8 we have

$$\Delta\gamma = -(\gamma_R - \gamma_L \cos \theta_1)(1 - \cos \theta_2) \tag{13}$$

by combining equations (4) and (11). The sign of $\Delta\gamma$, which may be regarded as a spreading coefficient, governs whether the rubber or the liquid wets the glass surface. If $\Delta\gamma$ is negative the rubber 'wets' the counterface and an appreciable force is required to separate the rubber from the glass. When $\Delta\gamma$ is positive the fluid wets the glass and there is no adhesion of the rubber to the counterface. This system has all the features of the system where two fluids are in contact with a solid surface and they both compete to wet the solid. The competitive wetting analogy may be taken further to provide a possible explanation for the difference between W_B and W_M at low peeling rates. If at low peeling rates $W_B \rightarrow \Delta\gamma_B$ and $W_M \rightarrow \Delta\gamma_M$ then adhesive hysteresis may arise because of hysteresis in the contact angle, θ_2. The contact angle, θ_2, is difficult to measure accurately and this hypothesis has not been tested. Measurements of θ_2, using interference techniques are unreliable, although angles in the range 40–70° have been estimated using the method. Experimental values of γ_R, γ_L, θ_1 give similar values of θ_2 when inserted into equation (13).

9. THE INFLUENCE OF SURFACE ROUGHNESS

Figure 10(a) shows the data of Fuller and Tabor for the influence of surface roughness on the adhesion of three spherical smooth rubber surfaces against various rough Perspex flats. The experimental method is that shown in Figure 1. It is seen that as the c.l.a. roughness is increased the adhesion is reduced and the reduction is proportionally greater for the rubbers of higher modulus. In contrast to the data obtained for smooth surfaces the measured adhesion is *not* a detectable function of the mutual radii of curvature of the contact. Fuller and Tabor[39] and Johnson[12] have analysed this problem theoretically by considering the balance between surface forces pulling the surface together and a few high asperities pushing the surface apart. Detailed analysis shows that the adhesion is a function of a dimensionless parameter, α, where

$$\alpha = \frac{4\sigma}{3} \left(\frac{4E}{3\pi\beta^{1/2} \Delta\gamma} \right)^{2/3} \tag{14}$$

and where σ and β are the standard deviation of asperity heights and the mean

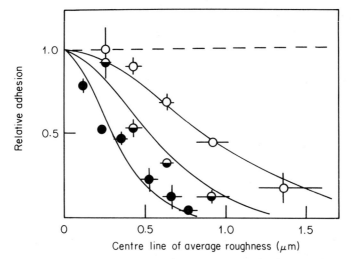

Figure 10. The adhesion of a smooth spherical rubber surface to a rough flat surface (experiment shown in Figure 1). The normalized pull-off force (pull-off force for a rough surface divided by the value for a smooth surface) as a function of the centre line average roughness of the flat surface. As the roughness increases the normalized adhesion decreases. The effect of roughness is more pronounced for the harder rubbers: open circles, $E = 2.2 \times 10^5$ Nm^{-2}; half filled circles $E = 6.8 \times 10^5$ Nm^{-2}; and filled circles $E = 2.4 \times 10^6$ Nm^{-2}. The lines are the computed curves taken from Fuller and Tabor[39] for the respective moduli. For this type of roughness the distribution of scale heights is Gaussian and the centre line average roughness is approximately equal to the standard deviation of the Gaussian distribution of height. In these experiments the surfaces were separated very slowly and thus the bulk viscoelastic contribution to pull-off force is very small. (Reproduced by permission of the Royal Society.)

radius of the asperities, respectively (Figure 10). The significance of equation (14) may be seen if we consider the relation

$$\alpha^{3/2} \propto \frac{E\sigma^{3/2}}{\beta^{1/2}\Delta\gamma} = \frac{E\sigma^{3/2}\beta^{1/2}}{\beta\Delta\gamma}$$

the denominator is a measure of the attractive adhesive force (see equation (3)) and numerator is, apart from a small numerical factor, the force required to push a sphere of a radius β on to elastomer to a depth σ. Figure 11 shows a comparison of this theory with the data of Briggs and Briscoe[35] on the rolling of roughened Perspex cylinders down smooth rubber inclines. Before considering the details of the comparison we note that by normalizing the data with respect to the smooth surfaces we can substantially remove any peeling rate, and beyond $\alpha = 3$ the adhesion is small.

By and large the agreement between theory and experiment is good, except that the enhanced adhesion at small roughnesses is not explained. Briggs and Briscoe[49] have been able to explain this effect in part by considering the force–extension curve derived by Johnson (Figure 1). For a single asperity contact the contact fails at E but for a multiple asperity contact some contacts fail at F. This enables more

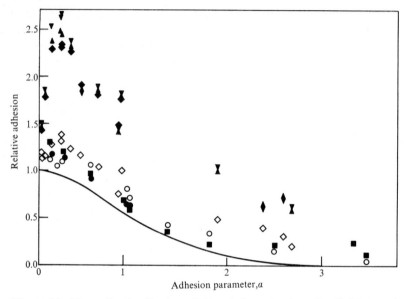

Figure 11. Normalized adhesion deduced from both pull-off (Figure 1) and rolling experiments (Figure 5(c)) as a function of the adhesion parameter (equation (13)). The data for each set of points were obtained over widely differing peeling velocities and normalized to the value for the nominally smooth surface. The pull-off data are shown as open symbols and the rolling data as closed symbols.

Pull-off data: $\circ$, RTV 602 ($E = 4.9 \times 10^5$ Pa); $\diamond$, Dow Corning XF-13-523 ($E = 6.3 \times 10^4$ Pa). In both cases the pull-off force for a smooth surface indicated a surface energy of $\Delta\gamma \simeq 100$ mJ m^{-2}.

Rolling resistance data for cylinders (radius 5 mm, length 10 mm) rolling down an inclined smooth rubber surface at the following speeds (mm s^{-1}): RTV 602: $\bullet$, 2; $\blacksquare$, 4; Dow Corning XF-13-523: $\blacklozenge$, 0.5; $\blacktriangle$, 1; $\blacktriangledown$, 2.

The net adhesion energy ($W_A = W_M - W_B$) is the net work required for a complete cycle of adhering together and then peeling apart unit area of interface and is deduced directly from the rolling resistance. It is markedly velocity dependent. The absolute values for nominally smooth surfaces, to which the above points are normalized, are respectively (mJ m^{-2}): 600, 970, 253, 384, and 612. Dusting the rubber surface with talcum powder to remove the adhesion reduced the relative rolling resistance to a very small value, indicating that bulk ploughing hysteretic losses were almost negligible in these experiments. Continuous line from theoretical treatment of Fuller and Tabor[39] and broken line the modified theoretical analysis of Briggs and Briscoe.[49] (Reproduced by permission of *Nature*.)

work to be dissipated in the failure process, and it turns out that this effect will be significant for $\alpha = \frac{1}{3}$ and that P_c (equation (3)) can now have a value which is almost 1.4 times greater than for smooth surfaces. This analysis was also extended to include a Rayleigh distribution of radii, the computed curve being shown in Figure 11. This approach accurately predicts the position of the adhesion maximum, but not the whole of the pronounced increase in adhesion. The agreement is less favourable for the softer rubbers. Apparently the interaction of surface forces is more complex than the simple model supposes. Thus, while the gross viscoelastic effects may be removed by normalization, significant viscoelastic effects seem to remain, particularly with the softest rubbers. A more serious difficulty also arises from the choice of $\Delta\gamma$ and not W_B (equation (3)) in equation (14) and in Figure 11. The analysis with W_B for $\Delta\gamma$ is intuitively the correct choice, but the comparison of theory and experiment is far better with the constant and much smaller value of $\Delta\gamma \approx 70$ m Jm^{-2}.

Apart from changing the absolute magnitude of the measured adhesion, surface roughness may also introduce significant differences in the apparent advancing and receding contact angle, θ_1 (Figure 8a). This is found to be the case for liquids on rough solid surfaces[50] and it is interesting to speculate whether this will increase the magnitude of the adhesive hysteresis, W_A.

10. FRICTION AND SCHALLAMACH WAVES

There is one aspect of the sliding friction of elastomers which is closely related to adhesion. This is when waves of detachment or Schallamach waves[36] traverse the contact (Figure 5(h)).[8,19,34] These waves resemble macro dislocations and no true sliding takes place in the contact. The passage of one wave may be thought of as the experiment shown in Figure 1 where a contact is broken and reformed. Thus, if F is the frictional force and V the imposed sliding velocity

$$FV = AW_A(v)f \tag{14}$$

where A is the area of contact and f the wave frequency. If we may measure f, A, F, and also the velocity, v, of the waves we may compute $W_A(v)$ and compare it with values obtained in experiments such as that shown in Figure 3. This is done in Figure 4 and we see that the agreement is good, indicating that most of the friction force may be accounted for in terms of adhesive losses. It turns out in practice that for small changes in V, F is almost independent of V; for as V increases, v increases also, and so does W_A, but the ratio f/v, the reciprocal of the Burgers vector, decreases almost exactly in proportion and F thus remains nearly constant. The studies of Barquins and Coutel[8] which cover a much wider range of temperature and velocity do, however, show that this mechanism can account for the viscoelastic nature of the sliding friction of rubbers. Generally, the frictional force is found to fit a 'master curve' similar to that shown in Figure 7 for adhesion.

11. CONCLUSIONS

It is clear that the adhesion of elastomers is governed by many interdependent factors. Surface forces are responsible for adhesion, but the magnitude of the adhesion is often governed by other factors such as the viscoelastic response of the rubber, the environment, and the topography of the solids. The behaviour of static contacts, both rough and smooth, seems to be quite well understood and theory and experiment are in general accord. The response of dynamic contacts is well documented but still awaits a satisfactory theoretical analysis. Indeed, the detailed interaction of adhesive forces and the bulk viscoelastic properties of the hinterland adjacent to the interface still presents the most important unresolved problem in the adhesion of elastomers.

ACKNOWLEDGEMENTS

The author is grateful to Professor D. Tabor, Professor K. Johnson, Dr. K. Kendall, Dr. J. Greenwood, and Dr. G. A. D. Briggs for valuable discussions. He also thanks the Ernest Oppenheimer Fund for the provision of a Research Fellowship.

REFERENCES

1. N. Gane, P. F. Pfaelzer, and D. Tabor, 'Adhesion between clean surfaces at light loads', *Proc. Roy. Soc. (London)*, **A340**, 495–517 (1974).
2. K. L. Johnson, K. Kendall, and A. D. Roberts, 'Surface energy and the contact of elastic solids', *Proc. Roy. Soc. (London)*, **A324**, 310–331 (1971).
3. S. Timoshenko and J. N. Goodier, *Theory of Elasticity*, McGraw-Hill, New York (1951).
4. D. Tabor, 'Surface forces and surface interactions', *J. Colloid and Interface Science*, **58**, 2–13 (1977).
5. S. Tomlinson, 'Molecular cohesion', *Phil. Mag.*, **6**, 695–712 (1928).
6. R. S. Bradley, 'The cohesive force between solid surfaces and the surface energy of solids', *Phil. Mag.*, **13**, 853–862 (1932).
7. B. V. Derjaguin, M. M. Müller, and P. Yu. Toporov, 'Effects of contact deformations on the adhesive particles', *J. Colloid and Interface Sci.*, **53**, 314–326 (1975).
8. M. Barquins and R. Courtel, 'Rubber friction and rheology of viscoelastic contacts', *Wear*, **32**, 133–150 (1975).
9. W. A. Zisman, 'Relation of equilibrium contact angle to the liquid and solid constitution', in *Contact Angle Wettability and Adhesion*, ed. F. Fowkes. A.C.S., *Advances in Chemistry*, series 43, pp. 1–52 (1964).
10. A. R. Savkoor, 'Adhesion and Friction of Elastic Solids in Contact', *Techn. Hogeschool Delft, Lab voor Voertigtechniek, Report*, p. 159 (1973).
11. A. R. Savkoor and G. A. D. Briggs, 'The effect of tangential force on the contact of elastic solids in adhesion', *Proc. Roy. Soc. (London)*, submitted for publication.
12. K. L. Johson, 'Non-Hertzian contact of elastic spheres', in *The Mechanics of the Contact between Deformable Bodies*, ed. A. D. de Pato and J. J. Kalker. Delft University Press, pp. 26–40 (1975).
13. A. D. Roberts, 'Nature of adhesive forces in rubber friction' in *Aspects of Adhesion*, to be published.

14. K. Davis, 'Surface charge and the contact of elastic solids', *J. Phys. D: Appl. Phys.*, **6**, 1017–1024 (1973).
15. B. V. Derjaguin *et al.*, 'Investigation on the Adhesion of polymer particles to the surface of a semiconductor', *J. Adhesion*, **4**, 65–71 (1972).
16. R. P. Campion, 'The influence of structure on autoadhesion and other forms of diffusion into polymers, *J. Adhesion*, **7**, 1–23 (1975).
17. F. Bueche, W. Cashin, and P. Debye, 'The measurement of self diffusion in Solid Polymers', *J. Chem. Phys.*, **20**, 1956–58 (1952).
18. A. Ahagon and A. N. Gent, 'Effect of interfacial bonding on the strength of adhesion', *J. Polymer Sci., Polymer Physics Edition*, **13**, 1285–1300 (1975).
19. G. A. D. Briggs and B. J. Briscoe, 'The dissipation of energy in the friction of rubber, *Wear*, **35**, 357–364 (1975).
20. G. A. D. Briggs, 'How rubber sticks and slides', Ph.D. Thesis, University of Cambridge, 1976.
21. K. Kendall, 'Shrinkage and peel strength of adhesive joints', *J. Phys. D: Appl. Phys.*, **6**, 1782–1787 (1973).
22. K. Kendall, 'The shapes of peeling solid films', *J. Adhesion*, **5**, 105–117 (1973).
23. K. Kendall, 'Rolling friction and adhesion between smooth solids', *Wear*, **33**, 351–358 (1975).
24. A. N. Gent and R. P. Petrich, 'Adhesion of viscoelastic materials to rigid substrates', *Proc. Roy. Soc.*, **A310**, 433–448 (1969).
25. A. N. Gent and J. Schultz, 'Equilibrium and non-equilibrium aspects of the strength of adhesion of viscoelastic molecules', *Proc. Int. Rubber Cong. Brighton*, I.R.I. (London), Paper Cl (1972).
26. A. N. Gent and A. J. Kinloch, 'Adhesion of viscoelastic materials to rigid substrates', *J. Polymer Sci.*, A2, **9** 659–668 (1971).
27. E. H. Andrews and A. J. Kinloch, 'Mechanics of adhesive failure', *Proc. Roy. Soc. (London)*, **A332**, 385–399 and 401–414 (1972).
28. E. H. Andrews and A. J. Kinloch, 'Mechanics of elastomeric adhesion', *J. Polymer Sci.*, Symposium No. 46, 1–14 (1974).
29. K. Kendall, 'Peel adhesion of solid films – the surface and bulk effects', *J. Adhesion*, **5**, 179–202 (1973).
30. A. J. Kinloch, W. A. Dukes, and R. A. Gledhill, 'Durability of adhesive joints', in *Adhesive Science and Technology*, Vol. **9B**, ed. L. H. Lee, Plenum Press, pp. 597–613 (1975).
31. M. L. Williams, 'The fracture threshold for an adhesive interlayer', *J. Appl. Polymer Sci.*, **14**, 1121–1126 (1970).
32. M. L. Williams, 'The relation of continuum mechanics to adhesive fracture', in Recent Advances in Adhesion, ed. L. H. Lee, Gordon and Breach, London (1973).
33. M. L. Williams, 'The continuum interpretation of fracture and adhesion', *J. Appl. Polymer Sci.*, **13**, 29–40 (1969).
34. A. D. Roberts and A. G. Thomas, 'Adhesion and friction of smooth rubber surfaces', *Wear*, **33**, 45–46 (1975).
35. G. A. D. Briggs and B. J. Briscoe, 'Effect of surface roughness of rolling friction and adhesion between elastic solids', *Nature*, **260**, 313–315 (1976).
36. A. Schallamach, 'How does rubber slide?', *Wear*, **17**, 301–314 (1971).
37. K. Kendall, 'Preparation and properties of rubber dislocations', *Nature*, **261**, 35–36 (1976).
38. A. D. Roberts and D. Tabor, 'The extrusion of liquids between highly elastic solids', *Proc. Roy. Soc. (London)*, **A325**, 323–345 (1971).
39. K. N. G. Fuller and D. Tabor, 'The effect of surface roughness on the adhesion of elastic solids', *Proc. Roy. Soc. (London)*, **A345**, 327–342 (1975).

40. A. D. Roberts and A. B. Othman, 'Rubber adhesion and the dwell time effect', *Wear*, **42**, 119–133 (1977).
41. K. Kendall, 'Effect of relaxation properties on the adhesion of rubber', *J. Polymer Sci., Polymer Physics Edition,* **12**, 295–301 (1974).
42. K. L. Johnson, private communication.
43. D. M. Rowson, 'An analysis of stick slip motion', *Wear*, **31**, 213–218 (1975).
44. C. A. Brockley and M. A. Green, 'Viscoelastic effects in boundary lubrication', *Nature,* **251**, 306–307 (1974).
45. See for example: D. Tabor, *Gases, Liquids and Solids*, Penguin Books London, p. 130 (1969).
46. B. J. Briscoe and C. R. McClune, 'The formation and geometry of entrapped liquid lenses at elastomer–glass interfaces', *J. Colloid and Interface Sci.*, to be published.
47. A. N. Gent and J. Schultz, 'Effect of wetting liquids on the strength of adhesion of viscoelastic materials', *J. Adhesion,* **3**, 281–294 (1972).
48. C. R. McClune, 'The properties of liquid films between highly elastic bodies', Ph.D. Thesis, University of Cambridge (1974).
49. G. A. D. Briggs and B. J. Briscoe, 'The effect of surface topography on the adhesion of elastic solid', *J. Phys. D: Appl. Phys.*, submitted for publication.
50. R. E. Johnson and R. H. Dettre, 'Contact angle hysteresis', in *Contact Angle Wettability and Adhesion*, ed. F. Fowkes. A.C.S., *Recent Advances in Chemistry*, series 43, pp. 112–145 (1964).

3

Surface Energetics and Adhesion

E. H. Andrews and N. E. King*

Queen Mary College, London

1. INTRODUCTION

When an adhesive bond is established between two materials the total surface energy of the system is diminished, this being the net effect of destroying two free surfaces and creating one new interface. If the atomic interactions across the new interface are secondary in nature (polar and dispersive only) the decrease in free energy is, by definition, the thermodynamic work of adhesion, w_A.

Larger decreases in free energy can occur if primary bonding is established across the interface or if entropy increases by the interdiffusion of the mating materials. On the other hand, if wetting is incomplete the average or apparent free-energy drop may be less than w_A. Further complications arise if the substrate is so 'rough' that mechanical 'keying' of the adhesive phase takes place.

Since adhesive failure results from the reverse process of debonding or separating the components of the adhesive joint, the minimum requirement is that the energy supplied to debond the system must exceed the free-energy drop that occurred on the establishment of the bond. This energy may be supplied mechanically, by the application of force, or physiocchemically by the arrival at the interface of migrating molecular species (e.g. water) which displace the adhesive molecules from the substrate. (This is only likely to occur for secondarily bonded interfaces, but is a potent threat to joints between hydrophobic and hydrophilic substances.)

In the light of this it was natural that attempts should be made to correlate adhesive strength, determined by mechanical tests, with w_A. Levine et al.[1] found a linear dependence of the tensile strength of adhesive joints on a somewhat different surface parameter, the critical surface tension γ_c.[2,3] The systems used were epoxy resins and polymeric substrates. Dahlquist[4] reported a direct relationship between the stress at the fracture front between adhesive and substrate (obtained from peel force data) and w_A, but similar work by Kaelble[5] using a rubber-like alkyl acrylate copolymer as the adhesive, failed to reveal a close correlation.

Correlations apart, the *magnitude* of the adhesive failure energy in mechanical tests is, almost invariably, very much larger than w_A. This simple fact has led some

*Now at Corporate Research and Development, Metal Box Ltd., Twyford Abbey Road, London, U.K.

authors to propose that adhesive bonds cannot be regarded merely as the result of close-range atomic interactions across a clearly defined interface. Voyutskii[6] therefore proposed interdiffusion of the bonding phases and Derjaguin[7] suggested electrostatic charge as the origins of joint strength. If these theories are true, the thermodynamic work of adhesion would be irrelevant to adhesive strength.

The first key to the role of w_A in determining adhesive strength was provided by Gent and Schultz in 1972.[8] Using a rubber-like adhesive bonded to a poly(ethylene terephthalate) substrate, they studied the effect on peel strength of immersion in various liquids. The difference Δ between w_A (in air) and w_{AL} (in a liquid) was calculated and plotted against the mechanical peeling energy as measured. A linear relationship was obtained and on the basis of this result they proposed that the adhesive failure energy could be factorized into two terms: one being the thermodynamic work of adhesion, w_A, and the other a factor arising from energy dissipation in the adhesive. The existence of the energy-dissipation element would explain both the fact that fracture energy always exceeds w_A and that (unlike w_A) it is strongly dependent on rate and temperature for many systems.

The remainder of this paper reviews, in some detail, the work carried out in the author's laboratory during the past five years and which has led to a theory of adhesion in which the roles of w_A and energy dissipation are explicitly defined. The theory has been well substantiated by experiment and the relevant results will be quoted.

2. THE DEFINITION OF SURFACE ENERGY

Classically, the surface energy of a solid is defined as half the energy required to separate unit area of two adjacent atomic planes and separate them to infinity in a vacuum. This definition, of course, assumes perfectly elastic behaviour, thermodynamic equilibrium, and the absence of any atmosphere. In practice none of these conditions normally apply, so that surface energies cannot be measured by fracture or cleavage experiments. We have already noted how energy-dissipation processes increase the energy required to produce new surfaces. A second factor is that atomic bond breakage is an activated process, as indicated in Figure 1. The equilibrium surface free energy is the quantity denoted ΔU on the diagram, but the energy required to activate the fracture process is the larger quantity U_{AB}. Although an amount of energy $(U_{AB} - \Delta U)$ is recovered as the system reverts to the state B (bond broken), this energy may not be returned to the strain field but appears instead as heat. From a mechanical point of view, therefore, the surface energy may be better defined by U_{AB} than by the equilibrium value ΔU.

Thirdly, of course, in most real situations surfaces are created in the presence of a fluid environment (gas, vapour, or liquid), and we are then concerned with interfacial energies rather than *in-vacuo* surface energies.

In the light of these comments we can now consider the surface energy parameters to be employed in this paper. The first of these is defined as the energy required to break the unit area of interatomic bonds across the fracture plane. We call this quantity the 'intrinsic fracture energy', $2\mathcal{T}_0$ if the fracture plane passes

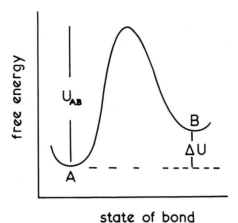

state of bond

Figure 1. Activated fracture of an interatomic bond. A, bond intact; B, bond
broken.

cohesively through a given solid. If the fracture occurs at an interface between two
solids we refer instead to the 'intrinsic adhesive failure energy' denoted θ_0.

The parameters $2\mathcal{T}_0$ (two, because two identical surfaces are produced by
fracture) and θ_0 are really identical in character and different symbols are only
used to provide differentiation, at a glance, between bulk fracture and adhesive
failure.

These intrinsic parameters are clearly of the nature of activation energies (like
U_{AB} in Figure 1) unless the fracture process is infinitely slow. They therefore have
minimum values equal to the thermodynamic equilibrium parameter (like ΔU in
Figure 1), but may exceed this value significantly. We can note at this point that
both U_{AB} and ΔU itself may be considerably smaller than for vacuum conditions if
the 'bond-broken' state B (Figure 1) is an oxidized or hydrated state as commonly
occurs in normal atmospheres. This is because the free-energy reduction caused by
reaction with environmental species lowers the free energy of state B by an equal
amount. Note also that Figure 1 applies whatever the nature of the interatomic
bond, i.e. primary or secondary, though, of course, ΔU and U_{AB} will be larger for
the stronger bonds.

Our second surface-energy parameter is that already defined as the thermodynamic work of adhesion. This quanitity was defined by Dupré[9] for a liquid (L)
on a solid surface (S) in a vapour (V) as

$$w_A = \gamma_S + \gamma_{SV} - \gamma_{SL} \tag{1}$$

where γ is the surface energy and the suffixes define the respective interfaces.

Extending this equation to an interface between two solids a and b, we have

$$w_A = \gamma_a + \gamma_b - \gamma_{ab} \tag{2}$$

However, the interfacial term γ_{ab} is given by theory (Owens and Wendt[10], Kaelble and Uy[11], Wu[12]) as

$$\gamma_{ab} = \gamma_a + \gamma_b - 2(\gamma_a^D \gamma_b^D)^{1/2} - 2(\gamma_a^P \gamma_b^P)^{1/2} \tag{3}$$

in which the dispersive (D) and polar (P) components of the surface energies are considered to interact only with their own kind. Combining equations (2) and (3) we thus obtain

$$w_A = 2(\gamma_a^D \gamma_b^D)^{1/2} + 2(\gamma_a^P \gamma_b^P)^{1/2} \tag{4}$$

Kloubek[13] examined values of w_A obtained from this equation (the 'extended Fowkes equation') and alternative theoretical equations and found no significant differences between the various predictions.

To obtain values for the dispersive and polar components of the solid surface energies, equation (3) is combined with the Young equation to give for a solid (S)/liquid (L) pair,

$$1 + \cos \alpha = \frac{2(\gamma_S^D \gamma_L^D)^{1/2}}{\gamma_L} + \frac{2(\gamma_S^P \gamma_L^P)^{1/2}}{\gamma_L} \tag{5}$$

where α is the contact angle.

If we now take a series of liquids $L_1, L_2, L_3 \ldots$ on one of our solid surfaces and measure $\alpha_1, \alpha_2, \alpha_3 \ldots$ we obtain a set of equations of the form (5) in which γ_S^D, γ_S^P are constants. If the dispersive and polar components of the surface energies of the various liquids are known (Fowkes[14] and Dann[15]), each pair of equations of the form of equation (5) yield solutions for γ_S^D, γ_S^P. These solutions are averaged after eliminating certain excessively large values (Kaelble[16]) and the process is repeated for the second solid involved in the adhesive joint. Then w_A can be calculated from equation (4).

This method is entirely suitable for organic solids whose surface energies are low (say $< 100 \text{ mJ m}^{-2}$), but lead to significant errors if applied to higher energy surfaces such as clean metals and oxides. This is because the analysis outlined above neglects the difference between the solid surface energy in, say, air or vacuum and the surface energy in the presence of the vapour of the test liquid. This difference (the so-called equilibrium spreading pressure, π_e) is negligible for organic surfaces but can be appreciable in the case of high-energy surfaces.

To take account of π_e requires modification of equation (5) to read

$$1 + \cos \alpha = \frac{2(\gamma_S^D \gamma_L^D)^{1/2}}{\gamma_L} + \frac{2(\gamma_S^P \gamma_L^P)^{1/2}}{\gamma_L} - \frac{\pi_e}{\gamma_L} \tag{6}$$

That is, the contact angle α increases with π_e. Equation (6) cannot be solved by the method outlined earlier if γ_S^D, γ_S^P, and π_e are all unknowns, but Harkins and Loesner[17] found that for organic liquids on metals the neglect of π_e gave results for w_A which were too low by 30–50% (true w_A values are obtained by simply adding the π_e term which is typically 30–40 mJ m^{-2} for organic vapour/metal systems).

To summarize, therefore, we have, for our present purposes, defined two surface-energy parameters.

(a) $\mathscr{T}_0$ or θ_0. The energy required to fracture the unit area of atomic bonds across a fracture plane. This quantity may represent the activation energy for bond rupture rather than an equilibrium 'surface' energy.

(b) w_A. The thermodynamic work of adhesion which is the equilibrium free-energy decrease (increase) on creating (separating) an interface between two substances which interact *only by secondary forces* (dispersive and polar).

It follows that θ_0 and w_A will only correspond if (i) there are no primary bonds across the fracture plane, (ii) the fracture is perfectly interfacial, and (iii) if θ_0 is measured under equilibrium conditions. Notice, however, that θ_0 remains a physically meaningful concept, even under conditions where it is not equivalent to w_A.

3. FRACTURE MECHANICS OF ADHESIVE FAILURE

A generalized theory of fracture mechanics proposed recently by the author,[18] gives the fracture energy (per unit area of newly created crack surface) as

$$\mathscr{T} = \mathscr{T}_0 \Phi(\dot{c}, T, \epsilon_0) \tag{7}$$

where $\mathscr{T}$ is the fracture energy (or critical apparent energy release rate) and Φ is a loss function dependent on crack velocity, temperature, and the strain, ϵ_0, at points remote from the crack. Furthermore, for a centre crack of length $2c$, in a tensile strain field,

$$\mathscr{T} = k_1(\epsilon_0)cW_{0(\text{crit})} \tag{8}$$

where W_0 is the applied (or input) energy density at points remote from the crack and $W_{0(\text{crit})}$ is the critical value of W_0 at which the crack propagates. The functions k_1 and Φ are further defined as follows:

$$k_1(\epsilon_0) = \sum_P q \left(x \frac{\partial f}{\partial x} + y \frac{\partial f}{\partial y} \right) \delta x \delta y \equiv \sum_P qg \, \delta x \delta y \tag{9}$$

where q is a negative constant (value -1 for linear solids), x, y are reduced cartesian coordinates, and f is the distribution function of energy density in the specimen. Summation is over all points P in the specimen. Also,

$$\Phi = k_1 / \left(k_1 - \sum_{PU} \beta g \, \delta x \delta y \right) \tag{10}$$

where β is the hysteresis ratio of the material at the point P (dependent on $\dot{c}$, T, and the local strain intensity). The summation 'PU' indicates summation only over points in the stress field which unload as the crack propagates.

The general validity of the theory has been demonstrated in a recent paper by Andrews and Fukahori[19] in which $\mathscr{T}$, $\mathscr{T}_0$, and Φ were evaluated independently and equation (7) shown to be correct.

The theory can be applied to the failure of an adhesive joint, as demonstrated by Andrews and Kinloch[20,21] and Andrews and King.[22] In this case the situation considered is shown in Figure 2 and the appropriate equation is

$$\theta = \theta_0 \, \Phi(\dot{c}, T, \epsilon_0) \tag{11}$$

$$\theta = k_1(\epsilon_0) c W_{0\,(\text{crit})} \tag{12}$$

where θ is the adhesive failure energy and the other terms are as previously defined. Note that equation (12) applies only to the centre-crack specimen shown in Figure 2. The adhesive failure energy θ can be measured using quite different test geometries and the appropriate equations (see Gent and Kinloch[23]). However, equation (11) is of general validity.

Equation (11) assumes that one of the solids comprising the joint (the solid referred to hereafter as the substrate), is rigid and elastic i.e. no energy is dissipated in this material by loading and unloading. Then Φ is the loss function for the other solid (i.e. the adhesive), and is the same quantity as appears in equation (7) for the cohesive fracture of this solid. Notice that we consider a single interface between substrate and adhesive, each of which is imagined to be present as a semi-infinite sheet.

From equations (7) and (11) we have,

$$\left.\begin{array}{l} \log 2 \, \mathcal{T} = \log 2 \, \mathcal{T}_0 + \log \Phi(\dot{c}, T, \epsilon_0) \\ \log \theta = \log \theta_0 \quad\; + \log \Phi(\dot{c}, T, \epsilon_0) \end{array}\right\} \tag{13}$$

and plots of $\log 2 \, \mathcal{T}$ and $\log \theta$ against $\dot{c}$ or T should thus be paralllel curves, i.e.

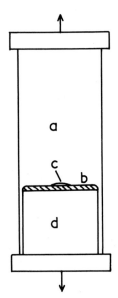

Figure 2. The simple-extension adhesive failure test specimen (a) adhesive sheet, (b) true substrate, (c) preformed crack, (d) rigid back-up substrate.

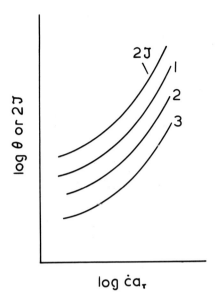

Figure 3. Predicted behaviour of cohesive and adhesive failure energies for the adhesive and a series of substrates 1,2,3. . . . , as functions of reduced rate of crack propagation (schematic).

identical curves mutually displaced along the log $2\mathscr{T}$ axis by an amount log $(2\,\mathscr{T}_0/\theta_0)$. The theory does not tell us the *form* of the curve unless we know β as a function of $\dot{c}$ or T. However, Φ must increase with increasing mechanical loss.

If the adhesive is a rheologically simple material, the time and temperature dependencies of Φ can be combined using the WLF superposition procedure,[24] i.e. by plotting log $2\mathscr{T}$ or log θ against log $\dot{c}a_T$ where a_T is the WLF shift factor. This procedure reduces data collected at different temperatures to a 'master curve' corresponding to a fixed temperature. In all the data reported below these 'master curves' are employed.

Returning to equation (13), we can now consider employing the same adhesive (thus the same Φ function) bonded to a variety of rigid substrates 1, 2, 3, etc. Then we should obtain a *series* of parallel curves (see Figure 3) with shifts, relative to the cohesive failure curve, of log $(2\,\mathscr{T}_0/\theta_{01})$, log $(2\,\mathscr{T}_0/\theta_{02})$, etc.

Clearly, if $\mathscr{T}_0$ is known (it can be measured or calculated for cross-linked elastomeric and thermosetting adhesives, see Andrews and Fukahori[19]), it is possible to calculate θ_{01}, θ_{02}, etc. from the measured vertical displacements of the curves.

In what follows, we show that the predicted behaviour is obtained experimentally and consider the significance of the θ_0 values derived from the data for a variety of adhesive systems.

4. EXPERIMENTAL OBSERVATIONS

These will be reported here only in summary form, since they have been published elsewhere in greater detail[20-22,25]

The systems examined are, firstly, a styrene–butadiene elastomeric adhesive simultaneously cured and bonded, in a heated press, to a variety of plastics as substrates. To maintain the requirement of rigidity in the substrate, the plastics were prepared as thin films, themselves bonded to a steel block. This arrangement satisfies the rigidity condition because of the very small volume of plastic material available for energy dissipation.

Measurements of $\mathcal{T}$ for the SBR, and of θ for the adhesive bonds, were made by recording the energy input to the system, up to the point at which the pre-formed crack or interface began to spread. This defined $W_{0\,(crit)}$ for the specimen shown in Figure 2, or the appropriate quantity for other types of specimen.[23] Here. $\mathcal{T}$ or θ is then evaluated from equation (12) or the appropriate equation for other types of

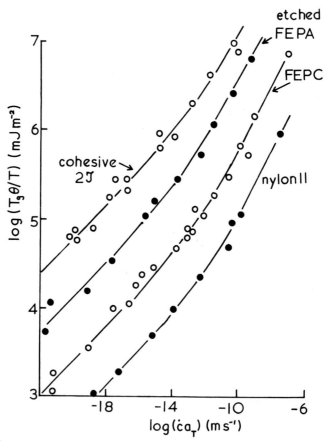

Figure 4. Actual behaviour of $2\mathcal{T}$ and θ for SBR and SBR bonded to a series of substrate films. Compare with Figure 3.

test piece. The value of $k_1(\epsilon_0)$ can be assessed independently and is, in any case, not very different from π for the systems studied.

Data for SBR and SBR-to-plastic bonds are displayed in Figure 4. The predicted parallel curves are obtained over a very wide range of adhesive joint strengths. In Figure 4 the log 2 $\mathscr{T}$ and log θ values have been 'normalized' by the factor (T_g/T), where T_g is the glass transition temperature of the SBR, as is normal in WLF master plots of mechanical data. This has little effect upon the points on a log–log plot.

A value of $2\mathscr{T}_0$ for the SBR adhesive was obtained, using the limiting fatigue method of Lake and Lindley,[26] as $8.32 \times 10^3 \mathrm{mJ\ m^{-2}}$. From this value, and equation (13), the θ_0 values displayed in Table 1 were then deduced for the substrate films detailed therein. In Table 1 also are included the thermodynamic works of adhesion, w_A, for the rubber–plastic pairs obtained, as outlined previously, from contact angle measurements.

It is immediately clear that, for the weaker adhesive joints which form the first group in Table 1, the values of w_A and θ_0 correspond very closely indeed. It was possible to demonstrate (see Andrews and Kinloch[27]) that for these joints the failure was truly interfacial and that no primary bonds were established at the interface. It would appear that these two conditions are sufficient to give the identity between θ_0 and w_A discussed previously.

The second group of joints in Table 1 refer to a single plastic, fluorinated ethylene-propylene film (FEPA), which had been subjected to surface treatment before bonding. These surface treatments included an undisclosed manufacturer's process and an etching treatment with a proprietary organic dispersion of sodium

Table 1 Comparison of θ_0 with the thermodynamic work of adhesion, w_A, for styrene–butadiene rubber bonded to various plastic film substrates

Substrate	$\theta_0 \mathrm{mJ\ m^{-2}}$	$w_A \mathrm{mJ\ m^{-2}}$
GROUP I		
Fluorinated ethylene-propylene (FEPA)	22	48
Poly(chlor-tri-fluor ethylene) (PCTFE)	75	63
Nylon 11	71	71
Poly(ethylene terephthalate)	79	72
Plasma-treated FEPA	69	57
GROUP II		
FEP C20 (proprietary surface treatment)	288	61
FEPA etched for (seconds):		
10	851	68
20	1170	70
60	1290	70
90	1620	71
120	1780	71
500	2420	72
1000	1990	72

naphthalene ('Tetra-etch' W. L. Gore and Associates). The etching time was varied as shown.

In this second group of joints, $\theta_0 \gg w_A$. It was also established that primary bonds were established across the interface during curing and bonding of the adhesive (etching introduces reactive double bonds into the substrate surface) and that failure was of a mixed-locus variety. That is, the failure was partly interfacial, but also partly cohesive in either the substrate or the adhesive.

By measuring, in a given fracture surface, the respective area fractions of interfacial failure (i), failure in adhesive (a) and failure in substrate (s), Andrews and Kinloch showed that θ_0 could be expressed as a weighted average of the three failure modes, thus,

$$\theta_0 = iI_0 + 2a\ \mathscr{T}_0 + 2sS_0 \tag{14}$$

where I_0, $\mathscr{T}_0$, and S_0 are the intrinsic failure energies for interfacial, cohesive-in-rubber, and cohesive-in-substrate failures, respectively, and, of course,

$$i + a + s = 1 \tag{15}$$

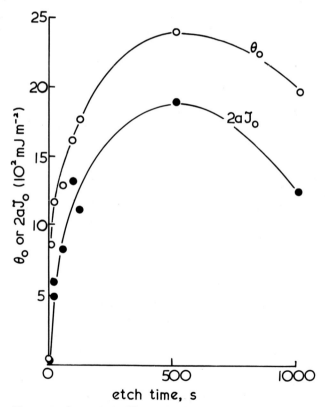

Figure 5. θ_0 and $2a\mathscr{T}_0$ as functions of etch time for SBR bonded to etched FEPA film, showing agreement between the two parameters to within 20%.

I_0 may simply equal w_A if no primary bonds are present in the interfacially debonding regions. Otherwise, $I_0 > w_A$. In the cases studied, the term $2a\mathcal{T}_0$ was found to account for 80% of the value of θ_0 and also for the major features of the dependence of θ_0 upon etching time (see Figure 5).

The concept of θ_0 as a weighted average can clearly be extended to cover other adhesion phenomena such as partial wetting and mechanical keying at the interface. Unwetted regions, u, would contribute nothing to θ_0 so that, for interfacial failure,

$$\theta_0 = I_0(1 - u) \tag{16}$$

In a similar way, mechanically keyed regions, k, would contribute to θ_0 by inducing fracture through the adhesive (assuming this is the weaker of the two solid phases). For otherwise interfacial failure, therefore:

$$\theta_0 = I_0(k - 1) + \mathcal{T}_0 k \tag{17}$$

What determines the locus of failure? In the authors' view this is normally the 'weakest link' in the highly stressed zone surrounding the propagating crack (or decohesion boundary). Thus, it depends entirely upon the relative bond strengths of the adhesive, interface, and substrate. Special exceptions to this simple rule have been discussed elsewhere,[22] but the weakest link concept is basically, and self-evidently, correct in most circumstances.

(i) Epoxy resin-to-metal bonds

A second series of investigations by Andrews and King employed similar methods to those already described to determine the failure characteristics of epoxy resin-to-metal bonds.[22] The epoxy was a diglycidyl ether of bisphenol A ('Shell 828') hardened by an amine/benzyl alcohol system ('Shell 114') and the resin-to-hardener ratio was varied. The metal substrates employed were aluminium, stainless steel, and gold, subjected to appropriate surface pre-treatments.[22] Some of the metal surfaces were further treated by ultraviolet (UV) radiation in the presence of ozone which has the effect of removing much of the surface contamination due to atmospheric hydrocarbons.[28]

The epoxy resin mixtures had glass transition temperatures in the range 40–90°C and joints were tested to failure both above and below the T_g.

Results above T_g, where the epoxy resin is a highly cross-linked rubber, were entirely consistent with those already described for SBR-to-plastic joints. This is illustrated in Figure 6 which shows log $2\mathcal{T}$ and log θ for the 5/2 resin/hardener composition, θ being determined for the Al substrate. These data show both the familiar behaviour of Φ with increasing reduced rate, and the parallel nature of the curves for $2\mathcal{T}$ and θ. Qualitatively similar data were obtained for each of the resin compositions employed and for all substrates, and θ_0 values were deduced as described earlier. These θ_0 values, along with w_A calculated for the various systems employed, are shown in Table 2. A number of interesting features arise from these results. Firstly, for all but the 5/1 composition (see later) the θ_0 values obtained are considerably higher than the measured w_A values, but are within a factor of about

58

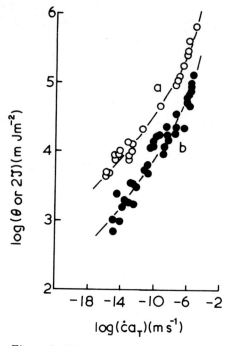

Figure 6. Cohesive and adhesive failure energies for (a) epoxy resin and (b) the same resin bonded to aluminium. Results for $T > T_g$.

Table 2 Comparison of θ_0 with the thermodynamic work of adhesion, w_A, for epoxy resins bonded to metal substrates. (Epoxies characterized by resin/hardener weight ratios.)

| Substrate | Epoxy | | | | | |
| | 5/1 (mJ m^{-2}) | | 5/2 (mJ m^{-2}) | | 5/3 (mJ m^{-2}) | |
	θ_0	w_A	θ_0	w_A	θ_0	w_A
Steel, etched	7820	—	700	—	715	—
UV/ozone treated	—	—	680	—	—	—
atomically clean	—	—	—	306	—	276
Al, etched	>4850	200[a]	580	206[a]	715	190[a]
atomically clean	—	333	—	306	—	276
Au, plated	1350	190[a]	800	200[a]	715	180[a]
UV/ozone treated	—	220[a]	600	220[a]	—	200[a]
atomically clean	—	396	—	360	—	325

[a]Approximate corrected values equal to double the values obtained experimentally by neglecting the equilibrium spreading pressure.

two of the w_A values calculated for atomically clean substrate surfaces (oxide or metal as the case may be). Secondly, UV/ozone treatment produces no increase in θ_0, despite the fact that it is known to be effective in purging such surfaces of organic contaminants. (This was confirmed by the large reductions in liquid contact angles resulting from the treatment.)

It would appear then, that the epoxy resin *itself* is capable of displacing organic contaminants from a 'clean' metal or oxide surface, no doubt aided by the elevated curing temperatures employed. The epoxy then bonds, effectively, to an atomically clean substrate, giving a relatively high w_A and thus θ_0. Although θ_0 remains a factor of two larger than w_A, the two values are sufficiently close to suggest that bonding is entirely secondary in nature (with the important exception of the 5/1 epoxy composition). Possible reasons for the discrepancy between θ_0 and w_A have been discussed elsewhere,[22] but it may simply be due to inaccuracies in the estimation of $\mathcal{T}_0$ which is more difficult in epoxy resins than in normal elastomers.

For the 5/1 resin, which contains a stoichiometric excess of epoxide groups, the θ_0 values reveal a startling difference from those obtained for all other compositions. For this resin, the interfacial starter crack normally diverted immediately into the resin phase, indicating $\theta_0 > 2 \mathcal{T}_0$. In a few cases of the steel substrate an interfacial failure *was* obtained at $\theta_0 \sim 8000 \text{ mJ m}^{-2}$, compared with $2 \mathcal{T}_0 \sim 4000 \text{ mJ m}^{-2}$. Again, the interfacial bonding is seen to be stronger than the cohesive bonding of the adhesive.

The observations apply to the *oxidized* surfaces (Al and steel). For gold, interfacial failure was still observed and $\theta_0 < 2 \mathcal{T}_0$. Nevertheless, even here θ_0 was three to four times larger than the atomically clean w_A values, indicating the undoubted incidence of primary bonding.

The picture thus emerged of some genuine primary bonding being established when there is an excess of epoxide rings in the resin which are not reacted with the amine hardener. Presumably the rings are opened by reaction with hydroxyl surface groups which are known to populate metal oxide surfaces and which are suspected of being present, even on gold. The result may be an ether-type linkage of the form, e.g.

$$\text{Al}-\text{O} \mid -\text{CH}_2-\underset{\underset{\text{OH}}{|}}{\text{CH}}-\text{R}$$

interface ↗

(ii) Epoxy resins below T_g

So far we have shown how successful the generalized fracture mechanics theory is in describing the failure characteristics of joints involving an adhesive in an elastomeric condition. Using this theory it is then possible to derive the parameter θ_0 which relates directly to the nature of the interfacial atomic bonding.

The picture is far less clear when the adhesive is, like the substrate, in a rigid condition. Figure 7 shows $\log \theta$ versus $\log \dot{c} a_T$ for the 5/2 epoxy bonded to steel over a wide temperature range, both above and below T_g. The data form a smooth

60

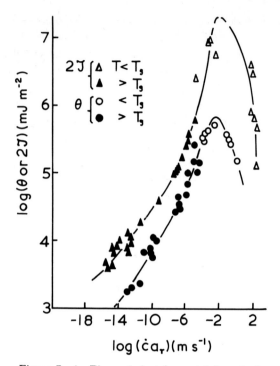

Figure 7. As Figure 6, but for a stainless steel substrate and temperatures above and below T_g.

curve with a maximum lying well within the glassy state region (i.e. the maximum does not represent the glass transition phenomenon itself).

On the same diagram are shown the data for $2\mathcal{T}$, i.e. cohesive fracture of the same resin. To the left of the maximum the adhesive and cohesive curves remain parallel, but as the peak is approached the curve for $2\mathcal{T}$ rises progressively higher than that for θ. Furthermore, on the right of the peak the curves are very widely separated and, possibly, no longer parallel.

Since the generalized theory is in no way limited to the rubber-like state, some explanation must be given for its failure to correlate the adhesive and cohesive data in the glassy range. Several possible reasons can be advanced.

Firstly, the three-dimensional strain field, and thus the function Φ, will be quite different for a crack tip at an interface with a rigid substrate and for one surrounded entirely by the polymer. This is because the rigid substrate inhibits contraction of the polymer 'sheet' in the direction normal to the plane of the sheet, creating a much larger degree of triaxiality in the stress field close to the crack tip.

As long as the adhesive is rubber-like, the highly strained zone at the crack is large and the major contributions to Φ arise from regions in the polymer sufficiently removed from the substrate to be unaffected by its presence. Thus, Φ is similar for both adhesive and cohesive cases.

When the epoxy resin becomes glassy, large deformations at the crack tip occur only in a small plastic zone. In the adhesive situation, therefore, the presence of the substrate exerts a much greater influence, reducing the plastic zone size and thus Φ. If Φ is much smaller in the adhesive situation than in the cohesive case, the qualitative trend of the results is explained.

5. THE EFFECTS OF ENVIRONMENT

Gent and Schultz[8] proposed that failure of an adhesive joint immersed in a liquid environment could be characterized by a modified thermodynamic work of adhesion w_{AL}. If w_A is the thermodynamic work of adhesion between two solids a and b *in vacuo* or an inert atmosphere, then

$$w_A = \gamma_a + \gamma_b - \gamma_{ab} \tag{2}$$

and

$$w_{AL} = \gamma_{aL} + \gamma_{bL} - \gamma_{ab} \tag{18}$$

where the suffix L denotes the presence of a liquid. Using the Young equation, it follows that

$$w_{AL} = w_A - \Delta \tag{19}$$

where

$$\Delta \equiv \gamma_L (\cos \alpha_{aL} + \cos \alpha_{bL}) \tag{20}$$

and α denotes the contact angle as appropriate.

Using Andrews' equation for the case of interfacial failure of secondarily bonded joints for which $\theta_0 \equiv w_A$, we find that the adhesive failure energy, θ_L, for failure under a liquid environment should be

$$\theta_L = w_{AL} \Phi \tag{21}$$

or

$$\theta_L / \theta = w_{AL} / w_A = 1 - \Delta / w_A \tag{22}$$

This prediction accords with the data of Gent and Schultz on joints of styrene–butadiene rubber bonded to Mylar film and reproduced in Figure 8. To fit equation (22) requires a value of 77 mJ m^{-2} for w_A which agrees very closely with the figure of 72.3 mJ m^{-2} obtained for the SBR–PET system by Andrews and Kinloch.[20]

An extreme case of the validity of equation (21) is illustrated by the work of Owens[29] who immersed polypropylene film, coated with a vinylidene chloride copolymer, in water and surfactants. In some cases the coating separated spontaneously from the film, and Owens was able to show that for these immersion liquids w_{AL} was negative. Provided w_{AL} remained positive, no separation was observed. Clearly equation (21) predicts zero adhesive strength for $w_{AL} \leqslant 0$.

Similarly, Gledhill and Kinloch[30] have shown that a negative value of w_{AL} of

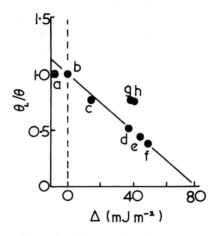

Figure 8. Data of Gent and Schultz[8] showing the ratio, θ_L/θ, of adhesive failure energies with and without liquid immersion, as a function of the parameter Δ. SBR bonded to Mylar film. (a) Water, (b) air, (c) 10% ethanol, (d) 50% methanol, (e) ethanol, (f) butanol, (g) glycol, (h) formamide. The high values for (g) and (h) are explained in Gent and Schultz.[8]

-255 mJ m^{-2} is to be expected for an epoxy resin bonded to mild steel if separation occurs under water (compared with $+291$ mJ m^{-2} in the absence of water). Spontaneous separation of such a joint is therefore expected from equilibrium considerations. However, for this to occur it is necessary for the water molecules to be transported to the interface, and this could occur by diffusion through the epoxy resin itself. Thus, joint failure was delayed in a manner controlled by the diffusion characteristic of the system, but its eventual occurrence could not be prevented.

6. SUMMARY

The work described and reviewed in this paper establishes unequivocally the relationship between the mechanical strength of adhesive joints (expressed in terms of the adhesive failure energy θ) and the energetics of the interface. In many cases the parameter θ_0, which appears in Andrews' equation.

$$\theta = \theta_0 \Phi$$

can be identified with the thermodynamic work of adhesion w_A. For non-interfacial failure, or where primary bonding occurs across the adhesive interface, $\theta_0 > w_A$ but can normally be interpreted as a weighted average over different modes of fracture. In the extreme case of wholly cohesive failure, θ_0 is equivalent to $2\mathcal{T}_0$, where $\mathcal{T}_0$ is the intrinsic failure energy of the cohesively failing material. In the important case of environmental failure, θ_0 is frequently identifiable with the modified work of adhesion w_{AL} (i.e. the work of adhesion in the presence of the liquid in question).

A number of problems remain, expecially that of relating cohesive and adhesive failure energies in a quantitative manner when both adhesive and substrate are rigid solids and the loss function Φ is modified in the presence of the boundary. The detailed interpretation of θ_0 as an activation energy for interatomic bond fracture, as opposed to an equilibrium surface energy, does not appear to be of primary importance, but clearly merits more detailed consideration.

REFERENCES

1. M. Levine, G. Ilkka, and P. Weiss, *J. Polym. Sci.*, **B2**, 915 (1964).
2. H. W. Fox and W. A. Zisman, *J. Coll. Inter. Sci.*, **5**, 514 (1950).
3. W. A. Zisman, *Advances in Chemistry*, series 43 (A.C.S., Washington D.C., 1964) p. 1.
4. C.A. Dahlquist, *Aspects of Adhesion*, Vol. 5, ed. D. J. Alner, (Univ. of London Press, 1969), p. 183.
5. D. H. Kaelble, *J. Adhesion*, **1**, 102 (1969).
6. S. S. Voyutskii, *Autohesion and Adhesion of High Polymers* (Interscience, New York, 1963).
7. B. V. Derjaguin, *Research*, **8**, 70 (1955).
8. A. N. Gent and J. Schultz, *J. Adhesion*, **3**, 281 (1972).
9. R. Houwink and G. Salomon, *Adhesion and Adhesives* (Elsevier, New York, 1962).
10. D. K. Owens and R. C. Wendt, *J. Appl. Polym. Sci.*, **13**, 1741 (1969).
11. D. H. Kaelble and K. C. Uy, *J. Adhesion*, **2**, 50 (1970).
12. S. Wu, *J. Adhesion*, **5**, 39 (1973).
13. J. Kloubek, *J. Adhesion*, **6**, 293 (1974).
14. F. M. Fowkes, *Treatise of Adhesion and Adhesives*, ed. R. L. Patrick (Marcel Dekker, New York, 1967) p. 325.
15. J. R. Dann, *J. Coll. Sci.*, **32**, 302 (1970).
16. D. H. Kaelble, *J. Adhesion*, **2**, 66 (1970).
17. W. D. Harkins and E. H. Loesner, *J. Chem. Phys.*, **18**, 556 (1950).
18. E. H. Andrews, *J. Mater. Sci.*, **9**, 887 (1974).
19. E. H. Andrews and Y. Fukahori *J. Mater Sci.*, **12**, 1307 (1977).
20. E. H. Andrews and A. J. Kinlock, *Proc. Roy. Soc. (Lond.)*, **A332**, 385 (1973).
21. E. H. Andrews and A. J. Kinloch, *J. Polym. Sci.*, **C**, **46**, 1 (1974).
22. E. H. Andrews and N. E. King, *J. Mater. Sci.*, **11**, 2004, (1976).
23. A. N. Gent and A. J Kinloch, *J. Polym. Sci.*, **A2**, **9**, 659 (1971).
24. M. L. Williams, R. F. Landel, and J. D. Ferry, *J. Amer. Chem. Soc.*, **77**, 3701 (1961).
25. N. E. King, Ph.D. Thesis (Univ. of London) 1976.
26. G. J. Lake and P. B. Lindley, *J. Appl. Polym. Sci.*, **9**, 1233 (1965).
27. E. H. Andrews and A. J. Kinloch, *Proc. Roy. Soc. (Lond.)*, **A332**, 401 (1973).
28. D. M. Mattox, *J. Vac. Sci. Tech.*, **11**, (1), 474 (1974).
29. D. K. Owens, *J. Appl. Polym. Sci.*, **14**, 1725 (1970).
30. R. A. Gledhill and A. J. Kinloch, *J. Adhesion*, **6**, 315 (1974).

4

The Movement of Electrical Charge Along Polymer Surfaces

T. J. Lewis

School of Electronic Engineering Science, University College of North Wales, Bangor

1. INTRODUCTION

The electrical properties of the surface of an organic polymer are likely to be quite different from those of the bulk. There will not only be structural differences but also chemical differences because of dangling bonds and oxidative and other surface reactions. Various adsorbed impurities, both neutral and charged from the surrounding atmosphere, will also be present, as will impurities or additives that have diffused to the surface from within the bulk. The electrical behaviour of a polymer surface is obviously important where the polymer is used as an insulant, but it is less obviously important in situations where charge on the surface affects other properties. Examples of the latter are the effects on appearance and wear of electrostatically attracted dust, the difficulty of printing where charge is present, and the triboelectric effects[1] when polymers are in rubbing contact with other solids.

Accurate measurements of the conductive properties of polymers are notoriously difficult to make, firstly because of the weakness of the currents that can be induced, even under considerable electric stress, secondly because of the time-dependence of the process, and thirdly because the previous history of the specimen affects both. The electrical properties of the surface regions are important in such measurements, whether currents normal or tangential to the surface are being studied.

In spite of the difficulties there have been many measurements of the normal component of current and there are many differing views about the mechanism of this process which have been reviewed recently by Lewis.[2] Most, if not all, of such measurements have been made by applying high electrical fields and so it is to be expected that the steady current ultimately reached, sometimes only after many hours, is injected via the electrodes on the polymer surfaces. As a consequence several authors (see, e.g., Taylor and Lewis[3]) have proposed that this current is controlled by the surface conditions at the electrode—metal contacts.

The second major way in which the surface is clearly important is in the many processes of contact charging with or without rubbing. Contact between a polymer surface and another surface will permit the transfer of electrons or ions and perhaps also the transfer of neutral mass. This may be encouraged by rubbing contact which generates heat and mechanical deformation. On separating the contact, charge can be left on the polymer surface, and in favourable circumstances may be great enough to cause electrical breakdown of the surrounding atmosphere. This has the effect of limiting surface-charge densities to less than 10^{-9} C cm^{-2} or less than about 6×10^9 charged sites per cm^2, which is small compared with the likely molecular packing density on a surface ($\sim 10^{15}$ cm^{-2}). The earlier work on contact electrification has been extensively reviewed by Harper[1] and there is a useful concise summary by Parkman.[4]

Concerning the tangential electrical properties of polymer surfaces the published literature is extremely sparse, although one suspects that much information of an empirical and proprietary nature probably exists in industry. The concept of surface resistivity is frequently employed, notably in dealing with the antistatic properties of polymers and values are often quoted, but there is little substantial evidence for the precise factors that control it. Although surface conduction is a measure of charge transport along the surface it does not directly give information about the nature of the charge carrier, whether it is ionic or electronic, positive or negative, or both. Some experiments are in progress, however, which should provide a much clearer picture of the surface conductive processes and these will be discussed below.

Although our primary interest here is in tangential properties of the surface, much information can be gained from a study of the ways in which the surface affects processes of conduction and charge transfer normal to the surface. Therefore, our attention will be directed initially to these normal properties. We shall begin by describing a model for the salient electrical properties of the surface and then we shall use this to discuss the experimental evidence.

2. ELECTRONIC PROPERTIES OF THE SURFACE

(i) The energy diagram

If, to a first approximation, a polymer can be considered as a molecular crystal then an appropriate diagram for the electronic energy states may be obtained by reference to that of crystalline anthracene[5] which is the ideal molecular crystal. The result is shown in Figure 1. We find narrow valence (bound state) and conduction (quasi-free) bands E_v and E_c because the intermolecular forces are weak and these are separated by a large energy gap. Within each band electron transport should be considered not as a *free* motion as in a normal covalent crystalline semiconductor, but rather as a *hopping* process between levels representing molecular groups separated both in space and energy. Hopping, as opposed to free or quasi-free motion, causes electrons or holes to be very much less mobile in polymers than in semiconductor solids and that, together with the wide band gap, is the reason for their very low conductivity.

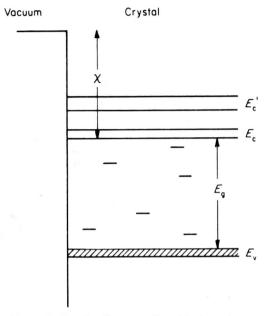

Figure 1. Band diagram for ideal polymer crystal with narrow valence and conduction bands E_v and E_c (0.1–0.5 eV wide) and a band gap E_g (4–8 eV). Also shown are localized states in the band gap. At the surface there is a potential energy step and the electron affinity is χ.

Above the conduction band E_c, a second conduction band E_c' is indicated which represents the states of extrinsic electrons introduced from outside into the polymer and corresponding in a gaseous state to negative molecular ions. Impurities or distortions in the crystal will introduce local energy states within the band gap just as in elemental covalent semiconductors. The surface of this ideal crystalline polymer will be represented by a step in potential energy and will define an electron affinity χ for the crystal as in Figure 1.

The simple crystalline model of Figure 1 will not be sufficient to represent a real polymer and its surface for a number of reasons. The most important of these are:

(a) Polymer solids are not 100% crystalline. Crystallite regions in some cases may occupy less than 50% of the volume. The non-crystalline or amorphous regions between the crystallites are not only structurally irregular but are also likely sinks for impurities. As a consequence a more likely band model taking account of amorphous regions might be Figure 2.

(b) The surface also represents an irregularity in the crystal structure where molecular folding, bond adjustment, the protrusion of side and end groups and topological changes will occur. All these disruptions will introduce a series of

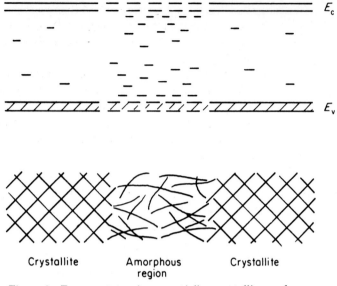

Figure 2. Energy states in a partially crystalline polymer.

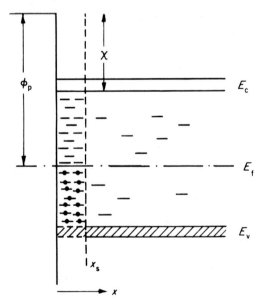

Figure 3. Localized states at a polymer surface, forming a surface region of depth x_s.
The number and energy distribution of the
states will depend on whether a crystallite or
an amorphous region extends to the surface
at the point in question. States up to the
Fermi energy E_f will be occupied.

surface states located in the band gap[6,7] and extending back into the bulk polymer by a molecular dimension or so (Figure 3). Moreover, where an amorphous region extends to the surface an even greater degree of destruction of the band structure can be expected.

(c) The surface will also hold impurities which have either come from the ambient medium or have diffused outward from the polymer bulk. These include catalyst residues, antioxidants, and processing agents. In fact the action of most antistatic additives is to produce an array of hydrophylic end-groups on an otherwise strongly hydrophobic polymer surface so that some enhanced degree of water adsorption is encouraged.

In equilibrium at a given temperature, the surface and bulk states may be expected to be occupied with a probability determined by the Fermi energy E_f (Figure 3). Unless the temperature is high, all states below E_f can be assumed to be occupied. When occupied, some of these states will be neutral and other charged, so to maintain an overall electrical neutrality band bending may be necessary, with the consequent acquisition of counter charges in states deeper into the polymer so that a surface double layer results (Figure 4). The existence of such layers is a familiar problem when considering the inorganic electronic solid state; see for example Henzler,[8] who discusses the problem with special reference to transport at surfaces. The location of a Fermi energy E_f also allows definition of a polymer work function ϕ_p which, as we see from Figure 4, is dependent on the surface states and may therefore vary from point to point on a polymer surface. Strong work-function patch effects may be expected.

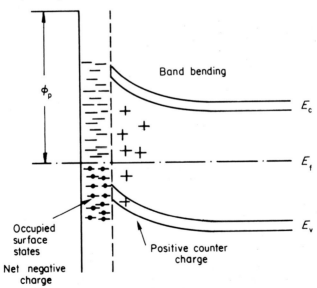

Figure 4. Double layer at surface of polymer set up by net negative charge in surface states and counter-positive charge in the adjacent bulk polymer, including some holes in the valence band.

(ii) Contact Phenomena

The concept of polymer work function gained acceptance after the work of Davies[9,10] and is very important in discussing all aspects of polymer contact phenomena. For example, on contact to a clean metal of work function ϕ_m electron transfer may be expected, the direction depending on the relative levels of the energy states and their donor or acceptor qualities. When equilibrium is reached, the charge transfer will establish a potential difference (the contact potential) causing band-bending and a common Fermi energy. In Figure 5 this situation is shown for $\phi_m <$ and $> \phi_p$, producing electron transfer from metal to polymer or vice versa. According to the work of Davies[10] and Fabish et al.[11,12] both situations are possible.

The existence of surface states in the band gap is very important in determining contact phenomena for two reasons. Firstly, without their presence in the band gap electron transfer from a metal to a polymer would be an unlikely occurrence, and secondly when it did occur it would have to be to states within the bulk, transfer to which would take times orders of magnitude greater than the usual contact times.[6] Even with surface states present it may be necessary in some instances to require bulk states to be ionized before equilibrium is finally reached. The time of charging is a useful indication of the existence and concentration of surface states.

From Figure 5, assuming that the equilibrium electron transfer can be accomplished in surface states alone, we have

$$\Phi_m - \phi_p = \tfrac{1}{2}e^2 N_s x_s^2 / \epsilon$$

where N_s is the density of charge in surface states extending to a depth x_s, e is the electronic charge, and ϵ is the permittivity of the polymer. The charge density in unit area of the polymer surface is

$$\sigma = eN_s x_s = 2\epsilon(\phi_m - \phi_p)/ex_s.$$

A similar treatment can be made when bulk states as well as surface states are involved, but the result is cumbersome. Such problems are well covered in textbooks on the electronic solid state. See for example Many et al.[13] An alternative view of the surface problem has been taken by Garton[14] who develops a classical model for the dielectric in which charge carriers become localized or trapped simply by the polarizing and ordering influence of the field of their own charge. Thus, there is no restriction on charge concentration as there would be if local polymer states had to exist to accommodate the charge carrier. Garton assumes that the carriers move into the polymer under the classical laws of diffusion and drift; Poisson's law is obeyed and in equilibrium the net current is zero. This treatment is attractive in that it does not require any specific properties of the polymer other than that it is a polarizable dielectric. Closer examination, however, will show that the model is not necessarily very different from the one discussed above, since it is very plausible that a surface region may be distinguished in which self-trapping could be different from that in the bulk. In fact Garton suggests that the trapping occurs within regions about 100 molecules thick.

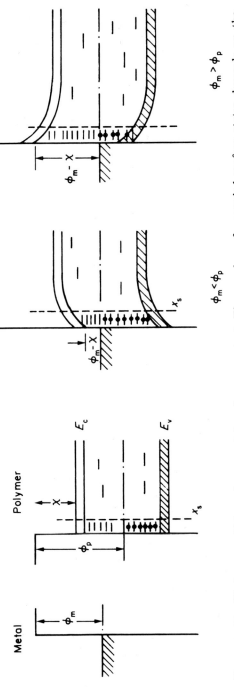

Figure 5. Equilibrium conditions for a metal–polymer contact. The number of occupied surface states depends on the relative magnitudes of ϕ_m and ϕ_p. If equilibrium cannot be reached by adjustment of surface-state occupation alone, then band-bending results.

The model described above receives experimental support in a variety of ways, and one of the most direct is that given by Davies[9,10] and Fabish *et al*[11,12] Charge σ is deposited on or in the surface of a polymer by contact with a metal of known work function which is then removed. The value of σ left behind can be determined and its subsequent behaviour followed using an electrostatic probe technique. The essential features of the method are shown in Figure 6 and are described, by Davies[15] among others. The probe consists of a shielded and insulated metal wire connected to a sensitive electrometer voltmeter. When passed across a charged surface, the probe system responds as a capacitative potential divider, the signal depending on geometrical factors such as the degree of screening which decides the effective area A seen by the probe, the distance d above the charged surface and the distance d_s from that surface to ground. Provided $C \gg C_p$ (see Figure 6) it can be shown that the charge density σ on the area of the polymer surface sensed by the probe is given by

$$\sigma = CV_p A^{-1}(1 + \epsilon_r d/d_s) \tag{1}$$

where ϵ_r is the relative permittivity of the polymer. The ratio C/A may easily be determined in a separate experiment in which the polymer surface is replaced by a metal one held at a known potential V. If the probe response is then V_p,

$$\frac{C}{A} = \frac{\epsilon_0}{d} \frac{V}{V_p}$$

in which ϵ_0 is the permittivity of free space.

Using this method to determine σ, Davies[10] found a correlation between charge transferred and metal work function and was able to deduce for a range of different polymers, each prepared by simple washing in isopropanol, that ϕ_p lay between 4.08 eV (nylon) and 4.85 eV (PVC). In an earlier paper,[9] using a slightly different technique, Davies found a work function of 4.7 eV for polyethylene and a penetration depth of about 1.8×10^{-6} m. The contact charge densities obtained

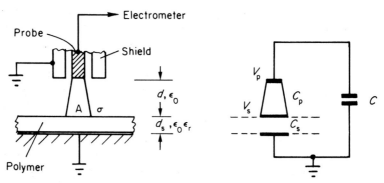

Figure 6. Induction-probe arrangement. Here C_p and C are the probe polymer surface and probe ground capacitances and C_s is the polymer sample capacitance for the effective area A investigated by the probe; V_p and V_s are the probe and local surface potentials, respectively.

were of the order $10^{-8} C \, cm^{-2}$ but, because the measurements were made under vacuum, gas discharges which would normally occur to reduce this value were avoided. There has been some subsequent discussion of the details of the model adopted by Davies, for which see Chowdry and Westgate[16] and Wintle,[17] but the essential features are not in question.

Fabish and coworkers[11,12,18] have recently extended the measurements and concepts put forward by Davies and have proposed that, on contact to a metal, electrons transfer to and from relaxed molecular ion states of the polymer which form broad acceptor and donor tails to the E_c and E_v bands, respectively, extending well into the band gap. The idea of localization by relaxation is similar to that proposed by Garton.[14]

3. DECAY OF SURFACE CHARGE INTO THE BULK

Charge placed on and in the surface region by contact, by frictional processes, or by using ion sources such as a corona discharge[19,20] or a Nernst filment[21] (both of which can provide ions of either sign) is free to move into and along the polymer surface. We shall first consider movement into the bulk, i.e. normal to the surface. Charge on thin films where the other surface is grounded is going to be encouraged strongly to move in this direction because of the remoteness of other grounded metal surfaces and the strength of the field of the charges and their images. The counter movement of charge from a grounded electrode on the opposite surface of a film is always a possibility and electrically yields the same result, namely a net decay of surface potential. We shall not consider this latter possibility here. The probe techniques discussed above which monitor the potential due to surface and bulk charges are ideally suited to the study of surface-charge decay, since they cause only a minor perturbation of the system.

The surface charge σ is found to decay at a rate controlled by temperature and by several other factors, as we shall see. The general form of the decay is shown in Figure 9 and specific examples for various polymers have been given by Davies[9] Ieda et al.[19,20] and Lewis et al.[22] The decay may sometimes continue for very long times, which is consistent with the idea that the charge is hell strongly in surface states and is also strongly trapped in the bulk where it has a very low mobility (possibly as low as $10^{-10} \, cm^2 \, V^{-1} \, s^{-1}$).

In some situations the decay will follow an exponential law with time, but sometimes the law is much more complex (see Baum et al.[23]). There may also be differences in the decay between positive and negative charge. A most intriguing result reported by Ieda et al.[19] was that the potential of a polyethylene surface, charged initially to a high value, decayed with time to cross rather than merge into decay curves which started from lower values. This surprising result has prompted much further study which sheds light on the importance of the surface states.

Many theoretical treatments of the problem exist, and that by Sonnonstine and Perlman[24] provides a unification of many of them. All the theories relate the decay of σ to processes in the bulk of the polymer, using the standard Poisson's and Ohm's laws and the condition that the sum of the displacement and conduction currents will be zero under the open-circuit conditions investigated. The various treatments differ in their assumptions as to whether or not charges move from the surface and become trapped in the bulk, and whether the charge carrier mobilities are constant or field-dependent. For our purposes here, the most interesting differences are in the conditions assumed to be imposed by the surface region where the charge resides initially. Thus, Batra and co-workers (Batra *et al.*[25,26] Seki and Batra[27]) assumed that the surface-charge region was negligibly thin and that part of the surface charge was injected instantaneously into the bulk. Wintle[28–30] assumed a finite initial penetration of charge. Sonnonstine and Perlman[24] offer two theories: the first involves Batra's assumption that there is an instantaneous injection of part of the charge; the second is more realistic in that injection is considered to be a time-dependent process so that the surface charge decays into the bulk exponentially with time. They also consider that the injection might be a field-dependent process which seems very plausible. In the earlier theory of Davies[9] it is postulated that the majority of injected charges remain at or near the surface so that there is little field distortion outside the thin surface region.

That there is no single theory appropriate to all cases of charge decay is not surprising when the role of the surface states is considered in detail. It is obvious that the various theoretical treatments differ essentially in their assignment of surface properties. Instantaneous injection, partial or otherwise, assumes that the surface states, or at least some of them, do not trap charge, whilst time-dependent injection assumes in effect that the surface space charge is activated from traps into the bulk where it becomes mobile. It is quite possible for a polymer surface to behave in either way depending on the manner in which it was first charged.

For example, contact charging is likely to transfer only electrons to the surface, and these will go into surface states which may be well inside the physical boundary of the polymer. On the other hand, corona charging places gaseous ions on the surface in the first instant and these may be one of several ionic species, depending on polarity and the gaseous environment. Thus, negative corona charging in air is likely to deposit CO_3^- ions and positive charging hydrated versions of H^+ NO^+ and NO_2^+.[23] These ions may remain as ionic entities burrowing into the gaseous surface monolayers or finding sites of appropriate configuration.[31] On the other hand, the charge may transfer from these ions to polymer surface states, much in the same way as occurs in contact charging with a metal, leaving a neutral gas molecule behind. An electron or a positive hole would thereby be introduced into the surface states. These possibilities are illustrated in Figures 7 and 8.

The possible need for some degree of activation energy is clear from the figures, but the amount is not only dependent on the relative surface-state and ion-state energies but also will be influenced by the field generated by the surface charges themselves. In many experiments the fields across the polymer films induced by surface charges exceed 10^7 V m^{-1} and such fields will serve to lower potential barriers to injection from surface states into the bulk. Charge may also be injected from the surface states by incident light, as it is in the photoconductive process with polymers such as PVK.

The recent experiments of Baum *et al.*[23] clearly illustrate the importance of

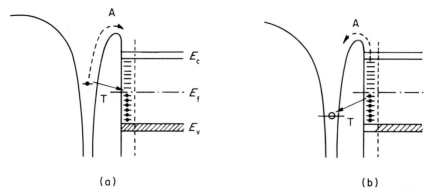

Figure 7. Illustrating transfer of ionic charge to surface states, leaving neutral gaseous molecules behind on surface. When the ions are close to the surface tunnelling (T) is most likely to occur, but activation (A) may also be possible, especially at higher temperatures or in the presence of radiation; (a) and (b) are for transfer of negative and positive ions, respectively.

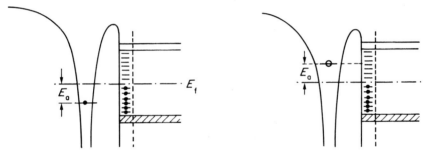

Figure 8. Situations where ion neutralization is difficult. Activation energy E_a must be provided to effect the charge transfer to a vacant state. Optical photons might provide the necessary energy in practice.

surface states. Samples of polyethylene without additives were charged negatively by exposure to a corona source in air. It was possible to regulate the amount and area of charging quite accurately. The subsequent decay of surface charge was then monitored by a high-resolution probe 80 μm in diameter. A typical result is shown in Figure 9 where it is seen that the central region decays much faster than the periphery. It was found that the reason for the different rates of decay was that only the central region was receiving corona light with an energy of several electronvolts and probably also excited gas molecules with equivalent excitation energies. Such energy would be capable of causing transfer of electrons, either from surface ions into the polymer surface states by processes illustrated in Figure 8 or, more likely in the present case, causing transfer of electrons in surface states into the bulk where they would move more readily. Another illustration of the same point is given in Figure 6 of the same paper by Baum *et al.* where exposure to corona light but not corona ions causes a marked increase in the decay of surface charge into the bulk. This enhanced decay could not be produced when the surface

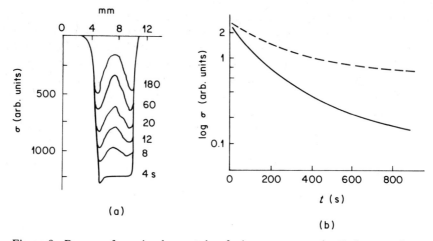

Figure 9. Decay of a circular patch of charge on a polyethylene surface, initially charged negatively by corona for 120 s. The rate of decay for the central region is much greater than for the periphery; (a) shows central line scans at various times after charging, and (b) the corresponding charge decay (Baum *et al.*[23]). (Reproduced by permission of the Institute of Physics.)

was positively charged, nor did it happen for either sign of charge on a polyethylene terephthalate surface, thus supporting the argument that surface states are important and specific to the particular ion and polymer. A reason for the cross-over phenomenon reported by Ieda[19] now becomes apparent. It is that the corona light, together with the high extracting field when the surface is highly charged, encourages efficient injection from surface states into the bulk. The theory of instantaneous partial injection[24] comes closest to explaining this situation.

The role that the field plays in encouraging injection from the surface states may also be important in more conventional studies of high-field conduction in polymers between metal electrodes. The space charge region is an interface between the metal and polymer which, at the cathode, could tend to prevent electron injection. The situation would be similar to that in Figure 5 ($\phi_m < \phi_p$), but with an electron-extracting potential energy gradient. Taylor and Lewis,[3] studying conduction in polyethylene and polyethylene terephthalate films, have proposed that such a surface space charge of electrons exists. They were able to deduce the likely shape of the surface barriers and obtained values of $\phi_m - \phi_p - \chi$ equal to 2.14 and 2.58 eV for polyethylene and polyethylene terephthalate, respectively, with aluminium electrodes.

4. MOVEMENT OF CHARGE ALONG THE SURFACE

The model developed so far for the surface region is applicable with little modification to charge transport processes tangential to the surface. There are two possible modes of charge transport along a practical polymer surface. One is a hopping transport between the surface states already described and taking place throughout the depth of the surface stratum, and the other is transport through a surface film of adsorbed material such as water. The latter is a likely mode when the humidity is high or the polymer has hydrophilic properties induced by surface-active additives. Literature on surface charge transport on polymers is scarce, probably because of the difficulties of accurate measurement, but surface resistivity is clearly an important technological parameter, especially in situations made hazardous by static discharge or where high-voltage flashover could occur.[4] Where surface-active agents are present and the humidity is high enough to allow the formation of a continuous water film the situation can be considered in terms of the double-layer theories adopted for colloids. This subject has been reviewed succinctly by Ottewill.[32] One half of the double layer will reside in electrolytic surface ions and thus will contribute to surface conduction.

(i) Surface conductivity

There are very few reliable measurements of the surface conductivity of polymers under conditions of controlled humidity where transport in the adsorbed water film is likely. Most work appears to have been done on glass, quartz and other inorganic surfaces.[1,33,34] Exceptions are the relatively brief studies by Sawa and Calderwood[35] and Awakuni and Calderwood[36] on oxidized low-density polyethylene and polytetrafluorethylene (Teflon) surfaces using parallel thin aluminium foil surface electrodes. In the case of absolutely dry oxidized polyethylene, currents were less than 10^{-15} A with average surface fields as large as 5 kV cm^{-1}. Even at humidities near the saturation point the surface conductivity is reported to be $<10^{-19}$ ohm^{-1}, in agreement with earlier reports quoted by them of the undetectable surface conductance of high-molecular-weight paraffins. Separate measurements indicated that there was little or no weight change of the polymer with humidity, confirming its hydrophobic nature. On oxidizing the surface by means of ultraviolet light, water sorption was increased and became quite marked for relative humidities greater than about 50%. There were corresponding increases in surface conductivity with water sorption (Figure 10(a)) and the current became non-ohmic at higher fields. Interpreting the non-linearity with the field as indicative of the presence of a space charge, Sawa and Calderwood used the well-known space-charge-controlled current density voltage relationship,[35]

$$J = \frac{9}{8}\mu\epsilon V^2/d^3$$

where J is the surface current density, V the applied surface voltage, and d the electrode spacing, to explain the behaviour. Assuming that the effective permittivity ϵ of the surface layer and its geometrical shape does not vary with humidity, they conclude that the effect of high humidity is to increase the mobility μ of the surface-charge carriers. However, lack of knowledge of the depth of the surface layer prevents a determination of the absolute magnitude of μ. The work shows clearly not only the role that water on the surface may play but also the important part played by the polymer itself. Awakuni and Calderwood[36] found that the

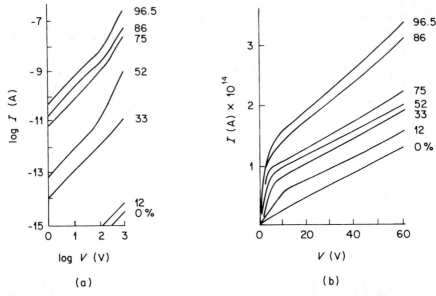

Figure 10. The surface–current voltage relationship at various relative humidities (%) for (a) oxidized polyethylene (Sawa and Calderwood[35]), and (b) PTFE. (Teflon) (Awakuni and Calderwood[36]). (Reproduced by permission of the Institute of Physics.)

behaviour of a Teflon surface was quite different (Figure 10(b)). The effects of humidity and field were much less than for oxidized polyethylene, but the field was not extended to such high values so the possibility of space-charge effects was not tested. The surface conductivity increased exponentially with water adsorption, unlike the case of quartz with which it was compared, where conductivity remained low until the first monolayer of water was complete. They concluded that, whereas on quartz, water molecules become mobile only when a first monolayer of water is complete, any single water molecule is mobile on Teflon.

Unfortunately, measurements of surface conductivity alone cannot indicate the nature or sign of the charge carrier. The assumption of space-charge limitation by Sawa and Calderwood implies injection from one electrode, and it would be useful to know which. Presumably, even if electron injection occurs at the cathode, transport is by negative ions on the water-adsorbed surface formed by electron attachment. In interpreting surface conduction measurements it may also be important to realize that the field between surface electrodes is almost certainly non-uniform unless the spacing is small. In the parallel electrode arrangement adopted for these studies probably no more than the central 50% of the gap is subjected to a uniform field. Near the electrode the field could be at least an order of magnitude larger, as calculations of the field and potential distributions will show.

(ii) Tangential charge transport

Returning now to situations where conduction in adsorbed films is not the major factor, the model for the surface states (Figures 3 and 4) will apply. Tangential charge transport in the surface stratum will be similar to normal transport already discussed and the charge is likely to move from one localized state to another. The fact that surface-charge densities are usually low (less than 1 site in 10^6 occupied) suggests that there will be no correlation restriction on hopping transport and that there will always be neighbouring empty sites into which a carrier may transfer. The fact that surface charge, once deposited, is difficult to remove other than by gross mechanical deformation suggests that it is normally held in states below the outer surface. It is also easy to see that it will tend to be confined to the surface stratum or channel, since the barriers to transfer from site to site will be lower in the tangential direction, where a higher density of localized states would be maintained, than in the normal direction where the localized state concentration will fall away and the potential barriers to hopping increase as the normal distance from the surface increases.

The problem of electronic transport via surface states is important in another sphere, namely in understanding semiconductor behaviour, expecially in respect of modern electronic devices. The subject has been extensively reviewed by Henzler[8] who concludes that the surface of a semiconductor will contain not only point defects but also grain boundaries, so that it is at best polycrystalline and strongly disturbed. He points out that the surface topology will also be important − a factor equally important in the case of polymer surfaces since a carrier may become localized not only by a potential energy well set up by atomic fields, but also by topographical wells where transverse motion will be prevented by physical intervention. Hopping, rather than extended state conductivity, can be expected with a reduced carrier mobility. So far surface-state conductivity has not been observed directly in semiconduction material such as silicon. What is more readily observed is the conductivity in the bulk space-charge layer or conductive channel induced just below the surface by the surface-charge field (field effect).[13] The possibility of field effects in insulating solids such as the polymers has not been explored, but there is no reason why a counter-charge should not appear, albeit slowly and accompanied by band-bending in the bulk polymer adjacent to a charged surface region as illustrated in Figure 4. Buried surface conduction might then be possible in this lower stratum.

Where surface-charge migration is likely to be relatively slow, the probe techniques discussed above may be employed to monitor the movement, provided the probe has good definition. For example, we may conclude from the results illustrated in Figure 9(a) that tangential charge migration was insignificant during the time of the experiment. This result is, of course, in keeping with the findings of Sawa and Calderwood.[35] More information concerning charge migration has been presented by Baum and Lewis.[21] In their experiments, charge of either sign was deposited gently on to a polyethylene terephthalate (PET) surface from a hot

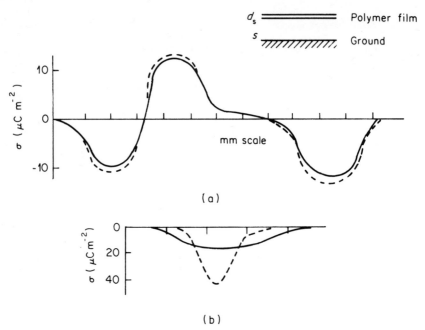

Figure 11. The stability of charge on PET surfaces. (a) Positive and negative charge on fresh untreated polymer, $-\,-\,-\,-$ initially and $\rule{1em}{0.4pt}$ after 24 h; $s = 1.6$ mm, $d_s = 25$ μm; (b) negative charge on a washed surface, $-\,-\,-\,-$ initially and $\rule{1em}{0.4pt}$ after 2.5 h, $s = 0$, $d_s = 25$ μm (Baum and Lewis[21]). (Reproduced by permission of the Institute of Physics.)

Nernst filament which is a useful and convenient source of low-energy ions.[37] On dry untreated film the surface migration of charge of either sign was practically non-existent at temperatures of about 21°C, even when charge patches of opposite sign were adjacent and the normal component of field had been reduced by making the grounded plane opposite the charge deposition remote (Figure 11(a)). When samples were washed in deionized water and dried, however, the charge was induced to spread even with the ground plane close to the surface (Figure 11(b)). Moreover, interactions between positive and negative charge could then be observed (Figure 12). The positive charge appears to be more mobile and has slowly moved to neutralize part of the negative charge.

Line scans of the type illustrated in Figures 11 and 12 do not give a complete picture, since any movement of charge at right angles to the scan line can be detected only as a fall in surface-charge magnitude on the scan line. Methods have now been developed which allow complete two-dimensional raster-mode probe scanning of a charge deposited,[22] the information being stored directly in a computer. The information can be recalled subsequently and displayed either as line scans or as two-dimensional plots of the charge density for any given time. It is also easy to determine from the computer store the total charge on the surface at any time.

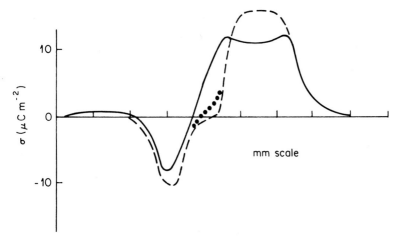

Figure 12. Charge interaction on washed PET film. The positive charge redistributes itself and neutralizes part of the negative charge. $-$ $-$ $-$ $-$ initially, $\cdots$ after 20 min, ——— after 24 h; $s = 1.6$ mm, $d_s = 25\,\mu$m (Baum and Lewis[21]). (Reproduced be permission of the Institute of Physics.)

An example of the results obtained in PET film at 165°C is shown in Figure 13. The film, 23 μm thick, is supported on a 3 mm thick, temperature-controlled block of PTFE on a grounded plane so that the field lines to ground are weak and tangential surface movement encouraged. The charge is deposited from a corona source. There is significant and very uniform spreading without loss of charge. Note how the high definition of the scanning probe allows charge-density contours to be separated. At lower temperatures the spreading is less rapid and at room temperature (21°C) it is non-existent. It is important to note that, without the detailed evidence provided by the two-dimensional maps, such as would be the situation arising with a poor definition probe and a single line scan, the conclusion would be that the surface charge decays with time by movement into the bulk rather than along the surface. It would then be natural but wrong to interpret the decay in terms of one of the theories such as were described earlier.

In fact the decay of the peak value of surface charge as in Figure 13, for example, follows closely the law for outward diffusion of particles from an initially uniform circular concentration as given by Crank.[38] In terms of the peak charge density σ_p at the centre of the circle, we find

$$\sigma_p(t) = \sigma_p(0)[1 - \exp(-r^2/4Dt)] \qquad (2)$$

where $\sigma_p(0)$ is the initial charge concentration in a patch of radius r, and D is the diffusion coefficient. Plots of $\log[1 - \sigma_p/\sigma_p(0)]$ or of the corresponding expression for surface potential V_p (equation (1)) versus t^{-1} should yield linear plots from which D may be determined, since r is also known from the probe measurements. Figure 14 shows such plots for the surface diffusion of negative charge. Similar results are obtained for positive charge. The diffusion law is accurately obeyed at

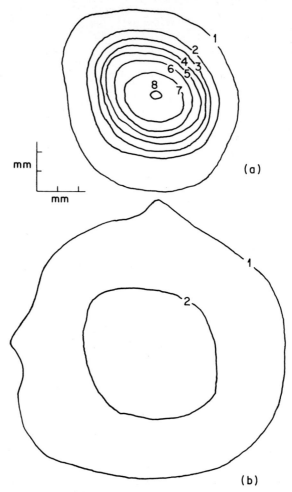

mm

mm

(a)

(b)

Figure 13. Computer-derived contour maps showing positive charge migration on a fresh PET surface
at 165°C. (a) 60 s, and (b) 120 s after charge deposition. Each contour interval is equivalent to a
charge density of 1.9×10^{-8} cm^{-2} (Lewis et al.[22]).

longer times, but as can be seen there are systematic deviations at shorter times
which become more marked as the temperature is lowered. The reason for this
deviation has still to be determined, but it is probably due to the fact that the
initial concentration of charge induces a field and therefore a drift force on the
charges which tends to overwhelm the diffusive processes which strictly apply to
uncharged particles.

In spite of these difficulties, D may be determined quite accurately and is found
to be in the range $10^{-4} - 10^{-6}$ cm^2 s^{-1} for both signs of charge in the temperature

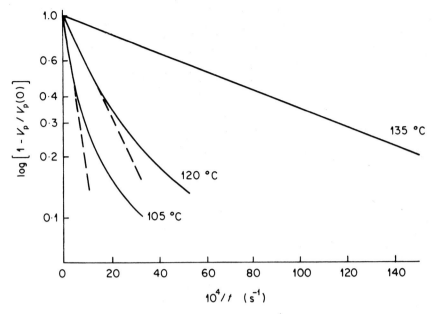

Figure 14. Plots of $\log[1 - V_p/V_p(0)]$ versus t^{-1} for a charge diffusing on a fresh PET surface at various temperatures. The theoretical law according to equation (21) (---) is obeyed at longer times.

range 135–98°C. Moreover, the results closely follow the Arrhenius law

$$D = D_0 \exp(-W/kT) \tag{3}$$

where the activation energy W is 1.51 and 1.44 eV for positive and negative charge diffusion, respectively. As stated earlier the likely ions deposited initially on the surface from the corona discharge are CO_3^- and hydrated H^+, NO^+, or NO_2^+, but the values of D are much greater than the coefficient of gaseous diffusion through bulk PET which are, for example, 3.84 and 5.37×10^{-8} cm^2 s^{-1} for CO_2 and N_2, respectively, at comparable temperatures.[39] This appears to be the first time that meaningful parameters have been established for charge movement on a polymer surface. The activation energy is comparable with that obtained from thermally stimulated current experiments on similar material.[21]

Here, D_0 (equation (2)) is found to have a value of approximately 5×10^{14} cm^2 s^{-1}. Now the diffusion process has been considered to be one of jumping between appropriate sites in the surface. These sites are assumed to be the localized states of the amorphous surface region, and the activation energy of 1.5 eV would then correspond to the depth of the corresponding energy wells. The process envisaged corresponds to vacancy diffusion in crystals, and thus we may write (see for example Van Beuren[40])

$$D_0 = \nu a^2 \alpha \exp(\Delta S/k)$$

where ν is an appropriate phonon frequency of the structure ($\sim 10^{13}$ s^{-1}), a the jump distance between localized states, α a geometrical factor which for the present level of precision can be taken as unity, ΔS the entropy change accompanying the jump, and k the Boltzmann constant. Assuming a to have the arbitrary but not unrealistic value of 25 nm we find $\Delta S = 2.6 \times 10^{-3} \text{eV K}^{-1}$. Although this figure has no great accuracy at the moment and more work is required to establish accurate values, the importance of the probe methods in obtaining meaningful data about surface conduction and diffusion processes is demonstrated. An activation energy of 1.5 eV suggests deep trapping states in the PET surface, and the entropy change indicates that a jump would be accomplished only after some local structural rearrangement involving increased disorder. There may be some significance in the fact that the entropy change is of the same order as that found from bulk conduction studies in polyethylene.[41]

The work is in its early stages and much more needs to be done to establish diffusion behaviour over a wider range of temperatures and under more controlled surface conditions. It will also be important to determine whether other polymers will behave in a similar way.

(iii) Charge transport adjacent to surface electrodes

Comment has already been made concerning the difficulty of obtaining detailed information from simple surface-conduction measurements. Much more information can be obtained from probe studies. For example, Baum and Lewis[21] have shown already how charge on a washed PET surface adjacent to an electrode held at a fixed potential with respect to ground may be neutralized by a charge of opposite sign emerging from the electrodes. Up to the present time little work has been done in this promising field, but there is enough to indicate the possibilities. An example, given by Lewis et al.[23] is shown in Figure 15.

Negative charge from a corona source was deposited initially adjacent to a 2 mm wide evaporated aluminium electrode on a fresh PET surface held at 130°C. With the electrode held at +100 V, charge on the surface adjacent to it becomes neutralized whilst at the same time charge remote from it diffuses away. This can be seen very clearly on line scans. That the positive charge is moving out from the electrode, or rather that the negative charge is moving in, is clearly seen by the patches of positive charge which appear at A in the figure. The opposite experiment with the positive charge deposited and the electrode at −100 V did not produce such a clear result; the charge was neutralized much less readily and there was little emission of negative charge from the electrode.

More striking experiments have been performed with a pair of parallel evaporated aluminium electrodes placed 11 mm apart on the initially uncharged PET surface and with a relatively high potential of 500 V between them. The sample of PET was mounted on a PTFE block as described earlier. Some of the results are depicted in Figure 16. Charge of either sign is seen to flow out readily from the electrodes across the polymer surface, progressing gradually with time and ultimately reaching a stable distribution. The electrodes, now with higher voltages on them than in the experiments described above, appear to be good charge

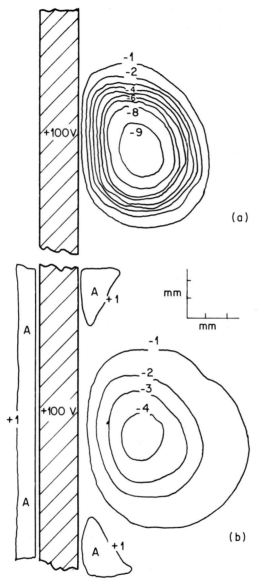

Figure 15. Behaviour of negative charge deposited adjacent to an electrode at +100V on a PET surface at 130°C. (a) 5 s, and (b) 5400 s after deposition. Note the emergence of positive charge from the electrode in (b). Charge contour intervals as for Figure 13 (Lewis *et al.*[22]).

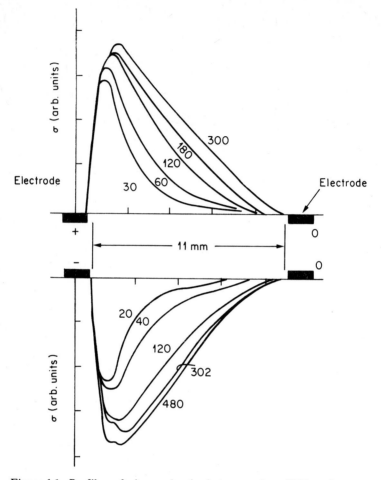

Figure 16. Profiles of charge developing on a clean PET surface at 135°C between parallel surface electrodes 11 mm apart. The lapsed time (s) is shown in the curves. The charge ultimately reaches a limiting distribution. Ordinate proportional to charge density.

emitters. Not only does the charge move out readily but it also collapses back quite easily into the electrode again when the electrode is grounded. The charge carriers are here electrode-generated and are therefore likely to be quite different in nature from charge carriers deposited as ions from a corona discharge. The most likely charge carriers in the present case will be electrons or holes produced in the polymer at the polymer–metal contacts (section 2(ii)). It is found, however, that once ejected into the surface the carrier motion may be described very well by diffusion laws. Following Crank[38] once more, it can be shown that the concentration of a substance diffusing out from a constant line source (in our case the concentration of surface charge σ diffusing from the electrode region) is given

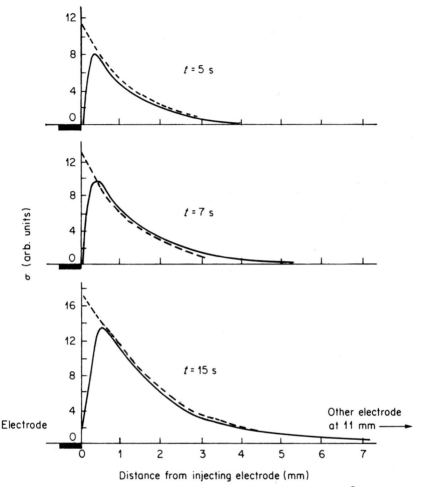

Figure 17. Diffusion of positive charge across a PET surface at 135°C from an electrode at a potential of +500 V with respect to a grounded electrode 11 mm away. The excellent agreement with theory (equation (4)) (– – –) can be seen, but note how some charge has already collapsed back into the electrode as soon as the potential is removed for recording purposes.

by

$$\sigma(x, t) = J \left\{ \left(\frac{2t}{\pi D_e} \right)^{1/2} \exp(-x^2/4D_e t) - \frac{x}{D_e} \operatorname{erf}\left[x/2\sqrt{(D_e t)}\right] \right\} \tag{4}$$

where $\sigma(x, t)$ is the concentration at a distance x from the electrode at time t, J the emission rate of the carriers from the electrode per unit length, and D_e the appropriate diffusion coefficient. This law is accurately obeyed, especially for the diffusion of positive charge as can be seen from Figure 17, and it is possible to estimate D_e to be 2.5×10^{-3} cm^2 s^{-1}, much greater than values for the self-diffusion of corona-deposited ions given above. We conclude that a different species of carrier is operating. The emission rate J may also be found from the results and has a value in the present experiments (where the electrode is held at ±500 V) of 4×10^{-12} A cm^{-1} at short times, but decreases with time as the diffusion proceeds and more charge appears on the surface. This suggests, as might be expected, that J is field-dependent. Experiments are proceeding to determine the factors that influence J and whether D_e follows a law similar to equation (3). The ready collapse of the charge back into the electrode, already commented upon, is clearly seen in Figure 17, where for small x the theoretical prediction of σ is greater than the experimental value, and quite considerable amounts of charge return to the electrode in the time it takes to remove the electrode voltage and scan the sample with the probe. By varying this time, information on charge collection at an electrode may also be gathered.

5. CONCLUSIONS

Although to date there has been relatively little direct evidence of charge migration on polymer surfaces, it is clear that the techniques now exist to study it in some detail. By control of temperature and surface preparation, and particularly from measurements under vacuum conditions, the roles of adsorbed layers, disordered surface layers, and the possible charge channels in the bulk immediately below the surface stratum should be separable. There is also now a distinct possibility of obtaining precise information about the behaviour of antistatic additives and the problems of surface conduction where high-voltage insulation is involved. A further dimension has also been added to the study of the metal—polymer interface and the nature of charge transitions across it. Although not stressed in the present report, the ability, via the computer to obtain two-dimensional charge maps is also important, since practical surfaces can be notoriously non-uniform in their conductive and antistatic properties.

ACKNOWLEDGEMENTS

The author acknowledges with pleasure the very considerable help and stimulation given by Dr. E. A. Baum and Mr. R. Toomer in the many aspects of this work.

REFERENCES

1. W. R. Harper, *Contact and Frictional Electrification*, Oxford, London (1967).
2. T. J. Lewis, *1976 Ann Rep. Conf. on Electrical Insulation and Dielectric Phenomena*, Nat. Acad. Sci. Washington (1977).
3. D. M. Taylor and T. J. Lewis, *J. Phys. D: Appl. Phys.*, **4**, 1346 (1971).
4. N. Parkman, 'Electrical properties of high polymers' in *Physics of Plastics*, ed. P. D. Ritchie, Iliffe, London (1965).
5. N. Karl, 'Organic semiconductors, Festkorperprobleme XIV', *Advances in Solid State Physics*, Pergamon—Vieweg, p. 261 (1974).
6. H. Krupp, *Static Electrification, 1971*, IPPS, Conf. Ser. No. 7, Inst. of Phys. London, p. 1 (1971).

7. H. Bauser, *Electrostatische Aufladung,* Dechema-Monographien, Band 72, No. 1370–1409, Verlag Chemie, Weinheim-Bergstrasse, p. 11 (1974).
8. M. Henzler, *Surface Physics of Materials,* ed. J. M. Blakely, Academic Press, London, p. 241 (1975).
9. D. K. Davies, *Static Electrification,* IPPS Conf. Ser. No. 4, Inst. of Phys., London, p. 29 (1967).
10. D. K. Davies, *Brit. J. Appl. Phys. (J. Phys. D),* 2, 1533 (1969).
11. T. J. Fabish, H. M. Saltsburg, and M. L. Hair, *J. Appl. Phys.,* 47, 930 (1976).
12. T. J. Fabish, H. M. Saltsburg, and M. L. Hair, *J. Appl. Phys.,* 47, 940 (1976).
13. A. Many, Y. Goldstein, and N. B. Grover, *Semiconductor Surfaces,* North-Holland, Amsterdam (1965).
14. C. G. Garton, *J. Phys. D: Appl. Phys.,* 7, 1814 (1974).
15. D. K. Davies, *J. Sci. Instrum.,* 44, 521 (1967).
16. A. Chowdry and C. R. Westgate, *J. Phys. D: Appl. Phys.,* 7, 713 (1974).
17. H. J. Wintle, *J. Phys. D: Appl. Phys.,* 7, 1128 (1974).
18. C. B. Duke and T. J. Fabish, *Phys. Rev. Lett.,* 37, 1075 (1976).
19. M. Ieda, G. Sawa, and U. Shinohara, *Electrical Engineering In Japan,* 88, 67 (1968).
20. M. Ieda, G. Sawa and U. Shinohara, *Japan J. Appl. Phys.,* 6, 793 (1968).
21. E. A. Baum and T. J. Lewis, *Static Electrification 1975,* IPPS Conf. Ser. No. 27, Institute of Physics, London, p. 130 (1975).
22. T. J. Lewis, E. A. Baum, and R. Toomer, *Proc. IIIrd Congress International de l'Electrostatique,* Grenoble, 1977, to be published.
23. E. A. Baum, T. J. Lewis, and R. Toomer, *J. Phys. D: Appl. Phys.,* 10, 487 (1977).
24. T. J. Sonnonstine and M. M. Perlman, *J. Appl. Phys.,* 46, 3975 (1975).
25. I. P. Batra, K. K. Kanazawa, and H. Seki, *J. Appl. Phys.,* 41, 3416 (1970).
26. I. P. Batra, K. K. Kanazawa, B. H. Schechtman, and H. Seki, *J. Appl. Phys.,* 42, 1124 (1971).
27. H. Seki and I. P. Batra, *J. Appl. Phys.,* 42, 2407 (1971).
28. H. J. Wintle, *J. Appl. Phys.,* 41, 4004 (1970).
29. H. J. Wintle, *Japan. J. Appl. Phys.,* 10, 659 (1971).
30. H. J. Wintle, *J. Appl. Phys.,* 43, 2927 (1972).
31. E. L. Zichy, *Advances in Static Electricity,* Vol. 1, European Fed. Chem. Engng, pp. 42–55, Vienna (1970).
32. R. H. Ottewill, *Static Electrification 1975,* IPPS Conf. Ser. No. 27, Institute of Physics, London, p. 56 (1975).
33. Z. Boksay, M. Varga, and A. Wikby, *J. Non-Cryst. Solids,* 17, 349 (1975).
34. D. Grant and E. C. Salthouse, *J. Phys. D: Appl. Phys.,* 10, 201 (1977).
35. G. Sawa and J. H. Calderwood, *J. Phys. C: Solid St. Phys.,* 4, 2313 (1971).
36. Y. Awakuni and J. H. Calderwood, *J. Phys. D: Appl. Phys.,* 5, 1038 (1972).
37. E. A. Baum, Ph.D. Thesis, University of Wales, U.K. (1975).
38. J. Crank, *The Mathematics of Diffusion,* Oxford, Clarendon Press, London (1967).
39. J. Crank and G. S. Park, *Diffusion in Polymers,* Academic Press, London (1968).
40. H. G. Van Beuren, *Imperfections in Crystals,* North-Holland, Amsterdam, p. 404 (1960).
41. C. G. Garton and N. Parkman, *Proc. I.E.E.,* 123, 271 (1976).

5

The Transport of Electrons from the Surface into the Bulk of Polystyrene

P. Keith Watson

Xerox Corporation, Rochester, N.Y.

1. INTRODUCTION

Highly insulating polymers such as polystyrene and polyethylene are not easily characterized from the electrical point of view. There is general agreement that their low conductivity is attributable to the small number of charge carriers of very low mobility, and that these properties in turn are associated with the high trap density in the polymer, but beyond that there is relatively little agreement.

Several experimental techniques have been used to study the electrical properties of polymers, and these have been largely concerned with the polymer bulk. The present work is based on the use of an electron-probe technique which enables one to study the motion of electrons as they drift from the surface into the polymer bulk. From this one is able to deduce properties such as electron range and the trapping parameters which effectively define the surface layer, which is related to other measurements such as contact charge exchange. We are not, therefore, concerned with the polymer surface *per se*, but with that part of the bulk which is in electrical contact with the polymer surface. Before turning to these measurements it is appropriate to review part of the literature on conduction in polymers in order to indicate what other measurement techniques have been used to study the electrical properties of these materials.

There are a large number of papers in the literature on electrical conduction of polymers between metal electrodes. This is by far the simplest measurement to perform on a polymer — hence, presumably, its popularity as a method. There are, however, problems in the interpretation of such measurements. In a homogeneous medium conduction is defined by the equation $\sigma = \Sigma \, n_i e \mu_i$. Clearly, a measurement of σ does not enable one to separate the number of charge carriers, their mobility, or their sign. Even so one can learn a great deal from such measurements, as shown by Barker.[1]

In his work on dc conductivity in polyethylene terephthalate Amborski[2] was able to show a marked structural dependence, field and temperature dependence of the conduction process; for example, crystallization reduced the conductivity by an order of magnitude. From these measurements Aborski deduced that the conduction in PET was ionic, not electronic in nature.

Saito *et al.*[3] studied dielectric relaxation and dc conductivity in PVC, PVAc, PCTFE, and amorphous PET as a function of temperature and applied pressure. The slope of the log σ versus $1/T$ plot changed at the glass transition temperature T_g and the change was associated with the increase in free volume in the polymer. The increase in free volume makes it easier for ionic charge carriers to move through the polymer (whereas it would decrease the range of electrons). On the basis of this and other evidence it was concluded that the charge carriers were ionic.

In the case of highly insulating polymers with a very low concentration of ionic charge carriers it is difficult to obtain useful information from low-field dc conductivity, though Miyamoto and Shibayama[4] have made conduction measurements on PS in the vicinity of T_g which indicate the presence of ionic charge carriers.

There is evidence that although the electrode—polymer interface acts as a blocking contact at low fields, carrier injection into the polymer can occur at high fields. Lengyel[5] studied electrical conductivity at high fields in PET and PVF and found that the current versus voltage curves followed a Schottky law over the range 20—200 kV/cm. From this he concluded that the currents were emission-limited and that the information obtained was about the metal—dielectric interface rather than about the polymer.

Taylor and Lewis[6] have used a similar approach to the study of high-field currents in PET and PE. Their results are not a good fit to a Schottky relationship, and they conclude that a more general type of injection barrier is needed to explain their results.

Lachish and Steiberger[7] have described transient conduction measurements on polystyrene films which indicate that the currents are predominantly along the surface rather than in the bulk of the material. To explain the fact that the charge carriers are confined to the surface layer of the polymer, Lachish and Steinberger introduce the idea that their electrodes are in contact with a surface layer having a conductivity which is higher than the polymer bulk.

This work raises an important question in the interpretation of those conduction measurements which rely upon high-field charge injection from electrodes into the polymer. This ambiguity of interpretation can be avoided by generating a charge carrier in the sample rather than relying upon electrode effects to create the carriers. Fowler[8] and others[9] have studied induced conductivity in polymers, using X-rays to generate the charge carriers. From the rate of decay of induced conductivity one may deduce values for the density of filled traps and the free-electron range. Hedvig[9] has summarized the work on radiation-induced conductivity in polymers.

An electron injection technique for the study of carrier transport in insulators was developed by Spear[10] and applied to polymers by Martin and Hirsch[11] and by Inuishi and his fellow workers.[12] In Spear's technique a thin sheet of electron—hole pairs is generated in the sample in a region adjacent to an electrode by a pulse of high-energy electrons or by light. Depending on the direction of the applied field, a pulse of electrons or holes can be extracted from the ionized region; these charge carriers drift across the sample under the influence of the applied field and the charge-carrier mobility is then deduced from the current transient.

There are three important requirements which must be met by a material for this time-of-flight measurement to work: the dielectric relaxation time must be longer than the transit time (otherwise the injected carriers are neutralized before they complete a transit); the carriers must not experience a wide distribution of trapping and release times in the material (otherwise the transit time becomes difficult to resolve); and the carriers must have a lifetime comparable to or greater than the transit time, so that sufficient carriers cross the sample without getting lost in deep traps.

These requirements are met by materials such as sulphur, selenium and a few other materials, but in the case of polymers, neither trapping nor lifetime requirements are easily met.

Martin and Hirsch used a modified form of this technique[11] and made observations of the current transients following the irradiating pulse of electrons. They concluded that one carrier species was very rapidly localized; that the mobile excess carrier decayed by bimolecular recombination (with a time constant in the order of 1 μsec); and that the mobile carriers could not complete a transit through the thickness of the material, but could be swept back into the bombarded electrode. From these measurements it was concluded that the mobile carriers are holes in polystyrene ($\mu = 1.0 \times 10^{-6}$ cm^2/volt sec at 20°C) and in polyethylene ($\mu = 4.5 \times 10^{-10}$ cm^2/volt sec at 80°C), and that electrons are mobile in PET ($\mu = 1.5 \times 10^{-6}$ cm^2/volt sec at 20°C).

Inuishi, et al.[12] have also measured carrier mobilities in PET and PS by the time-of-flight method, using an electron beam for carrier generation. By using very large values of applied field and relatively thin samples they were able to drive either sign of carriers across the sample and thus measure both electron and hole mobilities. In PET they obtained 2×10^{-5} cm^2/volt sec for the electron and 1×10^{-4} cm^2/volt sec for hole mobilities, with an activation energy of about 0.3 eV.

The migration and trapping of extrinsic charge carriers in polystyrene, polymethyl methacrylate and several other polymers have been studied by Reiser, et al.[13] They used a negative corona discharge to charge films of the polymers and from the rate of decay of surface potential, measured by an electrometer, they were able to deduce charge—carrier mobilities. Mobilities increased with solvent content of the polymer and with temperature. Charge trapping was indicated by persistent residual surface potentials; this was related to the injected charge density, film thickness, and carrier mobility. By adding substances that could act as trapping centres for the injected charges, they were able to increase the trapping.

Seiwatz and Richardson[14] used a uniform Townsend discharge to deposit the charge on to Mylar film and from measurements of charge deposited and surface potential they found that the depth of the injected space charge was of the order of 1 μm at a field strength of 10^6 volts/cm.

A different type of charge-injection process has been used by Davies[15,16] and by Fabish *et al.*[17,18] in their work on contact charge exchange. Electrons are injected into the polymer by a contacting metal, and from these measurements they infer that the charges have moved a short distance into the polymer.

Fabish *et al.*[17,18] have described measurements in which a series of metals was used in contact experiments on films of polystyrenes of various thicknesses. The results are interpreted in terms of an energy-selective injection process (each metal communicating with a discrete portion of the electronic state distribution in the polymer); the measurements indicate the presence of bulk states ($\sim 10^{14}/cm^3$) within a 2–4 μm surface or boundary layer, into which electrons can be injected from the Fermi level of the contacting metal.

The measurements of Davies and of Fabish *et al.*, taken in conjunction with those of Martin and Hirsch and of Inuishi *et al.*, indicate that the electron range in polymers such as PE and PS is a few micrometers in extent; this range is determined by the trapping properties of the polymer, and clearly, if one is to understand the transport of electrons from the surface into the bulk of a polymer, one needs to know more about this important property. The experimental programme described in section 2 was undertaken with this need in mind.

2. EXPERIMENTAL STUDY OF ELECTRON TRANSPORT IN POLYMERS

The aim of this work is to investigate the transport of electrons from the surface into the bulk of highly insulating polymers such as polystyrene and polyethylene. The experimental technique is based on the use of an electron beam to inject a space charge of electrons into the free surface of the polymer film under study. This space charge provides the electric field for the measurement and, in the case of highly insulating polymers, it provides the charge carriers for the conduction process. Thus, we apply a pulse of charge to the sample and this charge produces a voltage on the free surface of the polymer film; the experiment then consists of the observation of the decay of this surface potential.

By observing the surface potential as a function of time one may distinguish between charge decay due to ionic charge carriers within the polymer (giving rise to exponential charge decay) and charge decay due to motion of the injected carriers. In the latter case, if the electron range is long enough, one is able to deduce the mobility of the electrons and their range as they move into the polymer; one also obtains information about the electron trapping parameters of the material. Thus, we use the drifting electrons as a probe with which to study the transport properties of the polymer.

The apparatus is shown in Figure 1. Two electron beams are used in the experiment: one beam injects charge into the free surface of the polymer film, the other beam monitors the surface potential of the film.[19,20]

In the present form of the experiment the polymer films are charged with a 2 keV electron beam, which scans over a 4 x 4 cm^2 area of the sample surface. The writing gun is automatically turned off when the required amount of charge has been deposited on the sample. In these measurements charging times are typically about 1 sec, but in order to follow the relatively fast transport processes which may

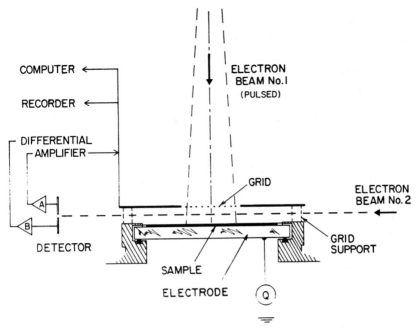

Figure 1. Schematic diagram of electron beam apparatus. Electron beam No. 1 supplies the pulse of electrons to charge up the free surface of the sample. Electron beam No. 2 monitors the surface potential.

occur in the surface layers of some polymers much shorter charging times are required.

The second electron beam is used to monitor the surface potential of the polymer film. This read-out beam passes between the sample surface and a reference grid situated above the charged area of the polymer film, as shown in Figure 1. The beam is collected on a split anode which is connected via a differential electrometer to a high-gain amplifier whose output is fed back to the reference grid. This feedback loop holds the read-out beam near its null position, and in so doing maintains the reference grid at the same potential as the sample surface, and in effect keeps the electric field equal to zero immediately above the sample. The grid potential is recorded and thus one obtains a direct reading of the surface potential of the sample and its variation with time.

(i) Charge decay from an ideal dielectric

The principle of the method can be seen most easily in the behaviour of a model dielectric, defined as one in which there are no deep traps, in which the charge carriers have a single value of mobility, and in which the injected carrier lifetime is much longer than the transit time (defined below).

Figure 2 shows the simple electrostatic picture of such a dielectric a short time after the injection of a pulse of charge Q, with the charge drifting under its own

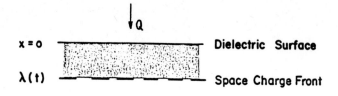

$$E(\lambda) = \frac{Q}{\epsilon}$$

$$\frac{d\lambda}{dt} = \mu E(\lambda)$$

$$V(t) = V_1 \left[1 - \frac{t}{2\tau} \right], \quad \tau = \frac{L^2}{\mu V_1}$$

Figure 2. Flow of charge in an ideal trap-free dielectric in which charge is drifting under the influence of its own space-charge field. The field rises linearly through the space charge, reaching a value Q/ϵ at the space-charge front.

space-charge field towards the substrate. The electric field above the dielectric is zero as noted previously and within the sample the electric field rises linearly through the space charge. The field at the space charge front λ is given by $E(\lambda) = Q/\epsilon$ and the space-charge density everywhere behind the carrier front is

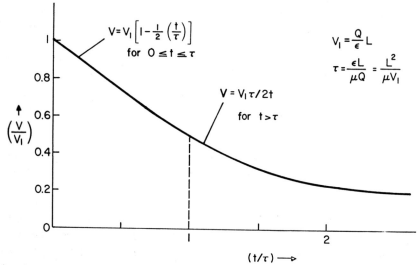

Figure 3. Decay of voltage from a pulse of charge in an ideal trap-free dielectric. In the time from zero to one transit time the voltage falls linearly from V_1 to $V_1/2$.

$q(x, t) = \epsilon/\mu t$, where μ is the carrier mobility and ϵ is the dielectric constant.[21] The velocity of the carrier front is given by

$$d\lambda/dt = \mu E\,(\lambda)$$

so that the carrier front arrives at the substrate in the transit time τ_1 given by $\tau_1 = \epsilon L/\mu Q = L^2/\mu V_1$, where L is the film thickness and V_1 is the initial potential due to charge Q. For this ideal case the velocity of the carrier front is constant for $t < \tau_1$. The decay of surface potential is shown in Figure 3, and is given by

$$V(t) = V_1\,[1 - t/2\tau_1] \quad \text{for } 0 \leqslant t \leqslant \tau_1$$

For t less than one transit time the surface potential decays linearly as the carrier front advances linearly across the gap. At time τ_1 the voltage has decayed to $V/2$, at which point the average carrier is halfway across the sample. Thus, during the initial part of the charge decay the voltage decrement gives an indication of the average position of the charge carriers in the film. (For $t > \tau$ the charge is collected at the substrate and the surface potential follows the hyperbolic law $V(t) = V_1 \tau_1/2t$. For present purposes, however, we are only concerned with times less than one transit time.)

The slope of the voltage decay curve is proportional to the carrier mobility:

$$dV/dt = V_1/2\tau_1 = \mu V_1^2/2L^2$$

This relationship provides a very useful test for the type of discharge:

(1) In the case of the ideal trap-free system postulated above, $1/V_1^2 \cdot dV/dt$ is constant for $0 < t < \tau_1$ (in some liquid dielectrics, for example, one observes this ideal behaviour).

(2) For a dielectric in which conduction is due to ionic species and where carrier lifetime is determined by ionic recombination processes within the material, charge decay is exponential, so that $1/V(t) \cdot dV/dt$ is constant and equal to the reciprocal of the dielectric relaxation time.

(3) In the case of highly insulating polymers, electron transport is limited by traps in the polymer and one finds that $1/V_1^2 \cdot dV/dt$ falls with time; the rate of this decrease is a measure of the trapping in the polymer.

(ii) Charge transport with trapping

The decay of charge from a highly insulating polymer differs significantly from the simple, ideal case discussed in section 2(i). Typical experimental results are shown in Figure 4. One finds that the rate of decay is not constant, but decreases with time long before the space-charge front has moved through the sample. Moreover, a high residual potential remains on the film; this indicates that charge carriers have become trapped in the polymer and thermal release rates are so slow that charges remain trapped indefinitely. Moreover, the electron mobility that one observes is extremely small and is temperature activated. For comparison, in well-defined organic crystals such as anthracene, tetracene, and phthalocyanine, charge—carrier mobilities are of order unity, and fall with increasing temperature.[22]

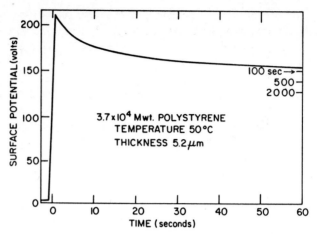

Figure 4. Decay of surface potential of polystyrene films. Typical decay curve for temperatures below T_g. The rate of decay is essentially zero for times longer than 1000 sec.

A simple trapping scheme which is consistent with the gross features of transport of this type is shown in Figure 5. The two sets of traps shown are characterized by their capture rates ω_1 and ω_2 and release rates r_1 and r_2. The shallow traps have a release rate $r_1 = \exp(-E_1/kT)$ and the deep traps have a release

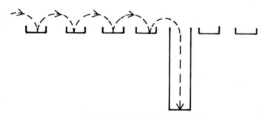

SIMPLE TRAPPING MODEL

SHALLOW TRAPS: CAPTURE RATE ω_1

RELEASE RATE $r_1 \sim e^{-E_1/kT}$

CARRIER MOBILITY $\simeq \mu_0 e^{-E_1/kT}$ for $\omega > r$

DEEP TRAPS: CAPTURE RATE ω_2

RELEASE RATE $r_2 = 0$

NUMBER OF CARRIERS $\simeq e^{-\omega_2 t}$

Figure 5. Trapping model for polystyrene. The shallow traps, 0.7 eV deep, determine the electron mobility. The deep traps, which determine carrier range, are too deep to be thermally ionized on this experimental time scale.

rate which approaches zero. If the capture rate for the deep traps is much less than for the shallow traps, then the carrier mobility is determined by the equilibrium of the charge carriers with the shallow traps, and the mobility has the form

$$\mu(T) \cong \mu_0 \exp(-E_1/kT)$$

where μ_0 defines a 'free mobility' of the carrier and E_1 is the depth of the shallow traps.

The effect of the deep traps is to reduce the total number of mobile charge carriers, and for a capture rate ω_2 the number of mobile carriers at time t is approximately given by

$$N(t) = N_0 \exp(-\omega_2 t)$$

Introducing these values into the decay equation discussed in section 2(i) one obtains a useful approximate solution to the decay equation

$$\frac{dV}{dt} \simeq \mu(T) \frac{V_1^2 \exp(-\omega_2 t)}{2L^2}$$

By extrapolating the results back to $t = 0$ one obtains the initial slope and hence the carrier mobility.

(iii) Multiple trapping

In the preceding section we have taken a heuristic approach to electron trapping in order to sketch out the general effects of traps on carrier transport. For a complete analysis of trapping one needs a more precise description of the process, involving more than one set of traps. Recently a number of papers have been written on multiple trapping by Schmidlin[23] and Noolandi.[24] It has been shown that one may obtain a relationship between the trapping parameters ω and r, and the transport equation is via the Laplace transformation.

Using this approach to multiple trapping, one can solve for any number of traps. For example Noolandi has analysed the experimental results of Pfister on the transport of holes in amorphous selenium over a wide range of temperatures, electric fields, and sample thicknesses. He finds that all the observed characteristics can be understood quantitatively in terms of the multiple-trapping model and that for such a model one need invoke only three sets of traps, whose capture and release times are obtained by computer fitting the analytical solution and the experimental results.

3. EXPERIMENTAL MEASUREMENTS ON POLYSTYRENE

The majority of our measurements have been made on monodisperse polystyrenes of 2×10^6, 1×10^5, 3.7×10^4 and 1×10^4 mol. wt. Some measurements have been made on a commercial polystyrene of broad molecular weight. The polymer films are cast from solution in benzene directly on to the surface of the metal substrate, with sample thicknesses in the range $2.5-20$ μm. The substrate consists of a copper electrode whose temperature is controlled by thermostat. Other substrate materials are plated or evaporated on to the copper surface. The metals used include nickel, gold, indium, and aluminium. The samples are baked-out in the vacuum system at 150°C for several hours and are then held overnight at 130°C before testing. Unless care is taken with the sample preparation, conduction due to residual solvent tends to dominate over the drift of the injected carriers. This is particularly evident with the higher molecular weight material because of its extremely high viscosity, even at 150°C, which inhibits the release of solvent. However, by following the bake-out procedure outlined above, we are able to remove virtually all the residual solvent and thus obtain a charge decay which is characteristic of the polymer.

(i) Experimental results below the glass transition temperature

We have studied the transport of electrons in polystyrene above and below the glass transition temperature ($T_g \sim 95-100°C$). Results above T_g are significantly different from those below T_g and will be dealt with separately.

A typical discharge curve for polystyrene is shown in Figure 4. The results were obtained on a $5.2\,\mu m$ sample 3.7×10^4 mol. wt. at $50°C$. The decay curve is characterized by an initial rate of discharge which decreases with time after the first few seconds and ceases altogether in some hundreds of seconds, leaving a high residual voltage on the films.

The high residual voltage indicates that the electrons are trapped within the polymer films, and this trapped charge can only be released by heating the sample above T_g.

The fact that the initial discharge is approximately linear suggests that when the electrons are first injected into the polymer they are able to move relatively freely before becoming deeply trapped. If this is correct, then, for a short time after being injected, the electrons will be drifting in their own space-charge field and the decay should exhibit space-charge-limited (SCL) behaviour, modified by the presence of traps in the polymer.

As already noted, in the case of an ideal dielectric in the trap-free limit the surface potential decays linearly to half its initial value in one transit time τ_1 and the rate of decay is given by $dV/dt = -\mu V_1^2/2L^2$.

If trapping occurs, these equations are modified and the departure from linear decay gives information about the trapping process. For example in the case of one set of deep traps with $r = 0$ (zero release rate) the number of free carriers decays exponentially and we have $N(t) = N_0 e^{-\omega t}$. The slope of the voltage decay curve then has the form $dV/dt = \mu V_1^2 e^{-\omega t}/2L^2$. We use this equation to test for trap-modified SCL conduction, and Figure 6 shows the charge decay data for a $5\,\mu m$ sample of polystyrene (3.7×10^4 mol. wt.) at 50, 70, and 90°C. These results agree reasonably well with the model at short times. At longer times, however, the simple model fails and one has to decide between the multiple-trapping model and a model in which trap density varies with distance into the polymer.

Extrapolating the results in Figure 6 to zero time gives the initial rate of decay,

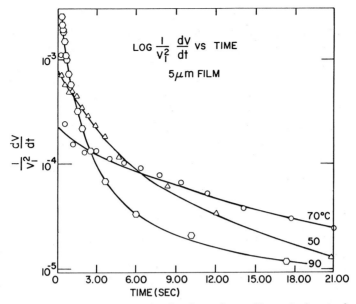

Figure 6. Normalized rate of voltage decay. For a single set of deep traps which are uniformly distributed in the polymer, the graphs of log $1/V_1^2 \cdot dV/dt$ should be linear in time.

and using this we can estimate the initial carrier mobility. A reasonable estimate for the electron mobility in the material at 50°C is between 0.8 and 1 x 10^{-10} cm²/-volt sec. At higher temperatures the carrier mobility increases rapidly. The results are subject to error because the charging time is comparable to the electron-trapping time, but mobilities at 70 and 90°C are of the following order of magnitude:

$$70°C \; \mu \sim 4 \times 10^{-10} \; cm^2/volt \; sec$$
$$90°C \; \mu \sim 2 \times 10^{-9} \; cm^2/volt \; sec$$

Values of electron mobility versus $1/T$ are plotted in Figure 7. From the slope of the curve one deduces an activation energy of 0.7 eV, and by extrapolation of the curve to $1/T = 0$ we obtain μ_0 of order unity.

We now turn to the question of carrier range. As noted in section 2(ii) the curve of $V(t)$ gives an approximate indication of the position of the injected carriers in the polymer; for example, when the surface potential has decayed, say, 10% from its initial value V_1, the average charge carrier will have moved 10% of the way through the sample. Clearly, this is accurate only for an ideal, trap-free dielectric, but it provides a useful indication of carrier position, even for a situation where trapping occurs, and a plot of decay rate versus voltage decrements enables one to estimate carrier range.

As an example of the use of this relationship we have plotted the experimental

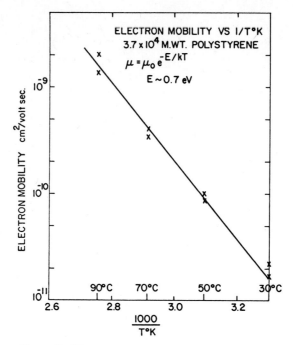

Figure 7. Electron mobility in polystyrene (3.7 × 10^4 mol. wt.). The activation energy is 0.7 eV and the free mobility is of order unity.

results from Figure 4 in the form

$$\frac{1}{V_1^2}\frac{dV}{dt} \quad \text{versus} \quad \frac{V_1 - V(t)}{V_1}$$

where $(V_1 - V(t))/V_1$ is the voltage decrement. This is shown in Figure 8. Other data from the same sample and same temperature, but different initial voltages are plotted on the same figure.

The graph is a rough profile of number carriers versus distance into polymer, and provides a useful measurement of carrier range in the material. For this particular material, the rate of voltage decay (and the number of carriers) has fallen to $1/e$ in about 0.5 μm. Beyond this point trapping becomes more rapid and about 95% of the injected carriers are trapped in 1 μm.

If now one repeats this experiment at various temperatures and if the multiple-trapping model is appropriate, with carriers trapped and released from traps of various depths, one would expect to see an increase in carrier range with increasing temperature. If, on the other hand, only very deep traps are involved, then the electron range should not increase significantly with increasing temperature. The results in Figure 9 show that the latter is true, suggesting that carrier range is dominated by one species of very deep trap and that negligible detrapping

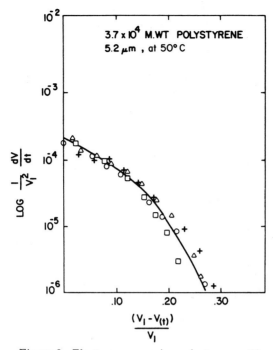

Figure 8. Electron range in polystyrene. The voltage decrement $(V_1 - V(t))/V_1$ is approximately proportional to the average carrier position in the film.

occurs, even at $90°C$. These results also suggest that trap density is relatively low near the sample surface and increases as one moves into the polymer bulk.

(ii) Experimental results above the glass transition temperature

Following the electron transport and trapping processes described above, the electrons end up frozen in deep traps in the polymer from which they are thermally released at a negligible rate; below T_g this means that virtually no charge decay occurs once the electrons have reached the end of their range, as shown in Figures 8 and 9. Above T_g, however, the situation is quite different: in the rubbery polymer the chain segments are free to move, and this motion of the molecules occurs by the cooperative movements of segments of the chain and the rate of this motion is a function of the free-volume fraction in the polymer.

Presumably, the cooperative motion of the chain segments gives rise to changes in the trapping environment in which the electrons are held. There is evidence that some of the traps are associated with microscopic voids in the polymer;[25] in such cases the traps will change and even disappear at a rate which depends on the motion of the polymer chain. Thus, electrons will be detrapped after some waiting period and then move on until they encounter another trap. This has a striking

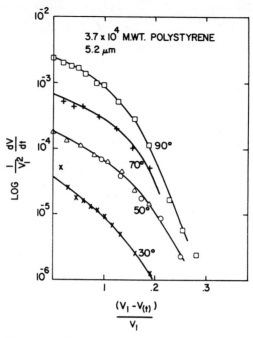

Figure 9. Electron range in polystyrene at 30, 50, 70, and 90°C. The range of the electrons is approximately independent of temperature.

effect on charge decay from the polymer, as may be seen from Figure 10, which shows curves of voltage versus time for temperatures above and below T_g.

The type of transport which we have outlined can be described in terms of a random-walk type of motion, with a distribution of waiting times determined by the micro-Brownian motion of the polymer chains. Moreover, because of the space-charge field which the injected electrons create in the polymer, the motion of the individual electrons will not be entirely random but will exhibit a field-dependent bias.

Scher and Montroll have developed a stochastic transport model for the motion of carriers undergoing a time-dependent random walk in the presence of a field-dependent bias.[26] The time dependence of the random walk is governed by a waiting time distribution $\psi(t)$. In any amorphous material there is a wide variation in the separation of nearest-neighbour sites and in the potential barrier between the sites, and both of these factors affect the waiting time (the time between successive carrier hops). This gives rise to waiting-time distributions with long tails, corresponding to long waiting times. A waiting-time distribution which has been examined in detail by Scher and Montroll has the form $\psi(t) \propto t^{-3/2}$. Based on this $\psi(t)$ they show that for long times the mean position of the carrier packet is $\langle l \rangle \propto \bar{l}\sqrt{t}$, where $\bar{l}$ is the mean displacement per step. Current, which is proportional

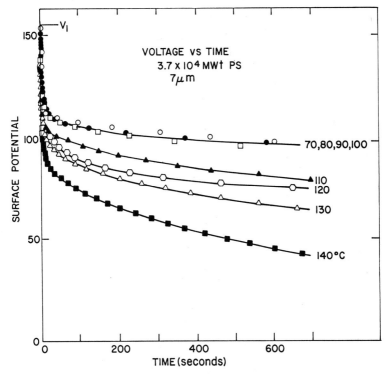

Figure 10. Decay of surface potential of polystyrene film at temperatures above and below T_g. The transition at T_g is clearly evident.

to the velocity of the carrier packet, is then given by $I(t) \propto d \langle l \rangle / dt \propto \bar{l}/\sqrt{t}$. Thus, current decreases with increasing time, even in the absence of a collecting electrode. This is because, as time increases, more and more carriers encounter long waiting times.

The mean displacement per step $\bar{l}$ is field dependent through the increased probability of hopping in the field direction. This can only be determined empirically. (For example, in As_2Se_3 it is found that the experimental data can be fitted by assuming a local asymmetry which is linear in field.)

In the present case we find that the result can be fitted by a weak field dependence of the form $\bar{l} \propto \sqrt{E}$. Then, $d \langle l \rangle / dt \propto \sqrt{(E/t)} \propto \sqrt{(V/Lt)}$. Current is given by the product of the number of carriers and their velocity. Hence, $I(t) \propto \epsilon V/L\sqrt{(V/Lt)}$. In our experiments we observe surface potential rather than current, but current is proportional to the rate of change of voltage, so we look for an experimental relationship between dV/dt and $\sqrt{(V^3/L^3 t)}$.

In Figure 11 we show a series of experimental results for polystyrene of molecular weight 2×10^6 plotted in this form. The result above T_g shows an excellent fit to the stochastic hopping model.

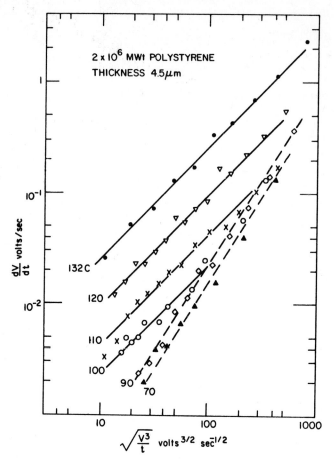

Figure 11. Rate of decay of surface potential versus the stochastic variable $\sqrt{(V^3/t)}$. In this logarithmic plot the slope is unity for temperatures above T_g.

4. CONCLUSIONS

The transport of electrons from the surface into the bulk of polystyrene can be characterized as a trap-modified, space-charge-limited process. We are able to observe both shallow and deep traps in polystyrene. The shallow traps determine the effective carrier mobility and the deep traps determine the penetration range of the electrons into the polymer. At temperatures below T_g, once the electrons have fallen into the deep traps in the polymer, they remain frozen into these traps indefinitely.

Above the glass transition temperature this initial stage in electron transport is followed by a much slower decay whose time dependence has the characteristic of stochastic hopping, and which we associate with segmental motion in the rubbery polymer.

REFERENCES

1. R. E. Barker Jr., *Pure and Appl. Chem.*, **46**, 157 (1976).
2. L. E. Amborski, *J. Polymer Sci.*, **62**, 331 (1962).
3. S. Saito, H. Sasabe, T. Nakajima, and K. Yada. *J. Polymer Sci.*, A2, **6**, 1297 (1968).
4. T. Miyamoto and K. Shibayama, *Polymer J.*, **6**, 79 (1974).
5. G. Lengyel, *J. Appl. Phys.*, **37**, 807 (1966).
6. D. M. Taylor and T. J. Lewis, *J. Phys. D: Appl. Phys.*, **4**, 1346 (1971).
7. U. Lachish and I. T. Steinberger, *J. Phys. D: Appl. Phys.*, **7**, 58 (1974).
8. J. F. Fowler, *Proc. Roy. Soc.*, **A236**, 464 (1956).
9. P. Hedvig, *Radiation Chemistry of Macromolecules*, Ch. 8, ed. M. Dole (Academic Press, New York 1972).
10. W. E. Spear, *J. Non-Cryst. Solids*, **1**, 197 (1969).
11. E. H. Martin and J. Hirsch, *J. Appl. Phys.*, **43**, 1001, 1008 (1972).
12. Y. Inuishi, K. Hayashi, and K. Yoshiro, *Japan, J. Appl. Phys.*, **14**, 39 (1975).
13. A. Reiser, M. W. B. Lock, and J. Knight, *Trans. Faraday Soc.*, **65**, 2168 (1969).
14. H. Seiwatz and D. E. Richardson, *I.E.E.E. Trans. Audio*, **AU-12**, 63 (1964).
15. D. K. Davies, *J. Phys. D: Appl. Phys.*, **2**, 1533 (1969).
16. D. K. Davies, *J. Phys. D: Appl. Phys.*, **5**, 162 (1972).
17. T. J. Fabish, H. M. Saltsburg, and M. L. Hair, *J. Appl. Phys.*, **47**, 930, 940 (1976).
18. T. J. Fabish and C. B. Duke, Chapter 6, this book.
19. P. K. Watson and T. M. Clancy, *Rev. Sci. Instr.*, **36**, 217 (1965).
20. P. K. Watson, J. M. Schneider and H. R. Till, *Physics of Fluids*, **13**, 1955 (1970).
21. J. M. Schneider and P. K. Watson, in *Proc. 1970 Conf. on Elec. Ins. and Diel. Phenom.*, p. 125 (N.A.S., Washington, 1971).
22. H. Meier, *Organic Semiconductors* (Verlag Chem., Weinheim, 1974).
23. F. W. Schmidlin, *Phys. Rev.*, B6 2362 (1977)
24. J. Noolandi, *Phys. Rev.*, to be published.
25. D. A. Copeland, N. R. Kestner, and J. Jortner, *J. Chem. Phys.*, **53**, 1189 (1970).
26. H. Scher and E. W. Montroll *Phys. Rev.*, **B12**, 2455 (1975).

6

Molecular Charge States in Polymers

T. J. Fabish and C. B. Duke

Xerox Corporation, New York

1. INTRODUCTION

Our purpose in this paper is to develop the hypotheses that charges injected into pendant-group polymers form molecular ion states and that these states are directly accessible experimentally via contact charge exchange of these polymers with various metals. The first hypothesis provides an alternative model to the traditional view[1-5] that, in so far as their electrical properties are concerned, polymers behave like band semiconductors with traps. In particular, our molecular-ion-state model[6-8] predicts that the low-energy charge states in polymers are both intrinsic and localized, in contrast to the semiconductor model in which the intrinsic states are regarded as delocalized (energy-band) states and the localized states active in conduction processes are regarded as extrinsic defect states. Indeed, the molecular-ion-state model predicts that the intrinsic charge-carrying states in polymers are more nearly analogous to ions in solution, although the detailed description of the conductivity differs in the two cases.[7,8] Consequently, the molecular-ion-state model differs from the traditional semiconductor models conceptually as well as in the technical details of its predictions.

The critical issue on which we focus our attention is the relationship between the triboelectric properties of polymers and the chemical structure of the macromolecules which comprise them. Recent studies[9,10] have established empirical correlations between these two quantities, although microscopic interpretations of these correlations were not advanced. More commonly, the triboelectric as well as the electrical properties of polymers are interpreted, *a posteriori*, in terms of the semiconductor model. In this model various 'trap' distributions,[11,12] whose microscopic character remains unspecified, are proposed to describe specific experiments[3,11] independent of the fact that it is known[8] that the origin and character of 'traps' in molecular solids and polymers are conceptually different than in (covalent) semiconductors. Specifically, in covalent solids, locally unsaturated ('dangling') bonds cause most traps, whereas in molecular solids polarization fluctuations are primarily responsible for them.[8] Traps of the sort common in covalent solids may occur in conjunction with chain terminations or possibly cross-linking in polymers. They are not, however, naturally associated with the character of the pendant groups in pendant-group polymers. Thus, the recently established correlations[9,10] between local chemical composition and triboelectric charging constitute strong if indirect evidence favouring the molecular-ion-state model over the traditional semiconductor model.

The substance of our presentation herein is based on a recent series of papers in which we reported measurements[13-15] of metal polymer contact charge exchange and injection, as well as the development of our molecular-ion-state model.[6-8] The molecular ion states are identified with relaxed, occupied, 'virtual' frontier orbitals (anion states) or with relaxed, empty, 'occupied' frontier orbitals (cation states), together with the accompanying electronic and atomic polarization cloud in the surrounding polymer medium. The state energies fluctuate by as much as $\Delta \sim 0.5$ eV from site to site because of thermal vibrations and local structural variations.[6] These fluctuations, in turn, cause[7,8] complete localization of the molecular ion states as well as their occurrence as broad distributions, evident in contact charging spectra and photoemission spectra obtained on thin solid films.[7] In other words, the spatial inhomogeneity inherent in polymer structures causes the energies of nominally identical molecular-ion states lying within a hopping distance of one another to differ by considerably more than the value of the intersite exchange integrals. In this event, a charge carrier at a specific molecular site possesses a vanishingly small probability density beyond its localization site. Furthermore, fluctuations in the local site energies of tenths of an electronvolt cause correspondingly broad distributions in the eigenvalues of the molecular ions, as revealed by spectroscopic probes of these distributions.

We endeavour to show how these concepts combine to provide an internally self-consistent, detailed framework for the discussion of charge states in polymeric insulators and of the role of these states in the phenomenon of contact charge exchange. We proceed by describing first the contact-charging experiments and then their interpretation in terms of a deduced distribution in energy of molecular ion states. We next develop a formalism by means of which we describe analytically the steady-state metal/polymer contact-charge exchange as a functional of the experimentally deduced distributions of insulator charge states. Finally, the model is tested by applying it to *predict* the contact-charge-exchange characteristic of a particular copolymer system.

2. EXPERIMENT

The apparatus, materials, contact-potential-difference measurements, and contact-charge-exchange experiments have been described previously.[13] Briefly, polymer films are solvent cast on flat disc electrodes ($\frac{11}{16}$ in in diameter), vacuum dried and installed on supports that are electrically insulating, but which allow the films to be heated (to return contact-charged films to a charge-neutral state for re-use). The contacting metals were flat polished discs ($\frac{9}{16}$ in in diameter) constructed from high-purity polycrystalline foils of indium, lead, gold, nickel, platinum, and tin. All measurements were made in a vacuum of about 10^{-4} Torr and with the polymers in their glassy state ($T \ll T_g$).

The film potential is the integral over the film thickness of the local electric field associated with the charge injected into the polymer film during contacts with one of the metals. It is measured by a standard technique.[16] Repeated contacts of a metal to a polymer film produce a steady-state value of film potential peculiar to the particular polymer film and metal. The object of the experiment is to investigate the dependencies of this steady-state film potential on the contacting metal, the film thickness, and the substrate metal.

Experimental results for five polymers are utilized in this work. They are poly(styrene) (PS), poly(methyl methacrylate) (PMMA), and three styrene/methyl methacrylate copolymers. The data for PS have been reported previously,[13] but are utilized here in a slightly different fashion in view of the new results for the other polymers.

3. RESULTS

The raw data obtained in our experiments are measurements of the steady-state potential of a polymer film as a function of its thickness. The observed[13] quadratic dependence reveals that the polymer states accessed by the injection process are 'bulk' states within a boundary layer 1—5 μm thick. We recall[13] that the charge injected by different metals is both additive and commutative, indicating that each metal injects charge into a different and unique set of polymer electronic states. This result is particularly vividly illustrated by PMMA which reversibly charges both positively and negatively, using Sn and In contacts, respectively.[14]

As described earlier,[13] the key to interpreting these data is the observation that the work functions of the metals establish an energy scale: larger work function

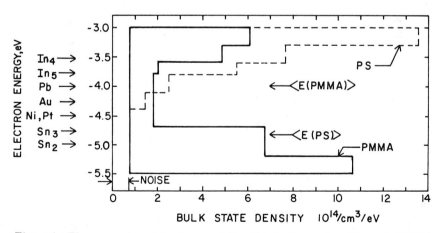

Figure 1. The experimental charge state distributions for PS and PMMA deduced by correlating the magnitudes of the charge injected by various contacting metals to the metallic Fermi levels as described in Fabish et al.[13] The Fermi level, defined as the enery level dividing (electron) acceptor states from donor states, is indicated by $\langle E(i) \rangle$ for PS and PMMA. The metals employed apparently inject charge into states extending from −3.5 to −5.0 eV.

metals injecting into lower energy polymer states. Thus, by arguing that each metal injects into a narrow band of states ($\Delta E \sim 0.4$ eV) just below (electron injection) or above (hole injection) its Fermi energy, one can convert the injected charge density measurements into distributions ('densities') of occupied charge states as a function of energy.[6-8,13,14] The fact that higher energy states do not decay quickly into lower energy ones (the commutativity property) is a consequence of the inhomogeneity of the energy distributions.[6-8] Thus, we refer to polymers as 'Fermi glasses'. The resulting charge-state distributions for PS and PMMA are shown in Figure 1.

The distributions of Figure 1 describe all of the features of the single and sequential metal contact experiments for PS and PMMA. In the case of PMMA, both (electron) acceptor and donor states appear in the energy range accessed by common metals, and a minimum occurs in the distribution separating the two types of charge states. The data for PS suggest that acceptor states in this material extend to energies below those probed by the highest work-function metal used in the contact experiments. The remainder of this paper is devoted to the construction and testing of a microscopic model of these polymer charge states.

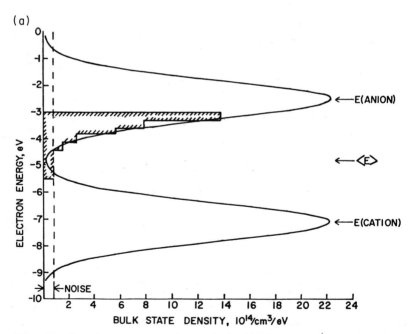

Figure 2. Gaussian representations of the solid-state anion and cation states. The measured state distributions for PS (a) and PMMA (b) are conveniently represented by double Gaussians to illustrate the displacement of the distribution centroids due to the difference in electronic structures of the charge localization sites in the two materials. The double Gaussian fits to the experimental contact charging data, which cover the

4. MODEL OF POLYMER CHARGE STATES

A molecular model for the charge states in polymers has been presented elsewhere.[6-8] An extension of that model is advanced herein that permits a particularly graphic description of the metal/polymer contact charge exchange.

To make the connection between the observed polymer charge states and molecular structure, it is envisioned that the thermodynamic drive (i.e. difference in electrochemical potentials) existent in a metal/polymer contact causes transfer of an electron residing at the Fermi level of the metal into the lowest virtual molecular orbital of the polymer structure, or the transfer of a hole into the highest valence state. For materials like PS and PMMA, these extremal eigenvalues can be identified as arising from the frontier orbitals associated with the π-electron states in the phenyl (PS) and ester (PMMA) groups.[6] Therefore, a direct correspondence exists between the molecular structure and the electronic (i.e. charge-exchange) properties of these materials. In the polymer state, however, large relaxation phenomena shift the molecular eigenvalues by several electronvolts (electrons move down in energy, holes up) relative to their gas-phase values.

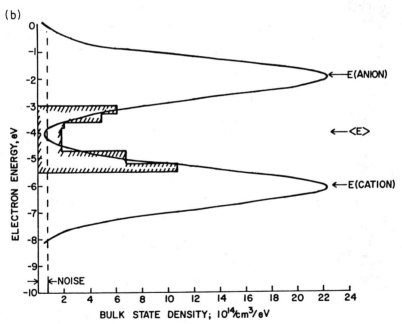

energy interval -3.5 to -5.0 eV as indicated by the hatched areas under the curves, are specified by their peak positions, halfwidth, and the normalization of the state density. The experimental results for PS and PMMA (in parentheses) are represented by Gaussians having E (acceptor peak) $= -2.5$ eV (-2.0 eV), E (donor peak) $= -7.1$ eV (-6.1 eV), half-width $= 0.845$ eV for both, and centroid $\langle E \rangle = -4.8$ eV (-4.1 eV). State densities are normalized to 4×10^{15} states/cm^3 in both cases.

Of equal importance to the magnitude of these relaxation energy shifts, however, is their explicit dependence on the local environment[6-8] and geometry[17] of the molecular moieties. It is the variation in these quantities from site to site in the polymer boundary layer that leads to a wide, inhomogeneously broadened distribution of charge states. In particular, the filled, fully relaxed molecular charge states can be represented by a double Gaussian, one peak of which is centred at the solid-state molecular anion energy (electron acceptor states), and the other at the solid-state cation energy (electron donor states).

Specific representations appropriate to PS and PMMA are given in Figure 2. Charge states lying in the extreme tails of the distributions govern the contact-charge-exchange characteristics of polymers. The centroid of the distribution of anion and cation states, $\langle E \rangle$, holds special significance in understanding the direction of the charge transfer in a contact; its position relative to the Fermi level of the contacting metal defines the thermodynamic drive.

5. ANALYTICAL DESCRIPTION OF METAL/POLYMER CHARGE TRANSFER

Knowledge of the distribution of the polymer charge states permits a concise analytical description of the metal/polymer contact charge exchange.

In order that a charge carrier be able to transfer from one material to another, a filled state must exist in the one and an empty state in the other, consistent with the energy and momentum constraints upon the microscopic transition process. In a proper theory of the process[18] its probability decays exponentially for energies deviating from the Fermi energy of the metal. Hence, each metal injects charge into polymer charge states within a narrow energy range, ΔE, of its Fermi energy, as indicated schematically (and referred to as the injection 'window') in Figure 3.

Unfortunately, the microscopic mechanism of contact charge exchange is not yet clearly established. Consequently, a detailed model of the rate of exchange is of little value. Nevertheless, the saturation value of the charge exchange (after many contacts) may be an intrinsic property of the polymer, independent of the details of the charge-exchange processes.

Presuming that such is the case, we may write an expression for the charge transfer in which it is limited by the availability of charge states in the polymer, i.e.

$$q_{MI} = \int_{E_F}^{E_F+\Delta E} dE \rho_D(E) f(E) - \int_{E_F-\Delta E}^{E_F} dE \rho_A(E)[1 - f(E)] \tag{1}$$

where q_{MI} is the charge density in insulator following steady-state contacts to metal, E_F the metallic Fermi level, ΔE the window for transfer from metal, $\rho_d(E)$ the density of insulator donor states, $\rho_A(E)$ the density of insulator acceptor states, $f(E)$ the probability that the insulator charge state is occupied, and $1 - f(E)$ the probability that the insulator charge state is not occupied.

Experimental results from metal/insulator contact experiments (Figure 1)

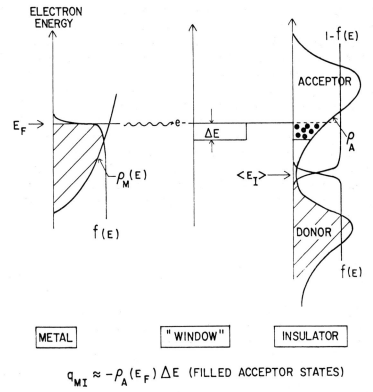

$$q_{MI} \approx -\rho_A(E_F)\, \Delta E \quad \text{(FILLED ACCEPTOR STATES)}$$

Figure 3. Schematic illustration of the energetics governing a metal/insulator contact for carrier injection into insulator acceptor states. The injection window is simplified in accordance with experimental and theoretical uncertainties. From experiment, $\Delta E \approx 0.4$ eV.

indicate that the transfer window ΔE is approximately 0.4 eV wide. It is consistent with the precision of the experiments and the level of understanding of the microscopic injection process to place the window at the metallic Fermi level and to take the densities of insulator states as slowly varying over ΔE, thereby reducing the integral to a product. These simplified energetics of metal/polymer contact charge exchange, indicated in Figure 3, constitute the definition of contact-charge-exchange spectroscopy.[6-8,13,14]

6. VALIDATION OF THE MODEL

A test of this interpretation of metal/polymer contact charge exchange can be achieved by combining different molecular sites into a composite material and determining its triboelectric response relative to those of the component moieties. The system examined here is a series of styrene/methyl methacrylate copolymers. According to the present model, the contact charging characteristics of the

116

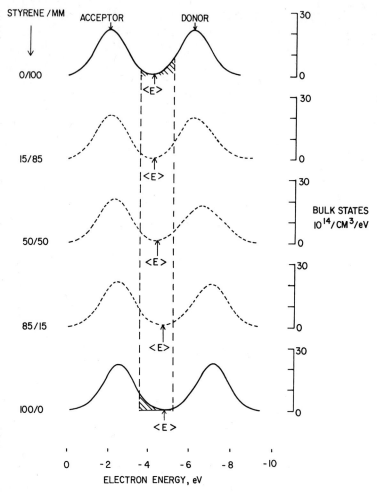

Figure 4. Prediction of the contact-charge-exchange properties of copolymers from those of the constituent moieties. The charge-state distributions for the copolymers (dashed curves) are calculated from the (measured) constituent distributions (solid curves) by weighting each constituent distribution by its mole fraction in the copolymer as specified in equation (2) of the text. The parameters for the Gaussian fits to the contact spectroscopic data for PS and PMMA were given in Figure 2. The calculated copolymer distributions show the continuous shift in the distribution centroids to higher (electron) energies with increasing mole fraction of methyl methacrylate.

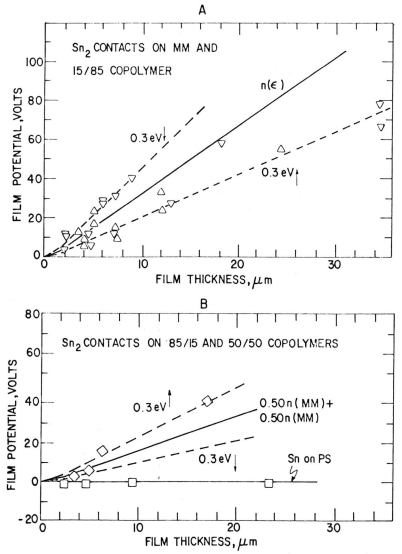

Figure 5. Comparison of the predicted charge exchange for styrene/methyl methacrylate copolymers with experiment. The 85/15 copolymer was found to be indistinguishable from 100/0, and 15/85 was indistinguishable from 0/100, as predicted. The steady-state film potentials for the 0/100 and 50/50 copolymer films studied through a metallic contact which accessed donor states in methyl methacrylate are displayed in (a) and (b). The solid curves are the predictions of the superposition model. The dashed curves correspond to an uncertainty of 0.3 eV in the position of the peak in the PMMA charge-state distribution Tin contacts on PS, horizontal solid line, and on the 85/15 copolymer, □, produced zero charge exchange (b). Hence, according to superposition, tin contacts on 50/50, ◇, should inject half the charge that is injected into 0/100 shown in (a). Within the intrinsic uncertainties of the experiment (±20%) and in the peak position, this prediction is seen in (b) to be consistent with the experimental results.

copolymer should be specified by its state distribution which is expected, in the first approximation, to be a simple linear superposition of the component state distributions.

$$\rho(E) = \chi \rho_{St}(E) + (1 - \chi)\rho_{MM}(E) \qquad (2)$$

where χ is the mole fraction styrene in copolymer and E the insulator charge state energy.

The copolymer distributions, calculated from the measured $\rho_i(E)$ for PS and PMMA, according to equation (2), are shown in Figure 4 for three mole fractions of styrene. The region of energy probed by the metals used in the contact-charge spectroscopy experiment also is identified in Figure 4. The calculated state distributions show that both PS and PMMA, and thus the copolymers, have significant numbers of acceptor states in the energy region probed by the metals, while only the copolymers having 50 or more parts methyl methacrylate contain detectable numbers of donor states. The major model prediction is, therefore, that the 85/15 (styrene/methyl methacrylate) copolymer ought to be indistinguishable from 100/0, whilst the 15/85 ought to be just like 0/100. The 50/50 copolymer represents the best test of equation (2) for contacts that utilize metals that access donor states.

The relevant results of the contact charge spectroscopy performed on the copolymers are given in Figure 5. The solid curve, $n(E)$, in Figure 5(a) gives the film potential values corresponding to the average density of states lying within 0.4 eV of the Fermi level of the contacting metal. (From Figure 1, $n_v = 2.7 \times 10^{14}/cm^3$ for tin contacts to PMMA; this applies to the 15/85 copolymer as well.) Since tin does not inject into PS (Figures 1 and 5(b)), according to equation (2), tin contacted to the 50/50 copolymer ought to access just half the number of states available in PMMA. The solid curve in Figure 5(b) is the result of this prediction, and is in satisfactory agreement with the results of the measurements.

7. SUMMARY

We reported herein several extensions of our earlier contact-charge-exchange measurements[1 3] and an expanded description of our theoretical model[6-8] used for their interpretation. Key features of the model include the identification of the polymer charge states with relaxed molecular ion states and the description of these states by inhomogeneously broadened (Gaussian) distributions. A consideration of the nature of the charge-transfer process led to our proposal of equation (1) for the description of the final saturation charge exchange. Our molecular-ion model leads directly to equation (2) describing the charge-state distributions for two-site insulators. Combining equations (1) and (2) leads to predictions of the charge-exchange spectra of two-site insulators (e.g. copolymers, mixtures, blends, etc.) in terms of those of their constituents. These predictions were shown to be consistent with the contact charge-exchange spectra for copolymers of PS and PMMA, thereby yielding a valuable verification of the internal consistency of our proposed models of charge exchange (equation (1)) and charge states in insulators.

ACKNOWLEDGEMENTS

The authors are indebted to Drs. J. J. O'Malley and J. B. Flannery for advice and encouragement, to Drs. P. Nielsen and H. Gibson for numerous helpful discussions, and to B. Fornalik and L. Kennedy for assistance.

REFERENCES

1. M. W. Williams, *J. Macromol. Sci. – Rev. Macromol. Chem.*, **C14**(2), 251 (1976)
2. C. G. Garton, *J. Phys. D: Appl. Phys.*, **7**, 1814 (1974).
3. A Chowdry and C. R. Westgate, *J. Phys. D: Appl. Phys.*, **7**, 713 (1974).
4. H. J. Wintle, *J. Appl. Phys.*, **44**, 2514 (1973).
5. H. J. Wintle, *J. Phys. D: Appl. Phys.*, **7**, L128 (1974).
6. C. B. Duke and T. J. Fabish, *Phys. Rev. Lett.*, **37**, 1075 (1976).
7. C. B. Duke, T. J. Fabish, and A. Paton, *Chem. Phys. Lett.*, **49**, 133 (1977).
8. C. B. Duke, *Surface Sci.*, Erwin Mueller Festschrift, 1977.
9. H. W. Gibson, *J. Am. Chem. Soc.*, **97**, 3832 (1975).
10. I. Shinohara, F. Yamomoto, H. Anzai, and S. Endo, *J. Electrostatics*, **2**, 99 (1976).
11. H. Krupp, *Inst. Phys. Conf. Ser.*, **11**, 1 (1971).
12. M. A. Lampert and P. Mark, *Current Injection in Solids* (Academic Press, New York, 1970), Ch. 1.
13. T. J. Fabish, H. M. Saltsburg, and M. L. Hair, *J. Appl. Phys.*, **47**, 930 (1976); **47**, 940 (1976).
14. T. J. Fabish and C. B. Duke, *J. Appl. Phys.*, **48**, 4256 (1977).
15. T. J. Fabish, H. M. Saltsburg, and M. L. Hair, submitted to *J. Appl. Phys.*
16. W. A. Zisman, *Rev. Sci. Instr.*, **3**, 367 (1932).
17. T. Kobayashi, K. Yokota, and S. Nagakura, *J. Electron Spec. and Related Phenon.*, **3**, 449 (1973).
18. C. B Duke, *Tunneling in Solids, Sold State Phys.*, Supp. 10 (Academic Press, New York, 1969).

7

Static Charges on Textile Surfaces

N. Wilson

Shirley Institute, Manchester

1. INTRODUCTION

Static charges are readily produced during the contact and separation of any two solid surfaces, the effect often being most marked when the materials are of a high electrical resistance and thereby not able to dissipate their charges rapidly. During contact, charge is transferred across the interface until an equilibrium is established, and on separating the surfaces a residual free charge of positive or negative polarity remains on one or other surface. It is this free charge which is the main cause of static problems.

The magnitude of the charge generated depends on several factors of which the more important are the relative affinities of the materials for charge of a given polarity, the electrical resistances of the materials — which can be strongly influenced by the relative humidity and temperature of the atmosphere in which they are conditioned — and the area of contact between the surfaces, which depends on the pressure between the materials and the degree of rubbing which may occur.

The nature of the charge carriers is, of course, important for an understanding of the mechanisms of the production and transfer of static charges and, as was noted by Harper,[1] opinions differ on whether the charging of insulators, of which many textiles are examples, is due to the transfer of ions or electrons or both. Despite much work carried out in the last decade on this topic there is still considerable disagreement.

However, most textile fibres contain impurities and surface finishes of one kind or another and for this reason the majority of the charge carriers are probably ions, in view of their existence in large numbers at or near the surface.

In recent years there has been a growing interest in static problems of a more applied nature where it is the magnitude of the charge produced and the consequences arising from it which matter rather than its type, polarity, or mode of generation. Static electricity has for many years been a problem to people in, for example, the textile-, paper-, and powder-manufacturing industries. The problem can be of a mechanical nature arising from the mutual attraction or repulsion of charged materials, as when fabric or paper sticks to rollers or charged filaments in a yarn 'balloon' out and become caught on protruding parts of machinery.

Fires due to incendiary spark discharges from highly charged materials have been a common occurrence in the manufacture of coated materials when flammable solvent vapours were present.

In remedying static problems during processing the manufacturer need take only temporary measures. Hydrophilic fibres, such as cotton or viscose rayon, can be static-prone in dry conditions because of their high electrical resistance, and a sufficient increase in the relative humidity (r.h.) of the atmosphere is all that is needed to solve the problem. However, it is more usual to render a troublesome fibre conductive by the application of an antistatic agent. Unfortunately, these agents are normally removed in subsequent processing and it is up to the finisher to apply a similar or more durable finish, the latter being preferable to the consumer.

It is only in the last two or three decades that static electricity has become meaningful to the majority of the public which is now familiar with 'static' as a nuisance and a hazard. For example, when walking over a carpet in a warm building, a charge can accumulate on the body and a shock may be felt on touching metal objects such as door handles and filing cabinets. The same may occur after rising from a chair, when a charge on the outer layer of the clothing induces a similar charge on the body.

The shock is accompanied by a small spark which often goes unnoticed, but in a situation where flammable gases, vapours, and powders are present, the spark may cause a fire or an explosion.

In this review a number of the more important investigations into the electrostatic behaviour of textile and polymeric materials, carried out during the last three decades, will be dealt with first before proceeding to the more applied type of problem outlined above. In particular, the problem of static charges on clothing and the hazards of spark discharges from the body will be dealt with at some length.

2. ELECTROSTATIC CHARGING OF TEXTILE AND POLYMER SURFACES

An account of the possible mechanisms of charge generation during the contact and rubbing of polymeric insulators with other materials is given by Henry.[2] The type of charge carrier is discussed and it is suggested that ions from the atmosphere which have previously been deposited on the surface, may play an important part in the charging behaviour of the material. A distinction is made between equilibrium electrification, when contact occurs only between the surfaces, and kinetic electrification owing to rubbing and other effects. It is argued that whatever the mechanisms of charge generation when equilibrium processes only are involved, it should be possible to arrange materials in an 'electrostatic series' such that any material in the series will become positively charged on coming into contact with another material below it; also, the charge densities produced at equilibrium for various pairs of materials would be expected to be additive.

Almost all attempts at determining an electrostatic series have been made by rubbing materials together rather than by contact. Lehmicke[3] measured both the charge and its polarity on rubbing pairs of different fibres together and established a triboelectric series for a wide range of textile materials. It was expected that the magnitudes of the charges generated would be related to the separation in the series of the materials concerned. This was found not to be the case. A closer, but not completely reliable correlation was observed between the charge generated on the fibre after rubbing against rubber and the moisture regain of the fibre. Fibres with a high moisture content tended to develop little charge and vice versa.

Ballou[4] investigated the sign of the charge produced by rubbing textile materials against various surfaces, including other textiles. Boiled and rinsed yarn samples were wrapped around insulating plastic formers 2 in square. The wound yarns were then rubbed against other samples in such a way that the direction of rubbing was along the yarn axis. Specimens were conditioned at 15% r.h. and, presumably, room temperature. By rubbing together every possible pair of a range of materials a consistent triboelectric series was established. It was noted that fibres containing amide groups, for example wool, nylon, and silk, are at the positive end of the series, and hydrocarbons and halogenated hydrocarbons, such as polyethylene and Saran (PVC) are at the negative end. The series is qualitative in that it ranks the materials according to the sign of the charge and does not allow a comparison of the magnitudes of the charges produced. It is interesting to note that wool, silk, and nylon have significantly higher dielectric constants that polyethylene and Saran, and that their charging behaviour is, to some degree, consistent with the 'law' first proposed by Coehn.[5] This states that the order of the materials in a triboelectric series will be that of the dielectric constants, i.e. on separating two materials after firm contact the one with the higher dielectric constant becomes positively charged. The 'law' was based on a comparison of published triboelectric series and of dielectric constants available at the time, and is nowadays considered useful only as a guide. Richards[6] found supporting evidence of Coehn's rule and derived an expression predicting the magnitude of the charge generated in terms of the dielectric constants of the materials being separated. Certain materials, for example ebonite and steel, behaved in a totally inconsistent manner to that predicted.

Referring back to the work of Ballou,[4] the question raised is how do fabrics made from blends of two fibres in the triboelectric series behave electrostatically? Clearly it would be expected to depend on the relative proportions of each fibre and their relative charging behaviour against a given material. Fabrics made from nylon and Dacron staple yarns, in various proportions were rubbed against a chrome-plated surface and a cotton surface — materials which occupy positions in the triboelectric series between the two components of the blend. The results showed that against the chrome-plated surface the net charge generated is practically zero when the percentage of nylon is between 40 and 50. When rubbed against cotton a rather higher proportion of nylon is required for neutrality, namely, about 75%. From these and other data it was concluded that a judicious choice of fibres from the triboelectric series, when blended together, can effectively reduce the troublesome effects of static charges in certain cases. The technique, however, is limited as the optimum blend depends on the nature of the rubbing material.

An often-quoted triboelectric series of 15 natural and synthetic fibres, believed to be one of the most reliable for practical purposes, is that due to Hersh and Montgomery.[7] Using mostly textile filaments the charge was generated by repeatedly rubbing one filament across a fixed one under controlled mechanical and ambient conditions. They adopted a symmetrical arrangement in which fibres were mounted at $+45°$ and $-45°$ with respect to the direction of motion in a horizontal plane, so that each fibre swept over the same length of the other. The magnitude of the charge produced was found to be proportional to the length of material rubbed and to the normal force between the fibres, although sometimes a limiting maximum value of charge was reached. The charge was found to be independent of the apparent area of contact between the fibres, and of the tension on the fibres, and the velocity of rubbing. The polarity of the charge was always reproducible and consistent and no difficulty was found in obtaining the triboelectric series given in Table 1. The magnitudes of the charges formed on rubbing pairs of materials together were, generally, fairly constant, i.e. materials well separated in the series gave charges similar to those obtained with materials adjacent to each other in the series.

Cunningham and Montgomery[8] used an apparatus similar to that of Hersh and Montgomery, with improvements to give better control of the mechanical variables, and to allow a variation of the ambient pressure. Whilst the triboelectric series of Hersh and Montgomery is confirmed, certain of the latter's results were not substantiated. For example, the charge produced on rubbing nylon on polyethylene has a square-root dependence on normal force rather than a linear dependence. Also, the charge is inversely proportional to the fibre diameter rather than independent of it as shown by Hersh and Montgomery.

Further improvements in the measuring techniques were made by Levy *et al.*[9] The apparatus allowed the measurement of frictional energy expended and the charge generated between two filaments during the $45°-45°$ mode of rubbing. The results revealed several differences from those obtained by the previous workers. With the various combinations of fibres tested the results showed that not several tens of rubs but thousands were necessary to reach a steady value of charge. The charge and the work of friction decreased with number of rubs, and interchanging the filaments produced practically the same equilibrium charge but of opposite polarity. The simultaneous fall in charge and work of friction is explained in terms of the stick–slip process. During sticking welded junctions are assumed to be

Table 1 Triboelectric series

Positive	Polyvinyl alcohol
Wool	Dacron (polyester)
Nylon	Orlon (acrylic)
Viscose (cellulose)	Polyvinyl chloride
Cotton	Dynel
Silk	Velon
Acetate	Polyethylene
Lucite (Perspex)	Teflon
	Negative

formed by asperities of the materials being rubbed. During slipping the junctions are broken and it is the material of lower shear resistance which shears, resulting in a lowering of the measured frictional work as asperities are broken off from one surface and not the other. An interesting behaviour was found when the same material was rubbed on itself. For polythene on polythene and nylon on nylon the frictional work at a given velocity of rubbing was independent of the number of rubs, and with the former material the average charge was practically zero. With nylon on nylon a constant, albeit small, charge was produced which was independent of the number of rubs. It was pointed out that with similar materials symmetrical welds are formed which probably break in the middle so that there is no preferential transfer of material in one direction or the other. In both cases the frictional work increased markedly with increase in the velocity of rubbing.

When different fibres in pairs were rubbed together at low, medium, and high velocities until a constant charge and frictional work was obtained at each velocity in turn, the results showed a fall in charge and, in most cases, an increase in the frictional work. The explanation of this behaviour given by Levy et al.[9] is as follows. As the velocity of rubbing increases, the number of plastic junctions (welds) increases owing to heating and more work is required to disrupt the junctions, resulting in a higher frictional work with increase in velocity. At the same time more material would be expected to transfer from one surface to the other, taking with it a charge of opposite polarity to that on the surface on which it settles. The effect would be partially to neutralise the charge on the surfaces, resulting in a fall in the net charge. This mass transfer of material appears to be the dominant factor in the charging behaviour of the materials, offsetting any effects of increase of excitation of charge carriers at the higher temperature. Microphotographs of a number of different fibres in pairs show that in each case the softer fibre is worn down and particles of one material are seen to be sticking to the other. Similarly, with identical fibres in pairs there is evidence of material transfer.

Henry[10] discusses two types of contact electrification, one requiring different materials and no rubbing (friction), the other requiring asymmetrical rubbing but not different materials. The first is suggestive of an 'equilibrium' effect, and the second a 'kinetic' process which depends on a type of rubbing such that the contact at one surface is over a relatively small area compared with that at the other surface. It is argued that if an equilibrium effect can be obtained for a series of materials without the interference of other effects, then the charges on separating the surfaces should be additive, i.e. for the same contact area and proximity of the surfaces, the charge on A after contact with C should be equal to the algebraic sum of the charges on A after contact with B, and on B after contact with C. This assumes, of course, that there is no loss of charge during separation of the surfaces because of discharges or conduction. In practice these conditions are difficult to fulfil without fairly elaborate equipment but, at least, it should be possible to arrange the materials in an electrostatic series which obeys the sign convention discussed earlier. To this end, Henry charged a number of materials, available in sheet form, in such a way as to minimize friction. The technique was to sandwich one sheet of material between two sheets of another material and pass the assembly between a pair of steel rollers, spring-loaded to provide a constant pressure. The 'sandwich' was then carefully opened and the charge on the inside material measured by transferring it to a Faraday cylinder connected to an electrometer, using forceps tipped with an insulator. This was repeated but using a 'sandwich' in which the outer and inner materials were reversed. The materials were ordinary commercial products in sheet form, consisting of various metals, polymers (including rubber) and filter paper. All the materials, except the filter paper, were given a suitable cleaning treatment after which they were handled only with the use of the forceps.

The charge on the specimens was determined from the mean of several readings and the true percentage scatter was about ±20%. In some instances a significant charge was obtained when the inner and outer materials were the same and it was necessary to 'correct' the charges obtained by contact of these materials with all the others by adding or subtracting a value sufficient to reduce the charges for like materials to zero. By this means the materials were arranged in a self-consistent electrostatic series. Henry does not claim anything of his results except that they are indicative of an equilibrium phenomenon. The results show that the magnitudes of the charges produced increase the further the separation of the materials in the series, but there are a number of exceptions to this which may be due to dissimilar true areas of contact between different kinds of material, and noticeable discharges through the air during the opening of many of the 'sandwiches'.

With reference to the 'kinetic effect', Henry considers the case of two rods of similar insulating material which are rubbed together in such a way that the contact is limited to one point on one rod and spread out over the length of the other rod. The charges on the rods, as measured using a Faraday cylinder, are opposite in polarity, and the charges are reversed when the asymmetry of the rubbing is reversed. Henry was of the opinion that local heating is primarily responsible for the observed effects. With asymmetric rubbing, friction would be concentrated on a smaller area of one surface than the other, and should result in a localized temperature gradient across the interface. This would cause the charge carriers, whatever their nature, to pass down the temperature gradient. If the carriers having a charge of one sign are either greater in number or lighter in weight than carriers of opposite sign, there will be a net charging of the two surfaces at the interface.

A more rapid but similar technique to the 'sandwich' method for arranging a number of polymeric materials (used in the photographic industry) as a triboelectric series has been used by Webers.[11] A number of reference materials 1 cm square were attached to the jaws of ordinary spring-loaded clothes-pegs, thereby providing the outer material of a 'sandwich'. A set of nine clamps were mounted in parallel and held open by means of a single spacer rod. The test sample was a strip of film about $\frac{3}{4}$ in wide held horizontally between the nine reference materials by clamps at each end. The reference and film-strip materials were conditioned for several hours at constant temperature and humidity and were then discharged by use of a radioactive (polonium) source. The spacer rod was then removed, thus allowing the reference materials to 'sandwich' the test strip. After 5 min the spacer rod was reinserted and the clamps removed. The film strip was immediately sprayed with a mixture of charged fluorescent powders, one yellow the other red. The powders were mixed in a roller mill for several hours, at the end of which the red powder was positively charged and the yellow powder negatively charged. After spraying, spots of red powder appeared where the film was negatively charged and yellow powder where the film was positively charged. The amount of powder deposited was an indication of the magnitude of the charge at that spot. As was noted by Henry[10] contact between a test strip and similar reference material did not necessarily give a zero charge, which behaviour was attributed to possible physical differences between the materials. Results indicate that the charging behaviour of a surface can be related with the nature and orientation of chemical groups in the polymer surface.

A study by Medley[12] of the charging behaviour of clean strips of nylon film (and of wool fibre) on rubbing against a metal wire has shown that the charge density on the polymer surface is critically dependent on the r.h. at which the material is conditioned. The film (or fibre) when attached to an insulating rod was drawn over a platinum wire and the charge on the latter was determined by the use of an electrometer. The effects of humidity, load, and speed of rubbing were examined. The charge density was found to be almost independent of humidity and not greatly dependent on the speed of rubbing at humidities below 70% r.h. At 70% r.h. the charge on the nylon reached a maximum and at higher humidities it fell rapidly to zero, because, it is thought, of conduction of charge through the polymer. The electrical conductivity of the material increased by about six orders of magnitude over the humidity range 0–70% r.h., but this apparently had little effect on the charging behaviour of the material. The factor controlling the magnitude of the residual charge on the polymer was considered by Medley to be the breakdown strength of the air. Experiments on a polythene film (hydrophobic) gave a similar maximum charge density with no steep fall at the higher humidities, thus supporting this view. Further evidence was obtained from charge measurements carried out in air saturated with carbon tetrachloride. This increases the dielectric strength of the atmosphere considerably and resulted in an increase in charge density on the nylon of 50%, and on the polythene of 100%.

The dissipation processes of electrical charges generated on textile products passing through rollers has been considered in detail by Medley.[13] For poorly conducting wool cloth or rovings it was shown that gaseous discharge between the charged product and the rollers limits the value of the charge retained by the product. This charge can further be reduced by placing earthed conductors close to the charged material.

A distinction is made between conduction of charge back through materials of finite conductance to the last point of contact with the rollers, and leakage of charge to the rollers via a surface contaminant at the roller nip. In the former case the critical condition for dissipation is found in terms of a dimensionless parameter which depends on the geometry of the system, the conductance of the material, and the speed of rubbing.

The conductance along an oiled roving is, however, low, and the critical value of the parameter discussed above would be too small to be attainable in practice. An alternative leakage mechanism suggested by Medley was that the surface agent acts as a leaky dielectric at the roller nip where the fibre—metal contact is broken.

The electrical resistance of textile materials as a measure of their antistatic properties has been studied by Wilson.[14] The objective was to find out whether a correlation between the resistance and rate of leakage of charge exists for a wide range of fabrics, including many treated with finishes, with resistances in the range $10^9 - 10^{15}$ ohm/cm^2. Resistances were measured longitudinally between opposite edges of a square of the material (longitudinal resistivity), and transversely through a 3 in square of fabric (transverse resistance). In determining the time constant of decay τ, i.e. the time taken for the charge to fall to $1/e$ of the original value, fabric samples were suspended vertically under tension from a horizontal metal rod mounted in paraffin wax supports and charged either by induction, when the fabrics were of lower resistivity ($< 10^{11}$ ohm/cm^2) or by stroking with an earthed metal rod when the fabric was of a higher resistivity. On earthing the specimen the rate of decay of charge was measured by use of a field meter. Leakage of charge transversely through the fabric was more difficult to measure and was achieved by bringing an earthed metal plate into contact with the fabric surface for a given time. This was repeated until the charge on the fabric had fallen to a suitably low value.

Plots of τ against either longitudinal resistivity or transverse resistance gave straight lines, indicating a simple relationship between the rate of leakage and the resistance.

From theoretical considerations, a plot of log τ against log resistance, was expected to give a straight line of slope equal to unity. The actual slope was 0.86, implying that charge decays faster than anticipated, but no explanation for this effect was given. A scale of antistatic ratings applicable to homogeneous fabrics, new fabric blends and fabrics treated with antistatic agents (Table 2; reproduced by permission of The Textile Institute) has been put forward.

However, as will be discussed later, the use of resistivity as an indicator of the static propensity of a material is limited. For example, coated fabrics in which one component (substrate or coating) may be more conducting than the other, the longitudinal resistivity may correspond to that of the more conducting material, and therefore does not represent the ability of the other component to dissipate charge. A second example is with fabrics containing small percentages of highly conducting fibre. The longitudinal resistivity of such a material can be very low, but

Table 2

Longitudinal resistivity (ohm/square)	Antistatic rating
$> 10^{13}$	Nil
$10^{12} - 10^{13}$	Poor
$10^{11} - 10^{12}$	Moderate
$10^{10} - 10^{11}$	Fairly good
$< 10^{10}$	Good

may be misleading as it does not correspond to the bulk property of the surface material.

Henry *et al.*[15] developed an elegant method of testing textile fabrics and other sheet materials for their liability to electrostatic charging and obtained experimental values of half decay times and corresponding surface resistivities for a wide range of textile materials. A cross-section and plan view of the sample holder are shown in Figures 1 and 2.

It consists of an outer electrode in the form of an alloy dish about 21 cm in diameter with a broad flange at the periphery (Figure 1). A thick polythene disc fills the cavity of the dish and the inner electrode in the form of an alloy disc with a screw hole at its centre is embedded into the polythene. The fabric specimen consists of a circular disc about 21 cm in diameter with a hole 11 mm in diameter at its centre. It rests on the dish supported by the polythene and is clamped by a recessed centre clamp, the rim of which makes firm contact with the specimen, and a metal ring which clamps the periphery of the specimen to the flange by means of Terry tool clips (not shown). The electrode system with the sample in position is mounted on a vertical shaft which is connected to a driving mechanism, and can be rotated at constant speeds within the range 25 rev sec^{-1} to 2 rev min.$^{-1}$ The specimen is charged by the contact of a shaped stainless steel roller which is mounted in miniature ball-bearings in a pivoted stirrup so that it may lie evenly on the sample. The pivots for the roller are supported by a piece of Perspex

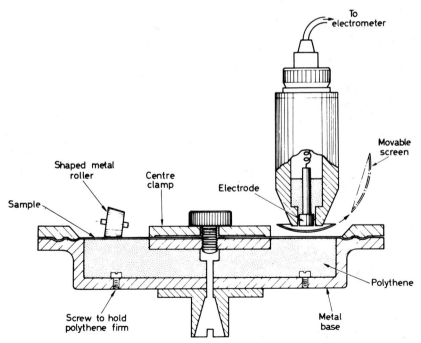

Figure 1. Vertical cross-section through sample holder and electrode (reproduced from Henry *et al.*[15] by permission of The Textile Institute).

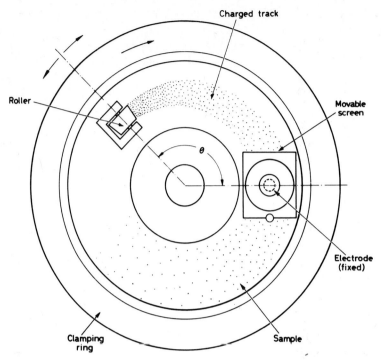

Figure 2. Plan view of sample holder, charging roller and electrode (reproduced from Henry et al.[15] by permission of The Textile Institute).

so that the roller may be insulated from earth when required. The whole of the roller assembly can be rotated about the axis of the sample holder to any desired angular separation from a detecting electrode. In the plan view (Figure 2) the roller is seen at an angle of θ to the detecting electrode, producing a charged track on the specimen as the latter rotates.

The detecting electrode consists of a highly insulated brass disc 1 cm in diameter, mounted in a larger aperture in the end of an earthed brass cylinder which acts as a screen. The electrode is connected via a length of insulated screened cable to a vibrating reed electrometer. A movable screen can be inserted between the electrode and sample and is used as a means of correcting disturbances such as those due to charging up of the electrode through the air by highly charged samples. Readings corresponding to the charge density on the sample were obtained from the difference in the readings with and without the screen in position.

The primary method of measuring the half decay time was to lower the roller on to the sample, uncover the detecting electrode, and adjust the speed of rotation of the sample to a value which is suitable to give a decay of charge round the track by a factor of 3 or 4. Readings were then taken with the electrode set at a number of different angular positions round the track. The logarithms of the corrected electrode readings were plotted against the angular positions and the best straight line drawn through the points. The half leak time $t_{1/2}$ was obtained by dividing the

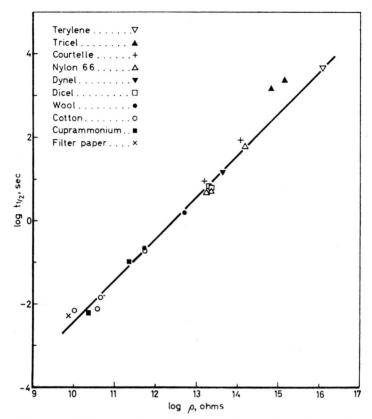

Figure 3. Half decay times plotted logarithmically against surface resistivities (reproduced by Henry et al.[15] by permission of The Textile Institute).

half leak angle by the speed of rotation. A plot of log half decay time against log surface resistivity for a number of plain weave fabrics is given in Figure 3, where it can be seen that a close proportionality exists between the half decay time and resistivity. It is also noted that the data are close to a line drawn to make angles of exactly 45° with the axes. In comparing these results with the findings of Wilson[14] in which the slope of the line was 0.86, Henry et al. considered the latter to be due to neutralization of charge following breakdown of the air at the edges of the samples where the electrostatic field would be greatest.

In experiments similar to those of Wilson, Shashoua[16] found that the rate of discharging was always greater than the rate of charging up at a given charge level and concluded that charge was 'radiating' to the air. In Henry's work there were no free edges of the specimens (i.e. edges exposed to air) and disturbances due to air breakdown were practically eliminated.

Arridge[17] investigated the way in which static charges decay on nylon filament free of antistatic finish. The filament was charged either by contact with a metal rod or by induction, using electrodes placed near to the filament. In the latter case the rate of build-up and decay of charge could be examined. The experiments were carried out at various relative humidities between 0 and 65%, presumably at room temperature. The decay of charge was observed by passing the filament through a Faraday cylinder attached to a vibrating-reed electrometer. By plotting the logarithm of the charge against either a linear or logarithmic time scale it was shown that at humidities below about 35% r.h. the charge remains localized and decays at approximately an exponential rate. Arridge explains the absence of spreading at low humidities by postulating a saturation current in which the charges are too few to be detected. For humidities above 35% r.h. the charges spread out along the filament and the decay of peak charge with time can be approximately represented by a $t^{-1/2}$ law. By applying a diffusion equation to the latter, values of the diffusion constant are obtained which are independent of (a) sign of charge, and (b) mode of charging (contact or polarization). The diffusion constant is shown to vary exponentially with r.h., supporting the view that the conduction process in nylon is ionic. It would also be expected from theoretical considerations that the conductivity of nylon should increase exponentially with r.h. The work of Hearle[18] has shown this to be the case.

Arridge[17] suggests that the similarity of the mode of decay for contact and induced charges indicates that the same carriers are involved in both and that they are therefore ionic.

In a study of the effect of moisture on the electrostatic charging behaviour of cotton, nylon, wool, and propylon fabrics, Sereda and Feldman[19] found that the charge produced reached a maximum at humidities intermediate between 0 and 65% r.h. Sets of specimens in triplicate were brought to equilibrium at 0, 20, 35, 50, and 65% r.h., either from the dry side (adsorption cycle) or the wet side (desorption cycle) during an interval of about a year. The charging experiments were done in a constant humidity box in which the apparatus could be handled without disturbing the conditions. The fabric specimen was charged by placing on a flat nickel-plated brass plate (4in x 8in) and passing over it 10 times a nickel-plated brass roller $2\frac{1}{2}$ in diameter and 5in long fitted with an insulating handle. The sample was then transferred to a Faraday cylinder with the use of insulating tongs and the charge determined. This procedure was repeated until a maximum charge was established. In addition the sorption isotherms of representative samples of each material were determined, using the apparatus described by Feldman and Sereda.[20]

Plots of the mean values of maximum potential (or charge) against r.h. for each of the fabrics showed an initial increase in potential to a maximum followed by a decrease. All the fabrics attained the same maximum charge within the limits of experimental accuracy. Using the method described by Brunauer et al.[21] the surface area exposed to water sorption was determined from the isotherms of each textile material, and hence an estimation of the r.h. condition when a monomolecular layer of water is formed during adsorption. The main characteristic of the charging curves was that the peaks occurred at the r.h.s at which a monomolecular layer of water is formed on the surface. It was suggested that the sorbed moisture on the fabric surfaces either contributes hydrogen ions to the metal or allows a more effective transfer of electrons from the metal, thus contributing to an increase in electrostatic charge up to a maximum. Beyond this point added moisture results in an increase in the conductivity of the surface and leakage of charge starts to limit the value of the charge which can be retained by the fabric. Sereda and Feldman[19] challenge Medley's[13] assertion that gaseous discharge combined with charge leakage controls the maximum charge, since, if this were the case, the maximum charge would occur at zero r.h. They do not, however, rule out the possibility that gaseous discharge has some influence.

It should be noted that for nylon, Medley[12] obtained a maximum charge at 70% r.h., a value considerably higher than Sereda and Feldman's[19] figure of 21% r.h. This would suggest a difference in the surface properties of the materials, because of either incomplete removal of surface impurities or, possibly, the unusually long period of conditioning used by Sereda and Feldman which would allow ample time for stray ions in the atmosphere to deposit on the samples. Such ions may be weakly bound to the surface and thereby able to transfer to other surfaces for which they have a greater affinity, resulting in charging.

Much of the early work on contact electrification was carried out with dissimilar metals and metals in contact with insulators. Harper[22] investigated the electrification of spherical specimens after light contact. The balls were mounted independently on triangular supports of stretched fibres of fused silica in horizontal planes one above the other. The balls were $\frac{1}{2}$in and $\frac{5}{12}$in in diameter. The larger, lower ball was sometimes of a conducting material, and when electrical contact with it was required it was supported on a similar suspension of phosphor-bronze wires. Specimens could be optically polished when of a suitable material. They were removed from their supports for cleaning with forceps having Perspex tips and a pneumatic device with a hard glass end shaped to fit the specimen. Contact between the balls was brought about by lifting the lower ball on its suspension into contact with the upper one and then lowering it. The upper ball was then lifted on its suspension into an inverted Faraday cylinder with three slots in its walls to accommodate the three quartz suspension fibres without touching them. Purified and dried argon was used to protect the specimens from atmospheric pollution. The argon could be ionized by passing over a suitable radioactive source and used to discharge the balls. The normal procedure was then to measure the cumulative charge for up to 10 successive contacts. The insulating materials which, after light contact with steel, chromium, and gold, showed negligible charging are as follows: amber, alkathene (polyethylene), Distrene (polystyrene), nylon, Perspex (polymethylmethacrylate), silicone on silica, Teflon (polytetrafluoroethylene), Caresin wax, pure cellulose, and Kel-F (polychlorotrifluoroethylene).

Two exceptions in the list were nylon and polyethylene in contact with steel. Both gave accumulated charges after 10 contacts. In the case of nylon the charge was attributed to possible electrophilic contamination, either on the steel or localized on that part of the nylon making contact with the steel. In the case of polyethylene there was a tendency for the polymer to adhere slightly to the metal and the observed charging might have been a result of material transfer.

In general, the experiments of Harper[22] showed that certain clean insulators are not able to become significantly charged on contact with metals. This indicates that no transfer of electrons takes place and that a number of synthetic polymers exhibit this electrophobic behaviour. However, under more realistic conditions where contaminants or surface finishes are present and where rubbing occurs, the charges produced on many of the materials listed above can be considerable.

The generation of static charge on polythene and glass after contact with various metals of known work function has been measured by Davies.[23] The effects of contamination were minimized, and discharges during the separation of the surfaces avoided, by carrying out the experiments under controlled vacuum conditions. Because of the difficulty in knowing the actual area of contact, owing to surface irregularities, multiple contacts were avoided. After preliminary measurements with various durations of contact of a metal and polythene a contact time of 15 min was found sufficient to produce a maximum charge. With glass the corresponding time was not more than 5 min.

The metal surfaces were polished to give the same degree of finish and the polythene and glass surfaces were cleaned ultrasonically and washed with isopropanol.

The surface charge density on the dielectrics was measured using an electro-meter-probe method. The probe was passed above the charged dielectric and the output displayed on an oscilloscope. A description of the probe system and its calibration is given by Davies.[24] The contact charge density was shown to increase linearly with the work function of the metal and is therefore consistent with a contact-potential effect. From the experimental data, values of about 4.7 and 4.3 eV for the work functions of polythene and glass were obtained. The calculated depths of penetration of the injected charge into these dielectrics was of the order of 10^4 Å which compares reasonably with results for semiconductors, but Davies notes that trapping and low electron mobility in insulators may reduce the values considerably.

Using the work functions and penetration depths obtained for polythene and glass from contacts with metals, an estimate of the charge transferred in a polythene–glass contact was made, assuming the surfaces have equal electron affinities. The value is about one-third of those obtained experimentally and it was concluded that the charging of dielectric surfaces by metal and dielectric contact is consistent with electron transfer. The charge densities may be adequately described by the electronic properties of the materials.

In the same paper Davies reports on the charge-decay behaviour for polythene and glass. Assuming the material is an electron conductor, a theoretical expression is derived for the decay of charge in terms of the initial charge density injected into the surface of the material, the density of the intrinsic charge carriers (thermally generated), and the thickness and permittivity of the specimen. Three regimes of charge decay were identified, corresponding to the intrinsic carrier density being negligible, significant, or dominant compared with the surface-charge density at finite time. Plots of the inverse of the charge density against time for glass at three different temperatures, 19, 54, and 104.5°C were found to be linear, corresponding to the case in which the intrinsic charge-carrier density is negligible.

Similar charge-decay measurements on polyethylene at temperatures of 74.5, 88.5, and 97.5°C showed a non-linear relationship between the inverse of the charge density and time at the higher temperatures. This behaviour is in excellent agreement with the theoretical prediction and corresponds to the case in which the intrinsic carrier density is significant. It was concluded that the decay of surface charge on insulators can be described in terms of electron (or hole) conduction in dielectrics, and that the retention of surface charge on insulators at low temperatures is caused by extremely low carrier mobilities.

The charging behaviour of various metals against a borosilicate glass under high vacuum has been investigated by Inculet and Wituschek.[25] A stainless steel carriage with four pins of the metal under test mounted in its base was allowed to slide down the inside of a glass tube, 3 cm in diameter and 37 cm long which was vibrated at 92 c/sec when inclined at an angle of 15° to the horizontal. The lower end of the glass tube is inserted into a Faraday cylinder used for measuring the charge collected by the trolley during its passage down the tube. The tube was meticulously cleaned before use and the metal pins were polished with fine alundum powder, rinsed in distilled water, and dried in air. Further cleaning of the metal surfaces was carried out by means of electric sputtering. The first series of data was obtained with Sn, Ag, Cu, and Au pins. The vacuum was held at 3×10^{-6} Torr for 30 min prior to sliding the carriage down the tube. The glass tube was cleaned once only for each metal and simply rotated for each subsequent experiment.

A plot of the charge against the work function of the metal showed no systematic relationship between the two.

In the second set of results, using metal pins of guaranteed purity, the vacuum was 3×10^{-7} Torr and the glass tube and pins were held in this vacuum for 24 hours before sliding the carriage. The pins were of Zr, Cu, Cu–Ni alloy, Ni, and Pt. On plotting charge against the work function of the metal an approximately linear relationship was found. From estimates of the area of glass rubbed by the pins, the charge divided by the area was calculated and plotted against the work function of the metal. The results showed an approximately linear relationship in which the charge per unit area decreased with increase in the metal work function. This is consistent with the charging behaviour of glass obtained by Davies.[23]

Further work by Davies[26] on the charging behaviour of various polymers (including nylon 66) *in vacuo* made use of rolling contact between film specimens and various metals. The principle of the operation is shown diagrammatically in Figure 4. The film specimens are mounted on an earthed metal drum and peripheral rolling contact is maintained between the specimens and an earthed contact wheel consisting of segments of cadmium, gold, zirconium, platinum, and aluminium. The charge density was measured by the shutter-probe system. Instead of the work functions of the metals, the contact-potential difference between each metal and a gold reference was used for correlation with the density of charge on the films. The apparatus was held in a vacuum of 10^{-6} Torr for two days in order to overcome adsorbed contaminant effects. The specimens were rotated slowly for five days until charge saturation was reached. Initially, the scatter in the results was large

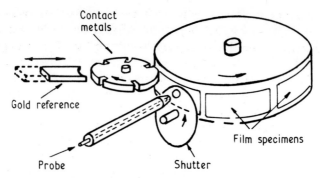

Figure 4. Diagram showing principle of operation.

because of non-uniformity in the polymer specimens brought about by an over-elaborate cleaning method. More consistent results were obtained after washing the samples with isopropanol.

Plots of charge density on a number of samples of nylon 66 and polycarbonate against the contact potential differences between the various metals and the gold reference are shown in Figure 5. The results indicate a linear dependence of charge density on the metal work function. Similar results have been obtained for PVC, PTFE, polyethylene terephthalate (PET), polyimide, and polystyrene. These results support Davies'[23] earlier findings that the charging behaviour of dielectric surfaces in contact with metals *in vacuo* may quantitatively be described in terms of the

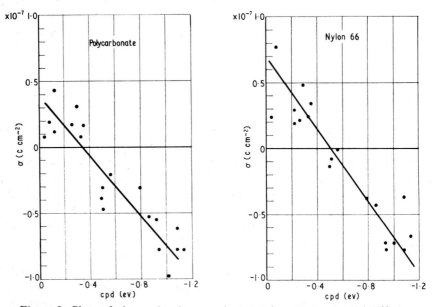

Figure 5. Plots of charge density σ against metal contact potential difference (c.p.d.) for polycarbonate and nylon 66 (reproduced from Davies;[26] copyright by the Institute of Physics).

electronic properties of the materials. Furthermore, the data provide a basis for the understanding of the charging behaviour of materials under ambient conditions.

These ideas are, to some extent, borne out by the work of Arridge[27] on the static electrification of nylon 66. The experiments consisted of charging a moving continuous thread by light contact with various metal surfaces mounted on a rotating wheel, in a manner such that there was zero relative velocity between the contacting surfaces. Contacts with a given metal could be made repeatedly at various points along the moving thread. The charge on the filament was measured by passing it through a Faraday cylinder, the output of which was fed to an oscilloscope via a vibrating-reed electrometer. Work functions of the metals were not used in this work as they are known to be greatly affected by the presence of oxide films, water, and other contaminants. Instead the 'contact potential' of a metal was measured immediately after contact with the fibre by rotating it to within a small distance (<0.02 mm) of a gold-plated reference electrode to which was connected a second electrometer. The output of the electrometer was displayed as a second trace on the oscilloscope used for the contact charges. The accuracy in the measurement of the contact potential was within 0.1 V and that of the charge down to 2.10^{-4} e.s.u.

Plots of 'contact potential' for 12 different metals against charge on the nylon filament, obtained in the room atmosphere and in dry nitrogen, showed good linear correlations with a zero charge corresponding to a 'contact potential' of 4.2 V for nylon.

A thread of polypropylene after contact with each metal surface acquired negative charges, implying that the 'contact potential' for this polymer was much higher than that of the metal with the highest 'contact potential' (4.7 V for gold).

Using an apparatus described earlier,[17] the rate of spread of charge on nylon 66 filaments with different diameters and draw ratios was examined. The diffusion coefficients for charge decay at 52% r.h. showed no clear change for fibres with diameters in the range 49–136 μm. This implies that there is no change in the mechanism of diffusion with fibre diameter as the coefficients are based on a one-dimensional diffusion equation.

The effect of draw ratios up to 3.8 was to reduce the diffusion coefficient considerably owing, it was thought, to the orientation of the molecular structure. The possible effect of an increase in crystallinity was discounted on the grounds that the drawn and undrawn fibres had the same isotropic refractive index.

Charge mobility on a drawn film of nylon 6 was observed after charging a small circular area by induction. A transverse scan of the charge distribution at a given radius showed a maximum intensity perpendicular to the drawing axis and a minima along the axis, indicating that the charge moves more easily across the molecular chain axes than along them.

The work of Davies,[24,26] Inculet and Wituschek,[25] and Arridge[27] on the charging behaviour of various insulators against metal, points to an electronic process of charging which can occur *in vacuo* or in ordinary atmosphere, and after single or multiple contact.

The role of adsorbed ions in contact charging between metals and insulators has been examined by Robins et al.,[28] using pyroelectric materials. These are readily polarized when heated and cooled in an electric field. Samples of polyvinyl fluoride (PVF) and tryglycine sulphate (TGS) so treated were mounted on a ground plate and coated with positive or negative ions (depending on the direction of the field during polarization) either by washing with methanol or by a stream of ionized argon. A metal sphere was supported in such a way that it could be brought repeatedly into contact with the pyroelectric insulator at the same spot. The charge transferred was obtained from measurements of the potential of the sphere before and after contact, using an electrometer. Charge transfer from virgin (unpolarized) samples was also measured. Most of the experiments were done in air or pure dry argon, but some on TGS *in vacuo* gave similar results to those in dry argon, indicating that charge density was not limited by air breakdown.

The results for both pyroelectrics showed the same polarity of charge on the sphere as that of the ions on the pyroelectric. The charge increased with number of contacts tending to become constant after a few dozen contacts. The charges for the virgin samples were comparatively small.

The results could be explained simply by supposing that ions at the insulator surface are neutralized by charge exchange with the metal, or alternatively that ions at the insulator surface are neutralized by charge exchange with the metal. Surface ions do appear to have a dominant effect on the charging of pyroelectric samples and there is no reason to suppose that the charging of ordinary insulators would be any different.

3. ELECTROSTATIC CHARGES ON CLOTHING

The problem which can arise from static charges on the clothing of a person insulated from earth by the shoes or the floor may be summarized by referring to Figure 6. It is seen that the charge on the clothing, here shown as positive, induces a positive charge on to the body which is available for dissipation, and causes a negative charge to come up to the surface of the skin where it is held captive by the positive charge on the adjacent clothing. Such distribution of charge on the clothing/person system can produce any of the effects listed below.

(1) On touching a large or earthed conductor the available charge is instantly dissipated and a spark produced, which in the presence of flammable gases, vapours, or powders, may cause an ignition.
(2) Similarly, when the charged clothing approaches a large or earthed conductor an incendiary spark may pass from the clothing to earth.
(3) Any dust particles coming within the influence of the electrostatic field near the clothing can be attracted to the latter, resulting in soiling.
(4) On touching a conductor, the dissipation of charge to earth can cause a shock to the body.
(5) The clothing may cling to the body due to the attraction of the charges of opposite polarities on the clothing and skin.

Of these (3), (4), and (5) are considered to be merely a nuisance while (1) and

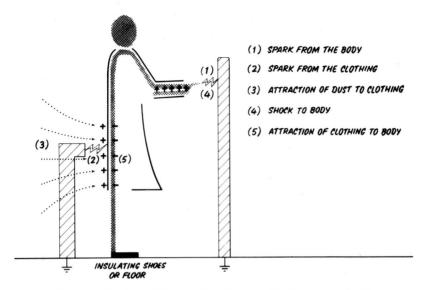

(1) SPARK FROM THE BODY

(2) SPARK FROM THE CLOTHING

(3) ATTRACTION OF DUST TO CLOTHING

(4) SHOCK TO BODY

(5) ATTRACTION OF CLOTHING TO BODY

INSULATING SHOES OR FLOOR

Figure 6. Various effects arising from static charges on clothing.

(2) can be hazardous. While there is little doubt that sparks direct from insulating surfaces can be of an incendiary nature, as, for example, demonstrated by Gibson and Lloyd,[29] it is well known that discharges from conductors are likely to be more energetic, as all the charge is dissipated at once, and are therefore potentially more dangerous. It is probably for this reason that the few investigations into the electrostatic charging of clothing which have been published are largely concerned with the hazardous effects of induced charge on the body.

The danger of static charges on textile surfaces in the form of clothing or on the body when insulated from earth, giving rise to incendiary sparks in locations where flammable gases, vapours, or powders are present, has for some years been a cause for concern by responsible people in occupations such as the armed forces, the petroleum and gas manufacturing industries, and in hospitals. An early attempt at understanding the problem of charge on the outer clothing was made by Guest *et al.*,[30] in work concerned with risks in the hospital owing to static electricity. They measured the voltage (or charge) induced on the body by that on the outer clothing on rising from a chair. This technique is realistic, representing an everyday bodily action which is likely to generate an amount of charge exceeded only by the quick removal of an outer garment.

In a report dealing with static electricity in the Apollo spacecraft, Potter and Baker[31] measured the voltage on a space-suited man, insulated from earth, after rubbing the suit with various materials. Results showed that the voltage induced on the man depended on the nature of the rubbing material. At 55% r.h. the highest voltage obtained was 1.6 kV after rubbing with a glass-fibre cloth. At 30% r.h. the value increased to 4.3 kV. The maximum electrical capacitance to earth of the man in the space suit was found to be about 200 pf, the same as that of the man's body.

The corresponding charge energy on the body was about 2.0 mJ. The minimum ignition energies of combustibles in the spacecraft are quoted as follows: flammable gas vapours in a 0.33 atmosphere of pure oxygen require 0.02 mJ, gas mists need about 1.0 mJ, and the ignition of combustible solids require between 10 and 100 mJ. A comparison of the stored charge energy on the body of the person with these ignition energies indicates that in theory there is enough available to cause the ignition of the gas vapours and mists.

The elimination of electrostatic charges from the person was conveniently achieved by the use of biomedical sensors attached to the body which allow the charge to pass quickly and safely to the spacecraft (earth). In examining other sources of high electrostatic charge in the spacecraft three were isolated as being potentially dangerous. A Teflon-coated couch cover was found to be highly charged after use by a crewman and could induce a high charge on to nearby conducting objects leading, possibly, to a spark. This problem was overcome by installing a grounded metal screen underneath the cover. Another potential hazard was identified during the withdrawal of lithium hydroxide canisters from storage lockers, when sparks were found to jump from the base of the canister to the top of the locker. The lockers are lined with pads of silicone rubber and Teflon pads, providing a snug fit for the canisters. From measurements of the capacitance of the canister relative to the locker and the voltage on the former, the theoretical spark energy was estimated to be 1.5 mJ. Grounding to the spacecraft during insertion and withdrawal from the lockers did away with the problem. The third problem was found on a Teflon belt and pulley system used to hoist lunar sample containers on to the lunar module. The belt became charged and was capable of attracting powdered minerals to its surface. It would therefore be possible for the belt to carry lunar dust, which may be toxic or flammable, into the lunar module. This was prevented by the insertion of a metal thread in a zigzag pattern into the belt which presumably neutralizes the static charge.

The risks of ignition because of static on outer clothing have been examined by Henry.[32] The charging behaviour and surface resistivities of various protective garments were measured after conditioning at various humidities in the range 23–85% r.h., at room temperature. The voltage on an operator wearing a garment and insulated from earth by wearing rubber-soled plimsolls was determined immediately on rising from a chair by touching the input of a Rothschild static voltmeter. Two types of chair cover were used, Rexine and PVC. Ten body voltage readings were taken using each chair cover and the mean of the 20 readings calculated. A correction was made to allow for the input capacitance of the instrument. The garments are of nylon, polyester-cotton, and cotton. The surface resistance of the materials was measured by using concentric ring electrodes placed on the surface of the outer material of the garment when supported horizontally by a sheet of insulating rubber. The resistance across the annulus of fabric between the electrodes was obtained by comparing with a standard resistor, and was converted to surface resistivity by multiplying by a constant determined from the radii of the electrodes. The surface resistivity of the material (ohm per square centimetre) is equivalent to the average resistance between opposite edges of a square in the surface of the material.

Plots of log body voltage against r.h. showed that the nylon garment gave higher voltages than any of the other garments over the whole range of humidity.

Two notable results were that an all-cotton garment gave higher voltages at low r.h.s than a new Terylene/cotton garment, and that a much worn and washed Terylene/cotton garment gave higher voltages than a new Terylene/cotton garment. The former implies that certain Terylene/cotton fabrics may give less of a static problem than cotton ones, but, as Henry points out, this might be altered when the contact is with seat coverings other than Rexine and PVC. The relative behaviour of the worn and new Terylene/cotton garments was attributed to the fact that with the much worn and washed garment the cotton component had been removed from its outer surface, leaving almost pure Terylene fibres behind.

Using the highest body voltage obtained from the 20 readings, a plot of the log maximum body voltage against log surface resistivity was made in order to estimate the critical value of the surface resistivity, applicable to homogeneous and new blended fabrics, below which dangerous levels of static would not be produced after contact with external surfaces in everyday usage. Knowing the capacitance of the operator (200 pf), the body voltage corresponding to 0.2 mJ, the minimum ignition energy of saturated hydrocarbon gases mixed with air was determined. Taking the most stringent material, cotton, which produced the highest body voltages at surface resistivities below 10^{13} ohm/cm^2, the upper limit of resistivity was found to be 10^{11} ohm/cm^2

Attempts at igniting a sensitive mixture of town gas and air by spark discharges from the body were made. The gas mixture was passed into an ignition chamber consisting of a Perspex box inside which was mounted an earthed metal sphere. The operator, insulated from earth, was then charged to a known voltage and a spark passed from a metal rod, held in the hand, to the metal sphere. One ignition in 10 trials was obtained when the charge energy on the body was between one and two orders of magnitude greater than the minimum ignition energy of the mixture. Henry suggested that an instantaneous loss of heat from the spark to the metal surfaces across which it passes (quenching) might, in part, account for the discrepancy.

The body voltage results showed that all the garments reached or exceeded the 0.2 mJ limit and that only the new nylon and all-cotton garments reached the energy limit at which ignitions of the town gas-air mixture were obtained.

In a study of the electrostatic hazards from insulated operators Bajinskis and Lott[33] measured the body voltage on an operator, insulated from earth by the shoes, immediately after rising from a chair. Various types of outer garment and seat materials were used and it was shown that voltages of 4 kV and higher were readily obtained. Similar experiments in which the operator wore shoes with soles of various electrical resistances were carried out, and the mean peak voltage and half decay time corresponding to each shoe sole was observed. A half decay time of more than 200 sec was obtained when wearing microcellular rubber shoes of a resistance exceeding 10^{13} ohm. The corresponding peak voltage was between 4 and 8 kV, depending on the nature of the outer garment and seat cover. With a half decay time of 0.1 sec, obtained when wearing conducting footwear, the corresponding peak voltage was 60 V. The capacitance of the operator to earth when wearing the conducting footwear was not measured and is likely to have been considerably higher than the value of 200 pf assumed by Bajimskis and Lott when estimating the energy of the charge stored on the body. However, this would probably not invalidate their conclusion that the hazard of static discharges from the body in the presence of highly sensitive materials such as, for example, lead styphnate (minimum ignition energy of 0.003 mJ), would be practically non-existent when the half decay time of charge leakage from the body is 0.1 sec. It was pointed out that this requirement would not apply to cases of extreme sensitivity, as, for example, with ultra-clean surfaces of primary detonators.

The effects of various types of carpet on the half decay time of charge from a person wearing conducting footwear and standing on a carpet was examined. At 30% r.h. only certain carpets treated with an antistatic agent or containing stainless steel fibres gave half decay times approaching 0.1 sec. However, at 50% r.h. certain wool carpets and carpets of nylon blended with a conducting fibre also met the latter requirement.

In discussing a number of potential and real accidents due to static electricity, an example is cited of an operator wearing microcellular soled shoes picking up a charged polythene container and walking to a blending vat whereupon an electrostatic discharge from the hand to the frame of the vat ignited the acetone solvent. Presumably the spark was due to an induced charge on the operator arising from that on the container.

The generation of charge on the body from that on the footwear and on the clothing has been measured by Wilson and Cavanagh[34] in assessing the electrostatic behaviour of some military clothing during wear. In ambient conditions of 10% r.h., 22°C, a standard walk on a polypropylene carpet was taken by an operator wearing different types of footwear. At the end of the walk, the charge on the body was measured by sharing with a standard capacitor while one foot was raised. Knowing the capacitance of the operator, the maximum energy of the stored charge on the body was found to be about 2.7 mJ — a value well above the minimum ignition energy of a fuel vapour—air mixture (0.2 mJ), and so considered to be hazardous.

The charge generated between adjacent layers of clothing during normal movement would be expected to have little inductive effect on the body, as equal quantities of charge of opposite polarities occur in close proximity to each other and are, therefore, mutually bound. In examining the charging behaviour of clothing, the outer garment was quickly removed, while both feet were firmly on the carpet, immediately following the standard walk. This action leaves the garment underneath the outer one in a charged condition and a similar charge is induced on the body. In this case, the total charge on the body arises from the repeated contact and separation of the shoes and carpet and the removal of the outer garment. The net charge was found to be similar to that obtained after a walk only, and was of opposite polarity. This implies that the charge energy on the body on removing the outer garment, without doing the walk, would be considerably higher than that after doing the walk alone, thus highlighting the particular danger of removing clothing in areas where flammable gases or vapours are present.

One of the means by which stored charge energy can safely be dissipated is by corona discharge from conductors in the form of sharp points. In examining this a man, insulated from earth, and holding a sharp metal needle in his fingers was charged to a potential of 20 kV. On holding the needle at a distance of 2 cm from an earthed metal plate a corona discharge occurred, leaving a residual voltage on the body of 2 kV. The corresponding voltage on holding the forefinger at the same distance from the plate was 18 kV. Thus, the needle was shown to be effective in greatly reducing the potential of the man safely when held at a small distance from the plate but, as noted by Wilson and Cavanagh, the residual charge energy on the man was still above the potentially dangerous level in the presence of petrol vapour of 0.2 mJ. Furthermore, the question remains as to whether or not an incendiary spark discharge would take place when the needle was rapidly brought up to the plate. Such an action could well prevent the safe discharge of much of the available energy on the body, leaving enough for the delivery of a single condensed spark of high energy.

In a report by Ormer[35] on the static propensity of navy clothing, a number of important factors relating to the hazards of static charges on the clothing/man system have been examined.

Experiments with capacitative sparks between various electrodes placed in an optimum mixture of propane and air at room temperature and pressure, showed that ignitions occurred when the energy of the charge on the capacitor was 0.56 mJ, when using pointed brass electrodes of radius 0.05 mm. With electrodes of radius 4.5 mm the minimum ignition energy increased to about 3.6 mJ. No explanation for this effect was given, but it is probably largely owing to differences in quenching of the spark. With the larger electrodes there is likely to be a greater loss of heat from the spark to the metal surfaces than with the pointed electrodes. The degree of quenching will also depend on the distance between the electrodes when the spark passes – the shorter the distance the greater the quenching. The separation of the pointed electrodes on passing a spark is likely to be greater than for the larger electrodes owing to the highly intense field at the points, and the spark should suffer less quenching.

These results are discussed in relation to an opinion held by Heidelberg[36] who has stated that flammable gases cannot be ignited by discharges from sharp points (i.e. radius not greater than 2 mm). Ormer believes that this opinion is based on a study of discharges from non-conducting surfaces to grounded electrodes and is not applicable to normal spark discharges between conducting electrodes.

Spark discharges between an earthed human finger and an electrode connected to a negatively charged 200 pf capacitor in an optimum propane/air atmosphere showed that the minimum ignition energy increased from about 3.5 to 4.5 mJ when electrodes of radii 1 and 5 mm, respectively, were used. The latter energy value is about 35 times the published minimum ignition energy of the mixture.

An experiment in which the charge was generated on the operator, whilst insulated from earth, by quickly removing an outer garment showed that the energy on the body required to cause the ignition of an optimum acetylene/air mixture by a spark from the fingertip to a brass electrode of 3 mm radius was 1.1 mJ. This is about 58 times the published minimum ignition energy of the gas mixture. No attempt was made to discover the cause of the discrepancies between the available charge energies on the body or the capacitor and the published minimum ignition energies of the propane/air and acetylene/air mixtures. The fact that the discrepancies for a discharge from the body are considerably greater than that from the capacitor implies that a larger proportion of the energy from the capacitor goes into the spark compared with that from the body.

Body voltages on various operators insulated from earth were measured immediately after the fast removal of an outer garment placed across the shoulders, drawn tightly around the body, and then snatched off. The garments were conditioned at 37% r.h., 21°C for at least 24 hours and the tests subsequently carried out under the same conditions. When a jacket with a nylon fleece lining was removed from a water-repellent treated, modacrylic poplin jacket worn by the operators, body voltages between 2.8 and 6.6 kV were obtained. The average value was 4.7 kV, corresponding to an energy value of 2.2 mJ. When a polyester/cotton material containing Brunsmet metallic fibres was removed from the same modacrylic jacket the maximum body voltage was 3.8 kV, equivalent to 1.44 mJ of energy. Thus, under appropriate conditions, when wearing either combination of these materials it should be possible to ignite an acetylene/air mixture by a spark from the fingertip.

As would be expected there was no significant charge generated on an operator insulated from earth while exercising for 15 sec. While copious charges would be produced on adjacent layers of clothing, owing to contact and rubbing, they would induce similar voltages of opposite polarity on the body, resulting in a low net voltage.

The voltage on the body can be affected by the rate at which an outer garment is removed. To examine this an operator insulated from earth and wearing a 50%/50% polyester/cotton jacket over a 50%/50% nylon/cotton jumper, removed the former in different time intervals under ambient conditions of 33% r.h., 20.5°C. The voltage on the operator after removing the jacket in 1 sec (snatched off) was 5.5 kV and after taking 8 sec was 2.6 kV. These results indicate that some neutralization of charge takes place when the garments are separated slowly. However, at lower humidities, or with more insulating fabrics the rate of charge dissipation may be so low that the slow removal of a garment will contribute little to safety. It is therefore advisable not to remove a garment in environments where sensitive gases, vapours, or powders are present.

Following the work of Henry[32] an extended study of the electrostatic behaviour of a wide range of clothing has been made by Wilson[37] in ambient conditions between 15 and 80% r.h. at about 21°C. The garments are of the type worn by military personnel. Some are of 100% cotton, Teklan, and Nomex, and there are various polyester/cellulosic fabrics. Also included are a number of coated fabrics with polymer coatings of polyurethane, Neoprene, and butyl rubber. Some of the garments were washed and rinsed before use. The properties examined were (a) surface electrical resistivity of the outer material of the garment, and (b) body voltage (or charge) induced on the body of an operator, when insulated from earth, on sliding out of a chair into a standing position. Various seat covers of the type used in military aircraft and armoured fighting vehicles were attached to the chair. Twenty voltage readings with each garment and chair cover were taken. The covers consisted of PVC-coated cotton, leather, cotton, and a lamb's-wool seat cover used by pilots.

The garments and chair covers were conditioned for about 20 hours, usually overnight, before starting the measurements.

The surface resistivities of all the clothing fabrics fell with increasing r.h. The most resistive fabrics were found to be Nomex, Teklan, and the nylon fabrics coated with polyurethane or butyl rubber.

The highest body voltages were obtained with the PVC-coated cotton and leather chair covers. Cotton garments were found to be the most static prone at humidities below 30% r.h. The Nomex fabric gave relatively low voltages against these chair covers.

For the purpose of assessing the potential hazard of charge induced on the body from that on the clothing, the maximum body voltages obtained with the PVC-coated cotton chair cover were used. A plot of log maximum body voltage against log surface resistivity is shown in Figure 7. It is seen that the maximum is not a univalued function of surface resistivity; different fabrics with the same surface resistivity produce widely different body voltages. The body voltage levels corresponding to the maximum ignition energies of synthetic gas and air (0.03 mJ) and natural gas and air (0.3 mJ), published by Sayer et al.[38] are shown to the right of the figure. Also included is the minimum ignition energy of a fuel vapour—air mixture (0.2 mJ) obtained by Lewis and Von Elbe.[39] The upper envelope of the

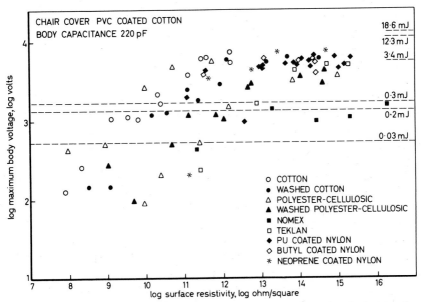

Figure 7. Plot of log maximum body voltage against log surface resistivity for various fabrics. Body capacitance 220 pf.

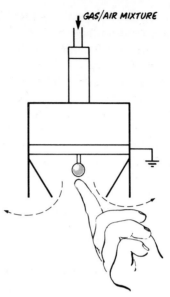

Figure 8. Simplified vertical section of the Perspex ignition chamber.

data represents the highest voltages obtainable in the chair experiment with fabrics of given surface resistivities, and it is seen that the 0.03 mJ and 0.2–0.3 mJ energy levels are reached with fabrics of resistivities of the order 10^8 and 10^9 ohm/cm^2, respectively. By textile standards these values are low and could only be achieved with fabrics containing hydrophilic fibres and those treated with antistatic agents, at high humidities, or with fabrics containing small percentages of steel or other highly conducting fibres. It is also seen that a wide range of fabrics can readily give rise to charge energies on the body sufficient, in theory, to cause the ignition of sensitive mixtures of various flammable gases and vapours mixed with air.

The question now arises as to what is the actual stored charge energy required on the body to cause the ignition of flammable gases by a spark discharge from the body to an earthed conductor. To answer this question we need to understand the nature of spark discharges from the body. In some preliminary measurements by Wilson[37] to find the energies of the stored charge on the body required to cause the ignition of stoichiometric mixtures of synthetic gas and air, and of natural gas and air, by spark discharges from the body, the same Perspex box with the earthed stainless steel ball mounted inside it, as described by Henry,[32] was used (Figure 8). The operator of capacitance 220 pf was insulated from earth and charged to a given voltage. He immediately discharged himself by bringing the finger up to the metal ball. This was done 100 times and if no ignition occurred the voltage was increased. The discharges were not from any particular part of the finger. The energies of the charge on the operator to produce at least one ignition in the 100 attempts were 3.4 and 18.6 mJ for the synthetic gas–air mixture and the natural gas–air mixture, respectively. Referring to Figure 7, it can be seen that the energy level of 3.4 mJ is reached by several fabrics, but none give voltages on the body equivalent to 18.6 mJ. The latter is about 62 times the published minimum ignition energy of the natural gas—air mixture, and assuming the same factor can be applied to a fuel vapour—air mixture the estimated energy of charge on the body for an ignition is 12.3 mJ. None of the fabrics gave voltages corresponding to this level of energy.

Clearly the discrepancies between the minimum ignition energies of the gas—air mixtures and the charge energies required on the body to produce ignitions are large. No attempt was made to measure the minimum ignition energies of the gas mixtures using capacitative sparks and electrodes of optimum dimensions and spacings, but some time was spent in using proportions of gas and air on either side of the stoichiometric values. For differences of a few per cent from the latter the number of ignitions in 50 attempts was greatest for mixtures which were 1 or 2% on the lean side of the stoichiometric value. This difference does not sensibly affect the minimum ignition energies of the gas—air mixtures and the results imply that the mixtures used in the ignition experiments were, for practical purposes, the most easily ignitable.

As noted by Henry[32] the loss of heat from the spark (quenching) to the electrodes across which the spark passes, is a factor probably contributing to the discrepancy, but Wilson[37] was of the opinion that this effect alone would not account for the discrepancy and suggested that the electrical resistance of the body might be an important factor to consider. Its effect would be to restrict the mobility of the charge leaving the body and to absorb part of the stored energy lost during the discharge as heat.

Accordingly, Wilson[39] began an investigation into the nature and incendiary behaviour of spark discharges from the body.

Ten volunteers took part. Their electrical resistances were obtained by measuring the current through each under an applied voltage of 30 V. Standing in bare feet on a metal plate they completed an electric circuit by touching a stainless steel ball with (a) the ball of the finger, and (b) the fingertips. The resistances measured through the ball of the finger were in reasonable agreement with each other and were invariably lower than those obtained through the fingertip. The values through the fingertip showed considerable variation between individuals, owing, it was thought, to differences in thickness and dryness of the skin.

The body capacitances of the operators were also obtained with the subjects standing in socks (or stockings) on a sheet of Perspex, by charging them up to a known voltage and sharing the charge with a standard capacitor. The voltage across the standard was measured and the capacitance of the operator calculated. The results showed variations between operators which were attributed to differences in their size, the average value being 256 pf. The capacitance of the operator who later did the ignition tests is 240 pf.

As the resistance through the ball of the finger was always less than through the tip it was considered likely that, for a given voltage on the body, sparks from the ball of the finger would more readily ignite a given gas mixture. In studying the nature of static discharges from the body it was therefore decided to observe the profiles (charge versus time) of sparks passed via the ball of the finger only.

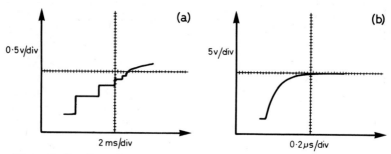

Figure 9. Profiles of spark discharges from the body at 5.0 kV; (a) all the sparks, (b) first spark.

The profiles were obtained by passing sparks to the stainless steel ball inside the Perspex box. The ball was connected to earth via a standard capacitor which is large compared with that of the body and which satisfies the time constant of the circuit. A high-speed storage oscilloscope (Tektronix, type 466) connected across the capacitor was used to record the rate of transfer of charge to the metal ball. The operator, with shoes removed, stood on the Perspex sheet and after being charged to a potential of 1.0 kV discharged himself by gently bringing the ball of the finger up to the metal ball. Data from five successive measurements were taken from the spark profiles displayed on the oscilloscope. The same procedure was followed with the subjects at potentials of 5.0 and 10.0 kV.

Typical profiles of voltage (or charge) against time during a spark discharge from the body at 5.0 kV are shown in Figure 9. In Figure 9(a) the steps indicate that a series of sparks was passed at intervals of a few milliseconds until the finger made contact with the metal ball. This behaviour was thought to be due to a fall in voltage at the finger, immediately following a discharge, to a value below the breakdown voltage of the spark gap. At this stage the charge ceased to flow, and as the finger approached the metal ball breakdown conditions were re-established and a second discharge occurred. This process continued until the finger touched the metal ball. It can be seen in Figure 9(a) that the charge transferred during the first spark is considerably greater than that in any of the succeeding sparks. A more detailed profile of this spark is shown in Figure 9(b) where it is seen that the charge is transferred within a few microseconds.

Knowing the initial voltage on the body, the body capacitance, and the charge lost from the body during the spark discharge, the corresponding loss in charge energy from the body was calculated. The average values of the charge energies lost from the 10 subjects during the first spark, and all the sparks, were plotted against the initial body voltage and are shown in Figure 10. The figure includes a curve showing the energy which would be lost if the body was completely discharged. At a given voltage the energy lost from the body during the first spark is more than half the total energy lost during the entire period of sparking. From this it was concluded that an ignition would probably be initiated by the first spark alone.

The distribution of the energy lost from the body during sparking should depend on the resistance of the body and of the spark gap at the instant the spark passes. In

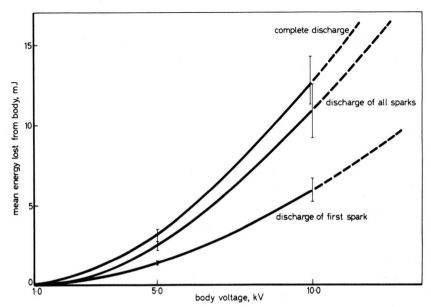

Figure 10. Plot of mean energy lost from the body against body voltage.

the present context the least conductive part of the body is the skin, which, if punctured at the point of discharge, may suffer an electrical breakdown. Should this be the case, the body resistance obtained under a low voltage might no longer apply during a spark discharge. Consequently, estimates of the body resistance during a spark discharge were determined from the initial slope (current) of the profile of the first spark, and the initial voltage on the body, assuming Ohm's law.

The discharge circuit was considered to be simply one in which the charge on the body was dissipated via the body resistance and the resistance of the spark gap measured at the instant of sparking. The circuit resistances obtained from the slope of the profile of the first spark (delivered by the same operator who later did the ignition experiments) are shown in Table 3. They correspond to different initial voltages on the body, and are a measure of the body resistance since, as was shown later, the resistance of the spark gap is small compared with that of the body. The data show that there is not breakdown in the body resistance during sparking,

Table 3

Body voltage (kV)	Circuit resistance (ohm)	Body resistance (ohm) at 30 V
1.0	1.4×10^4	3.1×10^4
5.0	0.7×10^4	
10.0	0.4×10^4	

Table 4

Estimated circuit resistance (ohm)	Circuit resistor (ohm)
1.3×10^5	1.0×10^5
1.4×10^4	1.1×10^4
1.6×10^3	1.5×10^3
800	470
360	47

although the values fall with increasing body voltage. It would be expected, therefore, that a large proportion of the stored charge energy will be lost to the body as heat.

The resistance of the spark gap was obtained in a similar manner by discharging a standard capacitor through various resistors of known value. The results are shown in Table 4. It is seen that there is fair agreement between the higher values of the circuit resistors and the estimated resistances of the circuit obtained from the slope data. However, at lower resistances, the latter indicate the presence of an additional resistance in the circuit of about 320 ohm. This was taken as representing the resistance of the spark gap at the instant the spark passes.

Ignition experiments similar to those done in the preliminary work were carried out by an operator with a body capacitance of 240 pf, standing without shoes on the Perspex sheet. The spark discharges were from the ball and the tip of the finger, and a rubber glove with part of the index finger removed was worn to protect the hand from burns.

The body voltages at which ignitions of the synthetic gas—air mixture and the natural gas—air mixture was first ignited by sparks from the ball of the finger are 4.3 and 12.5 kV, respectively, being equivalent to charge energies on the body of 2.2 and 18.8 mJ. The corresponding energies for discharges via the fingertip are 4.3 and 27.0 mJ.

The energy E_A lost from the body during the discharge of the first spark is taken to be absorbed as heat by the body resistance R and the resistance of the spark gap r, according to the following equation:

$$E_A = \int_0^t R \cdot i^2 \, dt + \int_0^t r \cdot i^2 \cdot dt \tag{1}$$

where i is the current at a time of t.

The energy absorbed by the spark gap E_G may be determined by numerical analysis, using data from the profiles of the first spark for discharges from the body at 1.0, 5.0, and 10.0 kV. However, a simpler treatment may be applied by assuming that R and r remain sensibly constant during a discharge, and can therefore be taken outside the integrals in equation (1). Now as $R \gg r$, the energy E_G lost in the spark gap is given approximately by

$$E_G = E_A \cdot r/R \tag{2}$$

The values of R were obtained by interpolation for the data given in Table 3, using the values of body voltage at which the synthetic gas—air and natural gas—air mixtures were ignited. The value of r is about 320 ohm. From data similar to that given in Figure 10, and applicable to the operator who did the ignition tests, the value of E_A corresponding to 4.3 kV is 1.08 mJ. Using equation (2) the estimated energy released in the spark gap when an ignition of the synthetic gas—air mixture occurs was 0.05 mJ. A similar calculation for the natural gas—air mixture gave 1.0 mJ. These values are reasonably close to the minimum ignition energies of 0.03 mJ for synthetic gas and air, and 0.3 mJ for natural gas and air. As might be expected from realistic experiments in which neither the electrodes nor the spark gap widths were chosen for optimum conditions of ignition, they are rather higher than the published values.

Attempts at igniting the gas—air mixtures by sparks from an operator via a sharply pointed metal electrode held in the hand showed that the synthetic gas—air mixture was ignited at a lower charge energy on the body (1.47 mJ) than when the ball of the finger was used. The result is in support of the work of Ormer[35] who ignited a propane gas—air mixture more readily by a spark from a pointed electrode than with a blunt one.

With the natural gas—air mixture ignitions were still possible with sparks via the pointed electrode, but were more difficult than with sparks from the ball of the finger. The reason is that the high body voltages required to produce ignitions cause much of the energy to be lost from the point in a corona discharge before the spark is passed.

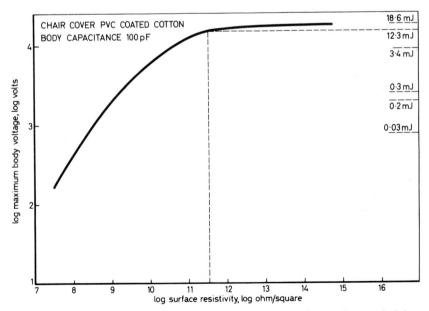

Figure 11. Plot of log maximum body voltage against log surface resistivity for various fabrics. Body capacitance 100 pf.

The last factor which was considered in order to ensure reasonable safety against incendiary spark discharges was the role of body capacitance. Taking a value of 100 pf on the likely minimum value for human beings and assuming the charge remains constant, the upper values of body voltage at various resistivities shown in Figure 7 were adjusted to take account of a change in capacitance of the operator from 220 to 100 pf. The results are given in Figure 11, from which can be deduced the maximum permissible surface resistivity of a fabric in order that a given energy level on the body is not exceeded when using the chair method of charging the body.

Taking the case of a fuel vapour—air mixture which for an ignition would require about 12.3 mJ of energy on the body, the maximum resistivity is about 3.0×10^{11} ohm/cm^2. This applies to homogeneous and new fabric blends which form the outermost material of the clothing.

REFERENCES

1. W. R. Harper, *Contact and Frictional Electrification*, Oxford Clarendon Press, 1967.
2. P. S. H. Henry, *J. Text. Inst.*, **48**, 1, P5, 1957.
3. D. J. Lehmicke, *Am. Dyes, Rptr.*, **38**, 24, 853, 1949.
4. J. W. Ballou, *Text. Res. J.*, **24**, 2, 146, 1954.
5. A. Coehn, *Ann. Phys.*, **64**, 217, 1898.
6. H. F. Richards, *Phys. Rev.*, **22**, 122, 1923.
7. S. P. Hersh and D. J. Montgomery, *Text. Res. J.*, **25**, 279, 1955.
8. R. G. Cunningham and D. J. Montgomery, *Text. Res. J.*, **28**, 12, 971, 1958.
9. J. B. Levy, J. H. Wakelin, W. J. Kauzmann, and J. H. Dillon, *Text. Res. J.*, **27**, 11, 897, 1958.
10. P. S. H. Henry, *Brit. J. Appl. Phys.*, **4**, Supplement No. 2, S31, 1953.
11. V. J. Webers, *J. Appl. Poly. Sci.*, **7**, 4, 1317, 1964.
12. J. A. Medley, *Nature*, **166**, 524, 1950.
13. J. A. Medley, *Brit. J. Appl. Phys.*, **4**, Supplement No. 2, S28, 1953.
14. D. Wilson, *J. Text. Inst.*, **54**, 3, T97, 1963.
15. P. S. H. Henry, R. G. Livesey, and A. M. Wood, *J. Text. Inst.*, **58**, 2, T55, 1967.
16. V. E. Shashoua, *J. Polymer Sci.*, **33**, 65, 1958.
17. R. G. C. Arridge, *Brit. J. Appl. Phys.*, **11**, 5, 202, 1960.
18. J. W. S. Hearle, *J. Text. Inst.*, **44**, 4, T117, 1953.
19. P. J. Sereda and R. F. Feldman, *J. Text. Inst.*, **55**, 4, T288, 1964.
20. R. F. Feldman and P. J. Sereda, *J. Amer. Concrete Inst.*, **58**, 203, 1961.
21. S. Brunauer, P. H. Emmett, and E. Teller, *J. Amer. Chem. Soc.*, **60**, 309, 1938.
22. W. R. Harper, *Proc. Roy. Soc.*, **A218**, 111, 1953.
23. D. K. Davies, *Static Electrification*, I.O.P. and Phys. Soc., Conf. Ser. No. 4, 29, 1967.
24. D. K. Davies, *J. Sci. Instrum.*, **44**, 521, 1967.
25. I. I. Inculet and E. P. Wituschek, *Static Electrification*, I.O.P. and Phys. Soc. Conf. Ser. No. 4, 29, 1967.
26. D. K. Davies, *Brit. J. Appl. Phys.*, Ser. 2, **2**, 1533, 1969.
27. R. G. C. Arridge, *Brit. J. Appl. Phys.*, **18**, 1311, 1967.
28. E. S. Robins, A. C. Rose-Innes, and J. Lowell, *Static Electrification*, I.O.P., Conf. Ser. No. 27, 115, 1975.
29. N. Gibson and F. C. Lloyd, *Brit. J. Appl. Phys.*, **16**, 1619, 1965.

30. P. G. Guest, V. W. Sikora, and B. Lewis, *Bureau of Mines Report 4833;* U.S. Department of the Interior, 1952.
31. A. E. Potter, Jr. and B. R. Baker, *NASA Technical Note TN-5579* Manned Spacecraft Center, Houston, Texas, U.S.A., 1969.
32. P. S. H. Henry, *Static Electrification* I.O.P., conf. Ser. No. 11, 212, 1971.
33. G. Bajinskis and S. A. Lott, *Report 521* Australian Defence Scientific Service, Defence Standards Laboratories, Maribyrnong, Victoria, Australia, 1972.
34. L. G. Wilson and P. Cavanagh, *Report No. 665*, Defence Research Establishment, Physics and Test Section, Ottawa, Canada.
35. G. M. Ormer, *Technical Report No. 109* Navy Clothing and Textile Research Unit, Natick, Massachusetts, U.S.A., 1974.
36. E. Heidelberg, *Static Electrification* I.O.P. and Phys. Soc., Conf. Ser. No. 4, 29, 1967.
37. N. Wilson, *Shirley Institute Report for Ministry of Defence*, SCRDE, 22/286/13. MOD Project No. A/70/GEN/10004, 1973.
38. J. F. Sayers, G. P. Tewari, and J. R. Wilson, *I.G.E. Journal*, May 1971.
39. Cf. N. Wilson, *Shirley Institute Report for Ministry of Defence,* SCRDE, 21/13/473. MOD Project No. A/78/CLO/47236/CB (CT) 4B, 1976.

8

Photopolymerization at Surfaces

A. N. Wright

General Electric Company, Schenectady, N.Y.

1. INTRODUCTION

White first reported, in 1961, that a monomer vapour, 1,3-butadiene, in contact with metallic substrates polymerizes under the influence of ultraviolet light.[1] In a brief note to the *Proceedings of the Chemical Society*, he indicated that on freshly prepared lead or tin films, a polymer deposit was formed selectively on illuminated areas at a rate dependent upon the nature of the underlying metal film, and with decreased rates at higher substrate temperatures. The rate of polymer deposition was essentially independent of monomer pressure over the range 5—100 Torr. Following earlier conclusions of Harris and Willard[2] and of McTigue and Buchanan[3] from studies of photo-activated exchanges of methyl iodide on glass and metal surfaces, respectively, White concluded that film formation was probably a consequence of selective photolysis on the surface in which decomposition occurs in the adsorbed layer of the monomer gas.

Indications of the potential for formation of thin polymers on surfaces under irradiation came from the earlier work of Ennos[4] who showed that the interaction of *electrons* with organic molecules adsorbed on the bombarded metallic surface was the origin of polymer films that contaminate specimens in the electron microscope. The organic monomer molecules were continually derived from residual vapours in the vacuum system, with hydrocarbon diffusion pump oil as the major source.[5]

Raising the temperature of the bombarded source also *decreased* the rate and extent of decontamination in this process. Poole had suggested that the mechanism of film formation, in the presence of apiezon grease, is the free-radical polymerization, under the action of the bombarding electrons, of organic molecules adsorbed on the surface and that the molecules are then cross-linked by the electron beam to form a solid polymer film.[6] Holland *et al.* reported that the films formed from silicone oil vapours could be *conducting* if formed at higher substrate temperatures such as occur during bake-out (~400°C) of the vacuum system.[7] (This rather paradoxical behaviour might well have provided an advance indication of the complexity of the processes taking place on surfaces when subjected to irradiation in the presence of polymerizable adsorbates!)

These early studies on surface photolysis were closely associated with a practical search for very thin, configurationally deposited, and strongly adherent insulating films for use in the microelectronic industry. They parallel many publications at about the same time describing the deposition of thin polymer films by electron bombardment, for example Christy reported that thin insulating films less than 100 Å thick could be produced by bombarding a substrate in the presence of silicone oil vapours;[8] Haller and White extensively described the polymerization of butadiene vapour on surfaces under low energy (250 eV) electron bombardment with an apparent *negative* activation energy of -6 ± 2 kcal/mole;[9] and Hill described electron-beam polymerization of insulating films from silicone pump fluid for use in multilayer microcircuits.[10] The role of organic polymers in thin-film electronics from a more chemistry-related viewpoint was reviewed at about that time by Allan and Stoddart.[11] Corrosion protection offered by polymer films deposited directly from the monomer on to metal surfaces by discharge and electrolytic techniques has been recently reviewed.[12]

Some more detailed information on the surface-photopolymerization process did become available in the mid-1960s. White reported further[13] on surface film formation from butadiene, describing a deposition rate from 100 to 200 Å/sec at monomer pressures about 2 Torr, a breakdown voltage of $5-6 \times 10^6$ volt/cm, and a dielectric constant of 2.65. He also indicated that the polymer film deposition rate on substrates did decrease with time unless the UV-absorbing film from the deposition on the inside of the quartz window to the vacuum system was removed between runs. The presence of oxygen did not appear to alter the rate in any way. Addition of hydrogen to the vacuum system increased the initial rate of polymer formation, which again decreased as the film on the window increased in thickness. Increase of initial monomer pressure beyond about 2 Torr led to decreased film thicknesses on substrates after irradiation times of 30 min, whilst the rate of polymer formation was proportional to the 0.86 power of the light intensity. There was some indication that similar protective/insulating films could be deposited outside a vacuum system in a mixture of flowing butadiene gas at a partial pressure of 2 Torr in an atmosphere of helium. Gregor briefly reported film deposition from methyl methacrylate at rates that increased continuously with pressure from ~1 to ~40 Å/min over the range ~0.2–4 Torr.[14] Gregor and McGee reported deposition of thin polymeric dielectric films (completely insulating at thicknesses > 500 Å) by UV irradiation of surfaces in contact with vapors of acrolein, methyl methacrylate, and divinylbenzene.[15] Rates of formation, followed by deposition on to an oscillating quartz crystal, showed negative temperature dependence for all three monomers. For acrolein, the rate of film formation reached a maximum of ~70 Å/min at a monomer pressure of ~5 Torr. Gregor briefly reviewed the early photodeposition work with emphasis on the dielectric properties of the thin insulating polymeric films.[16] It had then become apparent that film growth rate does probably not depend on the nature of the substrate material, since deposition after the first monolayer or so occurs on the surface of the deposited film.

Conduction measurements on photodeposited surface films ~1000 Å thick led Fabian to conclude that the high resistivity of the films from butadiene resulted from a structure with less than one carbon—carbon double bond per four carbon atoms.[17] Films from cyclo-octatetraene had still greater resistivity, suggesting that a conjugated, straight-chain polyvinylene is not formed in the UV process. The current carrying capacity (10 mA) of films photodeposited from acetylene was attributed to retention of considerable electron-rich double bonds. Fabian attributed an even higher current (50 mA) in films produced from cyanogen to the formation of resonance structures along with retention of considerable unsaturation.

Nevertheless, little detailed scientific information was available in the mid-1960s about the UV surface photopolymerization process. Much pertinent background data was known in the literature, for example the early work of Melville[18] and of Gee[19] on the gas-phase photopolymerization of methyl methacrylate and butadiene; a study of the photochemistry (with a medium mercury arc emitting strongly in the region (2000—2500 Å) of 1,3-butadiene at 4 Torr pressure in which Srinivasan also detected polymer formation as a 'by-product';[20] and an investigation of the primary photolysis and mechanism of gas-phase photopolymerization of 1,3-butadiene in which a decrease in all volatile products with increasing pressure led Haller and Srinivasan to speculate that all such products derive from a vibrationally excited, ground state, 1,3-butadiene molecule formed by intramolecular rearrangement of the first excited singlet state populated during the primary photolytic act.[21] Labelling experiments further identified three primary processes in the vapour-phase photolysis of butadiene,[22] and Srinivasan and Sonntag reported on UV-absorption of 1,3-butadiene, both near the absorption maximum of 2100 Å and at longer wavelengths.[23] Since polymerization in the gas phase exhibited an induction period, with a rate dependent on the square of the number of photons absorbed, Haller and Srinivasan suggested that polymer formation may involve photolysis of a volatile product of the primary photodecomposition of the butadiene.[21]

It was apparent by the mid-1960s, however, that the photo process could provide a selective method of depositing polymeric films on surfaces in a patterned manner, and with high integrity when very thin. Moreover, the photon energy of 4—5 eV in the UV offered the potential for retention of the properties of the starting material, since activation approximates the excitation energy for polymerization reactions as opposed to the excess energy usually involved in electron-beam[8-10] and glow-discharge[24-27] processes. These induce considerable fragmentation of the 'monomeric' molecule and — at least for the latter technique — can induce surface film formation from starting materials as 'non-monomeric' as methane and pentane by decomposition to such fundamental building units as CH and CH_2. As will be described, the surface-photopolymerization technique permits the controlled deposition of very thin films and coatings close in properties to the polymers formed from recognized monomers by conventional polymerization techniques. In addition, non-monomeric materials such as phenol[28] yield films different for every feedstock, yet retaining in solid form many of the characteristics/properties of the irradiated vapour.

The process reviewed in this chapter is limited to photochemical formation of polymer films on surfaces which involves interaction on the surface with absorbates from the vapour phase of the starting material. Although the mechanisms of polymer formation may be complex, involving in many cases, as will be seen, the interaction of UV radiation with *products* of gas-phase photolysis adsorbed on the surface, it is this coincidence of light- and surface-adsorbed species from the gas phase which leads to the major characteristics of the process: surface film formation patterned by the configuration of incident light, strong adherence to most substrates, and a negative film growth-rate dependence on substrate temperature. Thus, processes such as the radiation (in this case γ-rays) induced polymerization of monolayers of adsorbed liquids on clays,[29] polymerization induced on clay surfaces previously impregnated with free-radical species,[30] polymerization on catalytic surfaces such as cobalt or nickel halides which may be accelerated by UV light,[31] irradiation during evaporation processes to induce cross-linking in polymeric films,[32] and the well-known processes for grafting polymers by irradiation of liquid monomers (see for example, Tsunooka et al.[33]) will not be directly addressed. Neither will the *photosensitization* of graft polymerization from the vapour phase, for example acrylics on to cotton in the presence of diketone,[34] acrylonitrile on to Aerosil in the presence of $Me_2Si_2Cl_2$,[35] nor vinyl isocyanate on to polythylene in the presence of benzophone,[36] be reviewed. Similarly, photo-induced *solid-state* polymerization of monomers such as acetaldehyde[37] and divinyl compounds[38] will not be considered. (Hasegawa provides a review of the early work on solid-state photopolymerization.[39]) The intriguing concept whereby thin polymeric films *previously* photodeposited (from benzaldehyde) on the inside of glass reactors may be used to *photosensitize* in a heterogeneous manner the gas-phase dimerization[40] of molecules such as pipery-lenes and 1,3-cyclohexadiene, also involves chemistry beyond the scope of the present review.

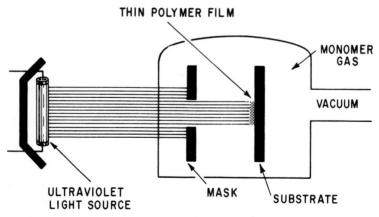

Figure 1. Schematic of UV surface-photopolymerization process. From Kunz et al.[41] (Reproduced by permission of the Society of Plastics Engineers.)

2. THE SURFACE-PHOTOPOLYMERIZATION PROCESS

The process is illustrated schematically[41] in Figure 1. Typical UV light sources that have been used by the author and his colleagues include a Hanovia 674 A, 700 W medium-pressure mercury lamp, and a General Electric H3T7 mercury lamp ballasted at 1000 W for optimum UV emission. Both lamps, usually positioned about 2 in from the substrate surface, emit significantly[42] at the wavelengths ~2000 Å required for deposition from many of the vapours, and give efficiencies[43] of 15–20% at the wavelength <3000 Å required for most of the polymerizations. Some of the earlier systems,[44] employing ultra-high-vacuum techniques in order to obtain very clean surfaces for mechanistic studies, are illustrated in Figure 2. In these examples, the entire deposition systems were constructed from fused quartz (GE type 204) and were bakeable to 300°C before monomer was introduced. The UV source in Figure 2(b) was a General Electric AH4 lamp with the Pyrex envelope removed and replaced with a slotted glass tube to permit passage of UV radiation from the inner bulb. The effect of light intensity on film growth rate was studied by varying the power input to the mercury lamps, by increasing distance between lamp and substrate, and by insertion of wire-mesh, neutral-density screens or glass filters into the light path and determining the relative intensity with phototube filter combinations. A calibrated thermopile was used for absolute intensity measurements in conjunction with experiments involving a Bausch and Lomb high-intensity monochromator covering the range 1800–4000 Å.

The window to the evacuated deposition system in the more generalized case represented by Figure 1 must be of quartz glass in order to transmit the lower wavelength UV, for example the General Electric Co. type 151 which transmits ~80% of the incident light at 1900 Å. In order to minimize heat input to the system, and to facilitate temperature control (discussed later) at the substrate surface, a jet of air was often blown between the lamp and window surface. However, care is required since differential cooling of either electrode areas of the lamp can induce variation in the spectral output.

The mask, necessary only for patterned deposition, was usually of a metallic material and permitted greater configurational control the closer it was to the substrate. However, complete contact of 'contact masks' evaporated on to thin quartz slides could greatly decrease or even inhibit polymer growth by restricting monomer access to the substrate.

The substrates that can be used in this process cover a wide range. Metal films evaporated on to glass microscope slides are particularly suitable to mechanistic studies susceptible to trace impurities, and can even be freshly deposited within the vacuum system just before polymerization begins, as illustrated in the left-hand side of Figure 2(b). Evaporated films about 0.5–1.0 μm thick from aluminium, lead, and gold were often used by the author. On the other hand, aluminium foil, metal coupons, plastics, and even paper[45] can be used as substrates when the main objective is formation of films for subsequent study or use. The substrates were normally prepared in advance and introduced into the system before it was evacuated.

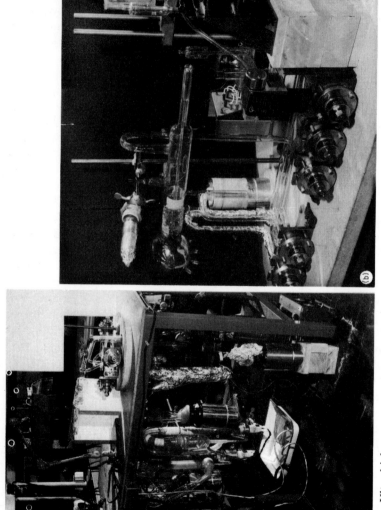

Figure 2. Ultra-high-vacuum systems for deposition of polymeric films: (a) simple system; (b) *in situ* deposition of substrates, mass spectrometer monitoring of gas-phase products. From Mimeault and Wright.[44] (Reproduced by permission of J. Wiley and Sons.)

Monomer vapours were maintained in the system at a pressure of $\leqslant 4$ Torr during the polymerization process. For example: butadiene[46] was maintained at a constant pressure of ~4 Torr simply by surrounding a reservoir of previously purified (bulb-to-bulb) liquid with a dry ice—acetone bath at $-78°C$; tetrafluoroethylene[46,47] was maintained at a suitable temperature in a lower temperature alcohol slurry after distillations at $-196°C$ to remove dissolved oxygen and a distillation at $-120°C$ to remove the polymerization inhibitor (α-pinene); hexachlorobutadiene[43,46] provided a vapour pressure of 0.135 Torr from purified (distillation on a spinning band column) liquid maintained at about room temperature ($18°C$); but *solid* starting materials such as phenol[28] and imides[48] had to be heated in shaded parts of the deposition system in order to provide vapour pressures in the range 0.1–3 Torr. Monomer vapour pressures were monitored by thermocouple and/or diaphragm gauges.

The surface-photopolymerization system was pre-evacuated to pressures of 10^{-5}–10^{-6} Torr, the valve to the pumps closed, valves to the monomer sources opened, and the deposition begun by exposure to the UV source which had been previously operating for at least 5 min in order to reach constant emission characteristics. The pressure during pumpdown was monitored by ionization gauges. In systems for mechanistic studies, such as shown in Figure 2, a prior bakeout to $300°C$ and evacuations by combinations of rotary and diffusion pumps (protected by liquid nitrogen) easily permitted pressures of 10^{-6} Torr. However, less stringent vacuum conditions did not appear to affect either the growth rates or property characteristics of the polymeric films (as long as water vapour and oxygen were reasonably removed) and many depositions were made in non-bakeable systems, even containing 'O'-rings, that were pre-evacuated to pressures ~10^{-5} Torr.

Temperature control at the substrate surface is important in these systems since the earliest reports[1] had indicated a strong and negative dependence of growth rate on temperature. Without attempts at cooling, the typical surface temperatures in systems illustrated by Figures 1 or 2 were about $210°C$ and a very slow film growth rate resulted. Air or nitrogen cooling external to the deposition system can permit temperatures about $100°C$. But the more precise temperature control needed for determination of activation energies for the polymerization process and reproducible growth rates for more routine deposition requires cooling of the substrate holder. Figure 3 illustrates temperature control by a copper block substrate holder that contains hollow tubes through which liquids can be circulated in a closed system regulated by a constant-temperature bath. The temperature of the bath, or the speed which the fluid (ethanol or Dowanol 33B) is pumped through the circulation system, ultimately determines the temperature of the substrate. As measured by thermocouples, this temperature control could be achieved over the range of about 25–$200°C$. This substrate temperature control was particularly important for deposition from C_2F_4[46,47] which only reproducibly participated in the surface-photopolymerization process at temperatures about $30°C$ (over the range 10–$60°C$, depending on the monomer pressure). Although non-bakeable, this system could be easily assembled, after loading the copper holder with substrates,

Figure 3. Deposition system with control of substrate temperature.

by bolting the flange at the end of the quartz tube, no glass blowing being necessary. A knife-edge and copper gasket arrangement between flanges permitted evacuation to approximately 10^{-5} Torr in less than 30 min without baking. Systems such as these were used for most of the experiments described in this chapter. For more routine deposition of films for property evaluation, systems such as that illustrated in Figure 4 were devised in which a quartz plate was attached to a removable metal box by silicone rubber, the box was connected to the vacuum system by 'O'-rings and simply lifted between depositions (Figure 4(a)), a large copper block providing temperature control as before. In the version illustrated in Figure 4(b), the temperature-controlled substrate holder was hinged and irradiation conveniently took place from below. Only rotary pumping was provided for pre-evacuation in these systems, and simple, non-bakeable valves led to the monomer sources.

For all types of systems, the inside of the quartz windows needed to be cleaned of polymeric deposit between runs in order to prevent decreases in growth rates on the substrates as films built up slowly, even at the high temperatures (~200°C) of the internal quartz surface. For longer experiments, substrates had to be moved *in situ* to 'fresh' window areas if decreases in growth rates were to be avoided.

In all cases, irradiations took place within a specially constructed metal box which automatically turned off the lamp if access doors were inadvertently opened. In addition to this protection against UV radiation, the ozone produced in the box was conducted by a circulating fan through an 'elephant trunk' to a fume-hood.

Figure 4. Metal systems for convenient surface – photopolymerization: (a) removable irradiation chamber, lamp above; (b) hinged substrate holder, irradiation from below.

3. METHODS OF FOLLOWING FILM GROWTH AND DETERMINING POLYMER PROPERTIES

The rate of film growth on irradiated surfaces is most conveniently followed by dielectric (capacitance) measurements on the insulating deposits. Crossed-film capacitors, such as illustrated in Figure 5 for deposition from 1,3-butadiene, may be formed by evaporation of the ground aluminium cross-strips on a glass microscope slide, deposition of a polymeric film in the middle while shading the ends of each cross-strip to ensure later electrical contact, and subsequent evaporation of a narrow aluminium strip down the middle of the polymer film. Subsequent capacitance measurements of 1 kHz with a General Radio Company type 1650A inductance bridge, and using silver paste for good electrical contact on the several cross-film capacitors on each side, gave very reproducible measurements of the

Figure 5. Crossed film capacitors from C_4H_6 deposition.

film's thickness. Interferometry proved to provide the best independent thickness measurement, especially when evaporated metal films were used to accentuate the edge of patterned depositions. (It might be cautioned here that evaporation of metals on to particularly thin (< 1000 Å) polymer films can lead to erroneous results owing to a metal deposit thinner on the polymer film than on the rest of the (glass or metal) substrate, presumably because of a different sticking coefficient for thin-film metal deposition on the polymeric surface.) Once the dielectric constants, a characterizing property itself, were determined for the films from each monomer then such capacitance measurements were routinely used for after-the-fact measurement of thicknesses and growth rates. For example, Figure 6 demonstrates the thickness of films deposited from 1,3-butadiene after irradiation of aluminium

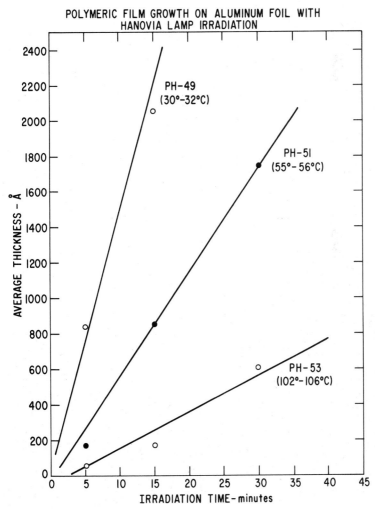

Figure 6. Growth of polymeric films from 1,3-butadiene.

substrates for different times at different substrate temperatures. The linear dependence on time permits deduction of growth rates which, when plotted in the typical Arrhenius fashion (Figure 7) permits calculation of activation energies, in this case −7 kcal/mole. Experiments such as these, incidentally, soon proved that the polymer film growth rate was indeed independent of the substrate material. More sensitive *in situ* measurement of growth rate, such as required for effect of changes in wavelength of irradiation during the deposition process, was obtained by direct deposition upon the surface of a quartz crystal microbalance.[47] The surface temperature could be controlled to ± 0.1°C and it was estimated that the sensitivity of the microbalance towards, for example, the polymer from C_2F_4 was of the order of a few ångström units.

The structural characteristics of the thin polymeric films were most conveniently studied by infrared absorption. The thinness of deposited films (usually < 1 μm) required development of a multiple-reflection technique. Polymer films about

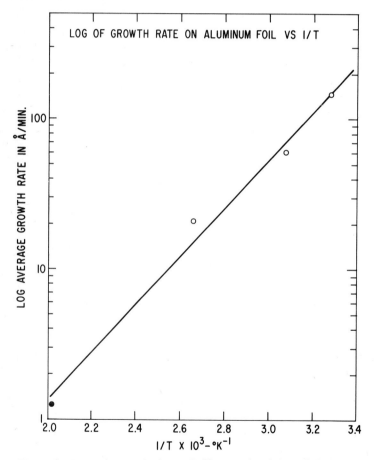

Figure 7. Log of growth from C_4H_6 on aluminium foil versus $1/T$.

6000 Å thick were deposited on two microscope slides freshly coated with evaporated aluminium films about 1 μm thick. These slides were arranged parallel and close together in a Beckman 1R12 so that the incident beam was subjected to multiple reflection during passage between the plates. As many as seven to eight reflections could be obtained from each slide, i.e. an equivalent film thickness as high as 560 000 Å. Five reflections from a film from butadiene about 20 000 Å thick gave clear evidence, for example, of *trans*-1,4 addition and the presence of a small amount of triple carbon–carbon bonds.[46]

Other analytical techniques that may be readily applied to the thin photo-polymerized films include elemental analysis, differential scanning calorimetry, optical microscopy, and X-ray. The films are uniquely suited to electron-microscopic studies of morphology, since the requirement for specimens sufficiently thin to permit electron diffraction may be directly addressed. Carbon replicas are not needed, and since the films exhibit considerable integrity at thicknesses about 300 Å after dissolution of the substrates on which they were deposited, they offer excellent sources for diffraction studies on specimens not disturbed by microtomy. However, the evidence for localized single-crystal structures reported by Wright from selected area electron-diffraction studies on films deposited, for example, from hexachlorobutadiene,[46] has never been fully explained. This type of film structure is not representative of the films' composition as a whole, and seems to occur in only a minute fraction of the surface area, and may either result from localized crystallization of largely carbonaceous by-products common to the monomers or may be an artefact of the system. It might be noted, however, that there have been several recent reports of supermolecular structures in *thin* polymeric films, for example spherulites in

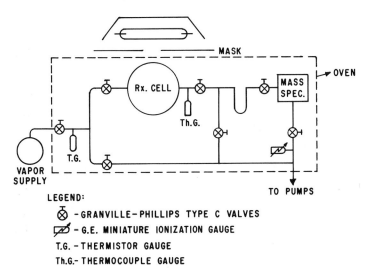

Figure 8. Schematic of deposition system with *in-situ* mass spectrometer. From Mimeault and Wright.[44] (Reproduced by permission of J. Wiley and Sons.)

surface films deposited from benzene in a glow discharge,[49] and extensive ordering at the metal interface in films deposited from trifluorochloroethylene.[50]

The gas-phase products of the surface-photopolymerization process were monitored most simply by mass spectrometer (CEC, model 21-104) analysis of aliquot samples obtained from 'break-seal' addends to the deposition system. For kinetic measurements of the gaseous products formed during the very initial stages of the irradiations a bakeable *in-situ* mass spectrometer was employed[44] for continuous monitoring, as shown schematically in Figure 8. In the case of polymerization from tetrafluorethylene,[47] infrared analysis proved convenient for detection of cyclic fluorocarbons among the gas-phase products.

4. SURFACE-PHOTOPOLYMERIZATION FROM VINYL AND DIENE MONOMERS

In 1967, Wright reported that surface photopolymerization from 1,3-butadiene showed an overall negative activation energy of approximately -8 kcal/mole.[46] Similarities in the structure of the films deposited from a variety of vinyl and diene monomers to those of the analogous polymers produced by conventional polymerization techniques confirmed that the low energies of the photo process do permit retention of many of characteristics of the individual monomers. Nevertheless, multiple infrared absorption data demonstrated conclusively that the films from C_4H_6 are not simply polybutadiene since they showed the presence of CH_3, CH_2, and aliphatic CH groups, and small amounts of carbon—carbon triple bonds. Absorption at $\sim$966 cm^{-1} suggested that the thermodynamically favoured *trans*-1,4 form of addition may be preferred under these reaction conditions. The complex nature of the photopolymerization process was further indicated by identification of at least 13 gas-phase photolysis products: H_2, CH_4, C_2H_2, C_2H_4, trace amounts of C_2H_6, C_3H_4, C_3H_6, C_3H_8, C_4H_8, C_5H_8, C_5H_{10}, and benzene and toluene. A slight increase in the C/H atomic ratio in the polymeric films by elemental analyses to 0.76, from the value of 0.66 for polybutadiene, paralleled a lower C/H ratio averaged over the major gas-phase products.[51] Differential scanning calorimetry showed no transition between -80 and 325°C, but endothermic behaviour above 130°C.

In some of the preliminary depositions from 1,3-butadiene it *appeared*[52] that although film growth was linear in time up to about 24 hours, there was an upper limit to the thickness of the films that could be deposited. For example, the data for deposition on *uncooled* aluminium substrates indicates negligible increase in breakdown voltages and hence thickness in films deposited over 24 and 48 hours. This reduced growth rate at longer radiation times was proved to be a consequence of polymer film growth (and hence UV absorption) on the inner side of the quartz reaction vessel. This was demonstrated in experiments involving substrate holding trays which could be repositioned to previously shaded sections of the quartz within the evacuated irradiation cylinder by the manipulation of external magnets. (Doepker, although primarily interested in C_2H_2 and C_2H_4 yields, also reported film formation on windows that drastically reduced the light intensity during the vacuum UV photolysis of 1,3-butadiene at 1236 and 1470 Å.[53]) Furthermore, in systems permitting substrate cooling, such as those illustrated in Figures 3 and 4, no upper limit (>1 μm) to film growth could be detected and difference in temperatures from the hot quartz window, and hence much faster growth rate on the substrate, made window cleaning mostly unnecessary. This capability to produce thicker films did lead to the observation[46] that, unique among all the films deposited in these laboratories from various monomers, those from 1,3-butadiene when thicker than ~5000 Å are unstable in the presence of air or moisture at high temperatures. They assume a 'crackled' appearance, often peel spontaneously from metallic substrates, and show birefringent behaviour. The 'crackled' films showed[52] no drastic chemical revision, for example no significant increase in C=O content, and appeared to involve an ordering process or phase change.

The optimum wavelengths for deposition from C_4H_6 seemed to be around 2700 Å, indicating that surface adsorption may relax the energy requirements for excitation to the first singlet state, although not permitting spin relaxation and hence excitation via the triplet at ~3500 Å.[52] Bukanaeva has also reported 2500–3000 Å as the most effective wavelength for surface-photopolymerization of films from butadiene (5–10 Torr) on aluminium or copper substrates at surface temperatures in the range −10–25°C.[54]

Growth rates at selected substrate temperatures, and representative dielectric properties are illustrated in Table 1 for deposition from 1,3-butadiene and other vinyl and diene monomers,[51] including ethylene.[55] Deposition from polar monomers such as acrylonitrile yielded thin films with higher dielectric constants such as are useful in passive electronic circuitry, but also showed higher loss factors.[51]

Imperforate thin films can also be configurationally deposited from *mixtures* of diene and vinyl monomers, for example butadiene plus styrene (69 Å/min at 81°C), butadiene plus acrylonitrile (500 Å/min at 44°C), butadiene plus methyl methacrylate 100 Å/min at 70°C), 1,5-hexadiene plus acrylonitrile (700 Å/min at 60°C), and 2,4-hexadiene + styrene (900 Å/min at 80°C) for equal pressures of monomers at about 4 Torr.[56] In addition to permitting a range of dielectric constants in the surface films, as determined by the choice of monomer pairs and their relative

Table 1
Polymeric films from olefinic monomers

Monomer	Measured growth rates (Å/min)	Surface polymerization temp. (°C)	Breakdown strength (volt/cm x 10^{-6})	Dielectric constant
1,3-butadiene	120	33	1.9	2.6
1,5-hexadiene	34	84	2.0	≥ 1.7
2,4-hexadiene	610	112	0.6	≥ 3.7
Ethylene	~ 3	−2	–	~ 2.0
Acrylonitrile	661	58	0.4	≥ 5.8
Styrene	247	108	1.0	2.8
Methyl-methacrylate	2	217 (uncooled)	1.7	≥ 1.6

concentration, the copolymerization technique introduced features such as cross-linking in the presence of dienes. For example, films from styrene were frequently non-insulating or showed low dielectric strengths at thicknesses of 4000 Å, yet mixtures of equal amounts of styrene and butadiene yielded films that were uniformly insulating at thicknesses down to 200 Å whilst still retaining the aromatic characteristic of the styrene feedstock. It might be noted that thin films formed from styrene by the arc glow-discharge technique have also been reported to show dielectric weak spots.[57] Kistiakowsky and Walter have reported slow deposition of a polymeric film on the surfaces of irradiation cells during photolysis of ketene at 2139 Å.[58] The plethora of products and complexity of the deposition of polymerization process from monomers such as 1,3-butadiene made difficult detailed mechanistic studies on the process, and these were reserved for the perhalogenated systems described in the following sections.

5. SURFACE-PHOTOPOLYMERIZATION FROM PERHALOGENATED VAPOURS

(i) C_2F_4

The report in 1967 of surface photopolymerization from C_2F_4 demonstrated for the first time that a polytetrafluoroethylene-like coating of high integrity when *very* thin could be deposited on substrates.[46] Although the high-performance properties of PTFE have been known for many years, the high molecular weight obtained from conventional polymerization and concomitant insolubility had precluded film formation by normal techniques such as casting from solutions. The surface-photopolymerization technique permitted, for example, direct deposition from the gas phase at about room temperature of films that could make substrates such as paper completely non-wettable (contact angles near $180°$) when the coatings were only ~100 Å thick.[45] It has now been established that relatively thin polymer films can also be deposited from C_2F_4 (or PTFE) by other processes, for example r.f. sputtering,[59] glow discharge,[60] hot-cathode discharge,[60] ion-beam etching,[61] and plasma polymerization.[62]

Surface photopolymerization from the monomer vapour at pressures of 1 Torr on to substrates maintained at a temperature ~30°C proceeded at growth rates of ~30 Å/min.[46] Although indistinguishable from PTFE by elemental analysis, slight structural differences were indicated by the presence of some CF_3 groups as revealed by infrared absorption at 980 cm^{-1}. The presence of a considerable amount of low molecular weight material in the photodeposited film was indicated by a loss in weight in air beginning at ~220°C, whilst differential scanning calorimetry could not detect the first-order crystal—crystal transition characteristic of PTFE at ~20°C.

Unlike deposition from most monomers, surface photopolymerization to produce thin, imperforate films of high integrity only occurred in the preferred substrate temperature range 0—60°C and at pressures $\leqslant 3$ Torr.[63] Photolysis at gas-phase pressures of C_2F_4 above ~8 Torr yielded a white 'floc' on all surfaces that was subsequently *fusible* (as opposed to sinterable) in air at 330°C to a clear deposit.[64] This product of (unsensitized) gas-phase photolysis at wavelengths ~2150 Å did show first-order transitions at 20–30°C and closely resembled PTFE in thermal stability, with no significant weight loss in air until approximately 500°C. Both the rate of 'deposition' on surfaces and the per cent retention on fusion increased linearly with monomer pressure over the range 8–760 Torr.

Mechanistic studies of the direct surface-photopolymerization process with C_2F_4 in the rather narrow operative regions of monomer pressure and substrate temperature were preceded by an investigation[44] of the interaction of perhalogenated molecules with clean (ultra-high-vacuum conditions) metallic surfaces in the presence and absence of UV irradiation. *In-situ* mass spectrometry demonstrated unexpectedly that, relative to surface photopolymerization of nonhalogenated monomers, CO is a significant product of the C_2F_4 surface photolysis. With clean aluminium or copper film substrates, prepared *in situ* by evaporation within the system (Figure 2(b)), the initial adsorption of C_2F_4 at room temperature as measured by resistivity changes was negligibly small and no decomposition products could be detected in the gas phase. During irradiation CO was produced at about twice the rate in the presence of clean aluminium than in the presence of the quartz glass surfaces alone. Although this study did demonstrate that large perturbations, such as dissociation adsorption, that might be induced by clean surfaces do not occur in this system in the absence of irradiation, the novel detection of CO generation did not greatly contribute to an elucidation of the mechanism of the surface photopolymerization. The CO may result from the interaction of some photo-excited carbon–fluorine species produced on or near the surface with oxygen found in the lattice as oxide, even under ultra-high-vacuum deposition conditions.

Detailed studies in which film growth was monitored by a very sensitive quartz crystal microbalance demonstrated that polymer deposition could not be initiated or maintained with light or wavelengths >2150 Å, the effective cut-off for absorption by C_2F_4 in the gas phase.[47] Adsorption on the surface, then, does not perturb the TFE molecule to the extent that the photolysis leading to polymer formation can be initiated by less energetic photons. The extent to which surface perturbation could lower the energy requirement for photolysis remained a pertinent question, since Shimizu *et al.* had reported that methane showed such effects when adsorbed on porous glass at 77 K.[65] In addition, the photodeposition from tetraethyl lead on surfaces to yield metallic films, which is controlled by surface reactions,[66] appeared to exhibit a wavelength sensitivity shift from 2540 to 2800–2900 Å when multilayers of TEL were adsorbed on to freshly prepared films of lead.[67]

At wavelengths of about 2100 Å the film growth rate depended on the 1.6 ± 0.4 power of the UV intensity, varied inversely with surface substrate temperature over the narrow range 10–60°C, and varied positively with monomer pressure from 0.5 to 10 Torr.[47] Film growth terminated immediately on removal of the light source, to commence again on re-irradiation. The surface-photopolymerization process was shown to be a rather inefficient one – and by no means a chain reaction – with about one monomer unit polymerized for each incident 2000 Å photon. Gas-phase by-products included $cyclo$-C_3F_6, C_3F_8, and higher molecular weight species of the type $(CF_2)_n$ up to C_9F_{18}, but there was no evidence for C_3F_6 or higher straight-chain unsaturated fluorocarbons.

A proposed[47] mechanism for the surface-photopolymerization process consistent with the experimental data invoked photodissociation of C_2F_4, followed by reactions of difluorocarbene and its higher homologues both in the gas phase and on the surface, i.e. direct photodissociation from a non-bonded excited state

$$C_2F_4 + h\nu \longrightarrow 2CF_2 \tag{1}$$

or

$$C_2F_4 + h\nu \longrightarrow CF_2 = CF\cdot + F\cdot \tag{2}$$

wherein reaction (2) may be favoured in the surface adsorbed state of the monomer over reaction (1), predominant in the gas phase. At the low monomer pressures deliberately chosen to favour surface photopolymerization, significant gas-phase polymerization via CF_2 is precluded. Nevertheless, it was concluded that the major source of radicals for surface-polymer growth must be in the gas phase *immediately* above the illuminated surface. The concentrating effect of surface adsorption then promotes the polymerization initiating reactions,

$$CF_2 + C_2F_4 \longrightarrow C_3F_6 \tag{3}$$

$$C_3F_6 + C_2F_4 \longrightarrow C_5F_{10}, \text{etc.} \longrightarrow \text{polymer} \tag{4}$$

Termination at each stage is possible, however, by ring-closure reactions such as

$$C_3F_6 \rightarrow cyclo\text{-}C_3F_6$$

or by recombination which would also apply to polymerization chains initiated by the monoradicals $F\cdot$ or $CF_2 = CF\cdot$, and hence contribute to low-molecular-weight surface products. The complex negative temperature dependence of film growth was attributed to the effect of surface temperature on the equilibrium between surface and gas phase, and also to its effect on the molecular-weight distribution of the adsorbed molecules. It was postulated that an equilibrium is established between the gas phase and the surface, with the latter accumulating the higher molecular weight material from reactions (4). Patterned deposition of films from TFE stable to temperature $>200°C$ probably also involves interaction of molecular species adsorbed on the surface, such as $CF_2 = CF(CF_2)_n - F$ derived from reactions (2), which can themselves significantly absorb lower wavelength UV and then participate within the films in molecular-weight-increasing reactions such as cross-linking. In the need, then, for repeated interaction with photons before stable polymeric film is produced, some analogy might be made with the direct polymerization of styrene as induced by laser irradiation in the liquid state at wavelengths >4000 Å.[68] In the latter case multiple-photon processes are necessary to *first* populate the molecular state that can photodissociate and subsequently participate in polymer-producing reactions.

It might be noted that under special deposition conditions, such as placing the substrate within 5 mm of the light source, surface photopolymerization to imperforate thin films can be made to occur in the higher pressure region 25–200 Torr to permit growth rates as high as 500 Å/min at substrate temperatures 0–70°C.[69] The distance CF_2 and other reactive intermediates travel in the photolysis zone before collision with the substrates can determine the extent of gas-phase photopolymerization and hence interference with the surface growth process.

Toy has shown, incidentally, that films can be surface polymerized from tetrafluorethylene on fluorine-activated substrates.[70] Although not involving photolysis, this two-step process wherein metal fluoride films are used as catalysts for surface polymerization does appear to offer an alternative approach to the deposition of thin, continuous, and adherent films from C_2F_4. The two steps involve gas-phase fluorination of the metal (copper, nickel, aluminium, titanium, steels) substrate followed by contact of these metal–fluoride surfaces at ambient to 100°C, with gaseous TFE at pressures between 2 and 10 atm. Toy postulated that the formation of polydifluoromethylene films is initiated by formation of an organo-metallic compound by way of high-valency fluorides serving as *in-situ* catalysts on the metal surfaces. Polymer growth might then proceed by an insertion polymerization mechanism.

Novel applications can derive from the extreme integrity, adherence, and resistance to most solvents of the very thin, configuratively deposited surface films from tetrafluorethylene. For example, the ability to *remove* them in a highly configurational manner by patterned UV irradiation in air (oxidative degradation) permits use as a 'positive' photoresist of very high resolution potential since they can resist solvents for metal removal, etc. at thicknesses <500 Å.[41]

(ii) Surface photopolymerization from other perfluorinated monomers

Hexafluorobutadiene (150 Å/min at substrate temperatures ~100°C) and hexafluoropropylene (9 Å/min at substrate temperatures 6–15°C) are also susceptible to the surface-photopolymerization process.[63] Millard has further demonstrated that higher molecular weight fluorocarbon monomers, i.e. perfluorocyclobutene, can be slowly surface photopolymerized to yield films of low surface energies.[71] Films 250–300 Å thick were reported on glass microscope slides after exposure for several days to UV in the presence of the monomer vapour at a pressure of a few tenths of a torr. The presence of mercury atoms (and hence the possibility of Hg-photosensitized reactions) in this study made mechanistic speculation complex, but it was concluded that both gas-phase and surface reactions were involved in the surface-photopolymerization process. Millard also reported surface films from hexafluoroisopropyl methacrylate, pentadecafluorooctyl acrylate, trifluoropropyl methyldichlorosilane, and (heptafluoroisopropoxy) proply-1-trichlorosilane.

(iii) Mechanistic studies of photopolymerization of C_4Cl_6

Wright described C_4Cl_6 as an extreme example of the ability of the surface-photopolymerization process to produce films from 'monomers' not subject to conventional forms of polymerization.[46] The steric hindrance effect makes this molecule quite inert in conventional polymerization and in most other reactions. Perchlorinated polymeric films were produced, however, on surfaces by this process at a rate of about 2 Å/min at (uncooled) substrate temperatures of $\sim177°C$, with rates increased to $\sim120°$ Å/min at temperatures near $30°C$. A negative activation energy of -18.5 kcal/mole was erroneously reported[46] as -8.5 kcal/mole in the first publication. Yet the surface films were by no means simply 'polyhexa-chlorobutadiene': the perchlorinated deposits showed a C/Cl atomic ratio of $\sim2/1$, while chlorine was identified as the principal gas-phase product. The highly carbonaceous surface films, roughly approximated by $(C_2Cl)_n$, were essentially transparent in the infrared and hence absorption measurements yielded little structural data. They were highly stable thermally with no oxidation detectable on exposure to steam at $100°C$. An exothermic reaction did occur in nitrogen at temperatures above $300°C$ and the films lost weight. As with surface photo-polymerization from C_2F_4, carbon monoxide was identified as a major gas-phase product during the early stages of the photolysis process.[44] For aluminium substrates, the CO yield was more than doubled over the relatively large (as compared to the C_2F_4) amounts detected in the presence of the surfaces of the quartz reactor system alone. For UV irradiation <2400 Å a linear dependence of growth rate for the highly stable film was observed over a threefold intensity change.[41]

Mechanistic studies indicated that efficient film growth from C_4Cl_6 occurs only at wavelengths <2400 Å, although spectral measurements show significant gas-phase absorption at wavelengths <2800 Å.[43] As indicated in Figure 9 an Arrhenius plot for film growth over the substrate temperature range $80-180°C$, based on the average of many depositions, yields a good straight line with an overall or 'apparent' activation energy of -18.5 kcal/mole. Measurement with several methods of varying light intensity confirmed the linear dependence of growth rate at wavelengths <2400 Å. Analysis of the *initial* reddish-brown, viscous oil surface product, as produced in a specially devised flow system, showed a chlorine content of 77.6%, or close to the ratio C_4Cl_5, while mass spectrometry indicated mostly C_8Cl_{10} and C_8Cl_8, with a small amount of C_8Cl_6. Ultraviolet absorption data indicated a trend to absorb more strongly and at longer wavelenths for the sequence monomer–initial surface product–polymeric film. The composition of the final film did depend on the combination of substrate temperature and UV intensity, for example a film deposited at $\sim170°C$ formed the very stable $(C_2Cl)_n$ previously reported, but at a surface temperature of $60°C$ a film that could be deposited at a faster rate showed an increased Cl content and was soluble in acetone. However, continued irradiation of this type of film in vacuum after deposition, and in the absence of monomer vapour, produced once again an acetone-insoluble film that appeared identical to those films formed directly at the higher substrate temperatures.

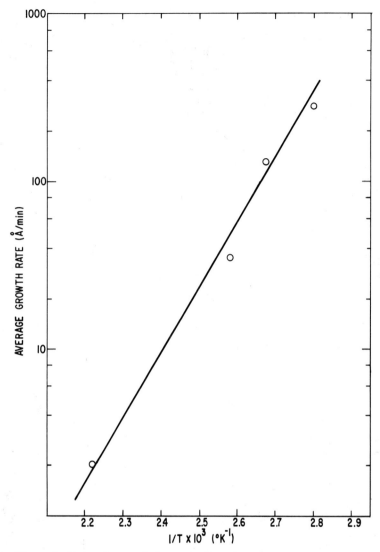

Figure 9. Log of growth from C_4Cl_6 evaporated aluminium films versus $1/T$. From Kunz and Wright.[43] (Reproduced by permission of The Chemical Society of London.)

The principal gas-phase products formed during film deposition were C_2Cl_4, C_2Cl_6, and CCl_4. The C_2Cl_4 and C_2Cl_6 yields reached plateau values after irradiation times of about 10 min, but the CCl_4 yield continued to increase linearly with irradiation times, for example for as long as 60 min. Measurements at short (0.5 min.) exposure times showed C_2Cl_4 to be the major initial product, and also confirmed the small CO yield previously reported. In the presence of *added* C_2H_6, the major products were HCl and C_4H_{10}, indicating the intermediate formation of

Cl atoms in the gas phase. No $C_2 Cl_4$, $C_2 Cl_6$, or CCl_4 could be detected under these conditions.

The mechanism suggested for the surface photopolymerization from hexachlorobutadiene emphasized the importance of gas-phase reactions that lead to reactive, higher molecular intermediates that can then react (often photolytically) on the surface to give the final polymeric films of high stability.[43] It was postulated that initiation involves formation of $C_4 Cl_5 \cdot$ radicals by photodissociation from $C_4 Cl_6$,

$$C_4 Cl_6 \xrightarrow{h\nu} C_4 Cl_5 \cdot + Cl \cdot \tag{5}$$

Gaseous products can then be explained by reactions $(6) \rightarrow (8)$,

$$Cl \cdot + C_4 Cl_6 \longrightarrow C_2 Cl_4 + C_2 Cl_3 \cdot \tag{6}$$

$$C_4 Cl_5 \cdot + Cl \cdot \longrightarrow C_4 Cl_6 \tag{7(a)}$$

$$\searrow C_2 Cl_4 + C_2 Cl_2 \tag{7(b)}$$

$$C_2 Cl_4 \xrightarrow[Cl\cdot]{h\nu} C_2 Cl_6 \xrightarrow[Cl\cdot]{h\nu} CCl_4 \tag{8}$$

The reaction (8) sequence, the progressive chlorination of $C_2 Cl_4$ to $C_2 Cl_6$ and finally to the more stable gaseous product CCl_4, could account for the induction period and other characteristics of the gaseous products. In the presence of both chlorine and UV light, the reaction,

$$Cl \cdot + C_2 Cl_6 \longrightarrow CCl_4 + CCl_3 \cdot \tag{9}$$

was suggested as the predominant mechanism for production of CCl_4. Combination of $C_4 Cl_5 \cdot$ radicals,

$$\nearrow C_8 Cl_{10} \tag{10(a)}$$

$$C_4 Cl_5 \cdot + C_4 Cl_5 \cdot$$

$$\searrow C_8 Cl_8 + Cl_2 \tag{10(b)}$$

was favoured over the reaction of $C_4 Cl_5 \cdot$ with $C_4 Cl_6$ molecules because of the linear relationship between UV intensity and film growth rate. Photodissociation of chlorine from C_8 compounds,

$$C_8 Cl_{10} \xrightarrow{h\nu} C_8 Cl_9 \cdot + Cl \cdot \tag{11}$$

could then yield radicals which combine with C_4 radicals

$$\nearrow C_{12} Cl_{14} \tag{12(a)}$$

$$C_4 Cl_5 \cdot + C_8 Cl_9 \cdot$$

$$\searrow C_{12} Cl_{12} + Cl_2 \tag{12(b)}$$

to form the C_{12} product. Reactions such as

$$C_4 Cl_5 \cdot + C_8 Cl_8 \longrightarrow C_{12} Cl_{13} \cdot \tag{13}$$

could occur between the radicals and the more unsaturated products. In this scheme, then, reactions (11)–(13) inclusive involve the photodissociation of chlorine from C_8 species followed by radical reactions to form C_{12} species. At this stage in the film-forming process, photodissociation could also occur from C_{12} species,

$$C_{12}Cl_{14} \xrightarrow{h\nu} C_{12}Cl_{13} \cdot + Cl \cdot \qquad (14)$$

and with the increasing carbon content of such species, and hence decreased vapour pressure, it was concluded that processes such as reaction (14) occur more frequently on the irradiated *surface* of the reaction system. In summary, radicals so formed on the surface will combine or react with less saturated molecules, forming large molecules with lower chlorine content, eventually leading to a stable polymeric film configurationally deposited in irradiated areas. The −18.5 kcal/mol value for the surface activation energy was recognized as an approximation for a complex process, involving surface residence time of a host of species among the photochemical products, and could be expected to be rate controlling over only a relatively restricted temperature range. It was concluded that the observed linear dependence of film growth rate on light intensity was consistent with a mechanism which essentially proposes formation of a product that can add to the surface and occurs by processes, such as recombination or termination of radicals, that are second order in radical concentration.

In addition to the ability of the very stable films from C_4Cl_6 to act as positive photoresists when as thin as 200 Å, the phenomenon whereby relatively unstable films formed at lower (<100°C) temperatures can be made acetone-insoluble by additional patterned irradiation in vacuum could be utilized to make possible a contact mask, *negative* photoresist system of high resolution when the films were first deposited at thicknesses ~500 Å.[41]

(iv) Chlorine and fluorine-containing monomers

Surface photopolymerization has also been demonstrated from monofluoro-trichloroethylene (~50 Å/min at substrate temperatures of ~115°C), and from trifluoromonochloroethylene (~230 Å/min at substrate temperature of ~117°C).[72] Gaseous mixtures of hexachlorobutadiene and tetrafluoethylene can also yield polymeric films, for example at rates of ~60 Å/min for substrate temperatures ~53°C.[63]

6. SURFACE PHOTOPOLYMERIZATION FROM OTHER VAPOURS

(i) Hydrocarbons without aliphatic unsaturation: OH-containing compounds

The surface-photopolymerization technique can be readily extended to starting materials which do not contain aliphatic unsaturation, although some, as solids, require heating to provide sufficient vapour pressures. Alkyl benzenes such as toluene and benzene had previously been reported to yield polymers on gas-phase photolysis at wavelengths $<$2537 Å.[73] Christopher and Wright demonstrated that vapours from the liquids ethylbenzene, toluene, 2,6-xylenol, and cyclohexanol slowly yielded patterned polymeric deposits at surface temperatures of $\sim$55°C.[28] The vapour from heated solid phenol yielded films at $\sim$180 Å/min for substrate temperatures of $\sim$100°C. The films from phenol and ethylbenzene retained aromaticity in the surface deposit, although the presence of CH_2 and CH_3 groups in the films from phenol indicated some ring opening and hydrogenation reactions. Nevertheless, the large amount of phenolic absorption in the film from phenol demonstrated some retention of the original phenolic C–O–H grouping. The presence of methyl groups in the film from cyclohexanol indicated ring-opening reactions.

(ii) Anhydrides and Dianhydrides

Phthalic anhydride heated to about 300°C in order to provide a vapour pressure of $\sim$1 Torr readily produced polymeric films on substrates at temperatures $\sim$200°C after irradiation times of 30 min.[74] Similarly, continuous, insulating, and strongly adhering films were obtained from pyromellitic dianhydride when heated to about 150°C to provide a vapour pressure of $\sim$0.1 Torr. If the film from pyromellitic dianhydride should be formed at very high substrate temperatures, for example on the surface of the ultraviolet lamp when placed *inside* the reaction chamber, the coating, although strongly adherent, showed some electrical conductivity with a resistance of 100 ohm/linear cm. This was the only surface-photopolymerized film studied by the author to show non-insulating characteristics.

(iii) Ketones

Thomas and Rodriguez have attributed an observed reduction in transmission in the 2000–3000 Å region of cell windows during the photolysis of cyclopropanone vapour (0.3–4 torr) to the formation of 'polycyclopropanone' on the window surface.[75] The polymer film appeared to be similar to that formed on thermal reaction and showed an activation energy of formation of $\sim$2 kcal/mole between 20 and 50°C.[76] Rodriguez *et al.* suggested that the polymer was formed by a process in which the wall stabilized an intermediate oxyallyl structure formed in the gas phase.[76]

(iv) Imides

The deposition of surface films from various imides at (uncooled) substrate temperatures of $\sim$200°C as reported by Christopher *et al.*[48] are summarized in Table 2. The solid 'monomers' were heated near the photolysis zone to provide vapour pressures of $\sim$0.2 Torr. The infrared absorption spectrum of the films from *N*-phenylmaleimide was remarkably similar to that of the monomer, except for disappearance of the carbon–carbon double bond, and elemental analysis showed

Table 2

Deposition from various imides

Starting compound	Growth Rate (Å/min)
N-phenylmaleimide	$\leqslant 1000$
p-tolylmaleimide	672
N-phenylphthalimide	47
N-phenyltetrahydro- phthalimide	220
N-vinylphthalimide	560
N-allylphthalimide	423
Phenyl imide of 5-norbornene-2,3 dicarboxylic anhydride	125

close agreement to a structure formed by polymerization at this unsaturation. However, identification of acetylene as a major gas-phase product suggested that photopolymerization may also proceed in part by species such as $C_6H_5N(CO)_2$ formed by elimination of C_2H_2. Formation of a 'copolymer' could decrease the tendency for the crystalline monomer to form crystalline polymer, and hence account for the absence of detectable first-order transitions over the temperature range $-100-400°C$. The films from N-phenylmaleimide were very highly temperature stable, with only a 2% weight loss in air until $390°C$. No structural changes could be detected in heating the films in air to $300°C$ for 15 hours. Dielectric loss measurements at room temperature were still $<0.20\%$ after a total of 22 hours exposure to $300°C$.

Phthalimide, succinimide, and pyromellitic diimide (10 Å/min) and several other imides are also susceptible to the surface-photopolymerization process.[77] Maleimide is, of course, susceptible to conventional polymerization by both anionic and free-radical propagation and Bamford *et al.* have reported that polymerization in the *liquid* phase can be photosensitized at 4358 Å by manganese carbonyl in the presence of carbon tetrachloride.[78] It has also been reported that maleimide from the gas phase (sublimation vapour) can be grafted on to ethyl cellulose and polyethlene films by UV irradiation from a high-pressure mercury lamp, as well as by X-ray irradiation.[79] In the photolysis process the polyethylene film gained weight immediately on irradiation without an induction period.[80] Although the rate of grafting from the maleimide vapour decreased with increasing amounts of added air, an induction period was still absent. In contrast, the liquid-phase photo process was completely inhibited by the presence of oxygen/air. Wavelengths in the region 2000–3800 Å were effective for grafting from the maleimide vapour which has an absorption peak at 2700 Å. Decreases in rate of grafting with reaction time was attributed to lack of growth on a surface completely covered with poly-maleimide. The amorphous nature of the deposited polymer, as opposed to the highly crystalline homopolymer formed by photolysis in the solid or liquid phase, led

Table 3

Deposition from silicon-containing monomers

Monomer	Monomer vapour pressure (Torr)	Substrate temp. (°C)	Growth rate (Å/min)
Vinyltriethoxysilane	0.5	160	5
	0.5	85	17
Divinyltetramethyl-disiloxane	3.5	100	77
Cyanoethyltriethoxy-silane	0.5	125	30

Hayakawa *et al.* to suggest that the grafting from the sublimed solid monomer may proceed by the same mechanism as that of the disordered solid-state polymerization.[80] Lack of crystalline structure could be attributed to rather short monomer sequences or obstruction by coexisting segments of the backbone polymer.

(v) Organosilicone Materials

Organosilanes and organosiloxanes also participate[52] in the surface-photopolymerization process as summarized in Table 3. The temperature stability of the films so produced was demonstrated by a dielectric loss factor for the films from vinyltriethoxy-silane of $<6.0\%$ after heating in air for 25 hours at 200°C.

(vi) Inorganic Materials

Carbon-rich films can be surface-photopolymerized from carbon suboxide, C_3O_2.[81] The major gas-phase product was CO and when the deposition was made at a slow rate (monomer pressure ~ 0.3 torr, substrate temperature $\sim 120°C$, film growth rate ~ 40 Å/min) the deposited films were well adhering and continuous. For deposition at monomer pressures of ~ 10 Torr and substrate temperatures of $-20 +25°C$ a somewhat powdery non-adhering film was obtained at growth rates as high as 10 000 Å/min. Deposition from this monomer resembles, then, surface photopolymerization from C_2F_4 in which a powdery form of polymer is deposited[64] from the gas phase outside well-defined photolysis conditions. Polymerization from C_3O_2 had been previously induced thermally,[82] with ionizing radiation,[83] and in the gas phase with UV light.[84] For gas-phase photolysis, it has been postulated that photodissociation of C_3O_2 at wavelengths of 2900–3100 Å produces C_2O as a ground-state triplet,[84] while irradiation at wavelengths <2900 Å may produce C_2O in an excited singlet state.[85] It might also be noted that Forchion and Willis reported the formation of polymer during the photolysis at 2537, 1470, and 1237 Å of carbon suboxide in the presence of hydrogen.[86] Gas-phase products were CO, CH_4, and C_2H_6, with a trace of C_4-compounds. They postulated that a polymer, as well as CO, may be produced by a reaction of C_3O_2 and C_2O. Bukowski and Porejko have shown that the thermal reaction of carbon suboxide with polyamides can produce block and graft copolymers.[87]

(vii) Complex starting materials — potential sources of non-thrombogenic coatings

The ability of the surface-photopolymerization process to retain many characteristics of the starting material in stable, surface-film form — even for unsaturated 'monomers' as exemplified by phenol[28] — led to attempts in our laboratories to localize on a surface materials of known or suspected anti-blood-clotting behaviour. It was demonstrated that films deposited (~150 Å/min) from the vapours of sodium heparin on substrates at temperatures of ~235°C did increase the time of first fibrin formation, and then of gross clotting, for freshly drawn human venous blood.[88] Although it is obvious that considerable photo- and thermal decomposition must have occurred at the reaction temperatures required to obtain vapour pressures of ~40 μ some biological activity seemed to be retained in the thin, strongly adhering, and continuous polymer coatings localized from the heparin. Similar results were obtained with films surface-deposited from the vapours of salicylic acid, p-aminobenzoic acid, 2-phenyl-1,3-indanedione, and 2-p-phenyl-sulfonyl-phenylindanedione-1,3 — all chosen for their possible effect on the role of platelets in thrombus formation or for effects on the fibrinolysis process which may help dissolve blood clots.[89] Detailed experiments involved in-vivo evaluation at the Royal Victoria Hospital, Montreal, of these and many other (a total of 15) films from platelet-influencing starting materials by insertion of a coated quartz ring intravascular thrombogenic model in mongrel dogs.[90] Although scatter in the data made it difficult quantitatively to assess the effectiveness of the different coatings in inhibition of thrombus formation, films from N-phenylmaleimide, N-dimethyl-amino-maleimide, and 1,2-diphenyl-3, 5-dioxo-4-allylprazolidine appeared to show the most promise. Infrared absorption studies confirmed that many of the functional groupings of the starting materials could be retained on the surface, although in several instances the use of an unsaturated analogue of the known platelet-interacting material helped minimize decomposition of the 'monomer', for example synthesis of the allyl derivative of ASA, allyl acetylsalicylate. There have been indications that negatively charged polymeric films formed by polymerization in a glow discharge may also offer non-thrombogenic properties.[91]

7. SUMMARY

In summary, the surface-photopolymerization process is capable of configurationally depositing, from a variety of starting vapours, stable, and strongly adherent, polymeric films of high integrity when very (<500 Å) thin. Almost any substrate is suitable for this in-vacuo deposition process, and the gaseous starting materials range from conventional monomers, such as ethylene and butadiene, through tetrafluorethylene to materials not subject to conventional forms of polymerization such as hexachlorobutadiene. Although considerable preliminary photodecomposition may occur in molecules without obvious polymerization routes, in most cases the polymeric deposit retains many of the structural and property characteristics of the starting material. The photolysis process is not a simple one and frequently involves photolysis on the surface of adsorbates previously formed as products of gas-phase reactions, either as primary photolysis products or from subsequent gas-phase reactions which may themselves involve photoreactions. Surface adsorption may slightly lower the energy of photo-excitation of some monomers, but the primary role of the surface seems to be kinetic rather than thermodynamic, i.e. a concentrating effect favouring polymer-forming reactions.

REFERENCES

1. P. White, *Proc. Chem. Soc.*, 337 (1961).
2. G. M. Harris and J. E. Willard, *J. Am. Chem. Soc.*, **76**, 4678 (1954).
3. P. T. McTigue and A. S. Buchanan, *Trans. Faraday Soc.*, **55**, 153 (1959).
4. A. E. Ennos, *Brit. J. Appl. Phys.*, **4**, 101 (1953).
5. A. E. Ennos, *Brit. J. Appl. Phys.*, **5**, 27 (1954).
6. K. M. Poole, *Proc. Phys. Soc. (London)*, **B66**, 542 (1953).
7. L. Holland, L. Laureson, and C. Priestland, *Rev. Sci. Instr.*, **34**, 377 (1963).
8. R. W. Christy, *J. Appl. Phys.*, **31**, 1680 (1960).
9. I. Haller and P. White, *J. Phys. Chem.*, **67**, 1784 (1963); P. White, *ibid.*, **67** 2493 (1963).
10. G. W. Hill, *Microelectronics and Reliability*, **4**, 109 (1965).
11. D. S. Allan and C. T. H. Stoddart, *Chem. in Britain*, 410 (1967).
12. K. Siewert and B. Jonach, *Korrosion* (Dresden), **6**, 23 (1975).
13. P. White, *Electronics Reliability and Microminiaturization*, **2**, 161 (1963); *Insulation*, 57 (1963); *Electrochem. Tech.*, **4**, 468 (1966).
14. L. V. Gregor, in 'The materials of thin-film devices', *Electro-Technology*, 108 (1963).
15. L. V. Gregor and H. L. McGee, *Proc. Fifth Annual Electron Beam Symposium* (Alloyd Corp.), p. 211, Cambridge, Mass. (1963).
16. L. V. Gregor, in *Physics of Thin Films*, ed. G. Hass and R. E. Thun, p. 131, Academic Press, New York (1966).
17. M. E. Fabian, *J. Mater. Sci.*, **2**, 424 (1967).
18. H. W. Melville, *Proc. Roy. Soc. (London)*, **A163**, 511 (1937); *ibid.*, **A167**, 99 (1938).
19. G. Gee, *Trans. Faraday Soc.*, **34**, 712 (1938).
20. R. Srinivasan, *J. Am. Chem. Soc.*, **82**, 5063 (1960).
21. I. Haller and R. Srinivasan, *J. Chem. Phys.*, **40**, 1992 (1964).
22. I. Haller and R. Srinivasan, *J. Am. Chem. Soc.*, **88**, 3694 (1966).
23. R. Srinivasan and F. I. Sonntag, *J. Am. Chem. Soc.*, **87**, 3778 (1965).
24. L. Holland, *Brit. J. Appl. Phys.*, **9**, 410 (1958).
25. A. Bradley and J. P. Hammes, *J. Electrochem. Soc.*, **110**, 15 (1963).
26. K. Jesch, J. E. Bloor, and P. L. Kronick, *J. Poly Sci.*, **A4**, 1487 (1966).
27. P. L. Spedding, *Nature*, **214**, 124 (1967).
28. A. Christopher and A. N. Wright, *J. Appl. Poly. Sci.*, **16**, 1057 (1972).
29. A. Blumstein, *J. Poly Sci.*, **A3**, 2665 (1965).
30. H. G. G. Dekking, *J. Poly Sci.*, **11**, 23 (1967).
31. W. S. Anderson, *J. Poly Sci.*, **A1**, 429 (1967).
32. M. White and P. Luff, U.S. Patent 3686022 (1972).

183

33. M. Tsunooka, M. Tanaka, and N. Murata, *Kogyo Kagaku Zasshi,* **72,** 1208 (1968).
34. H. L. Needles and R. P. Seiber, U.S. Patent 3 933 607 (1976)
35. L. A. Negievich, *Ukr. Khim. Zh.,* **41,** 1 (1975); L. A. Negievich, and A. A. Kachan, *ibid.,* **42,** 608 (1976).
36. A. A. Kachan, Yu. G. Lebo, and V. A. Shrubovich, *Vysokomol. Soedin.,* **A12,** 214 (1970).
37. E. I. Finkel'shtein, *Vysokomol. Soedin.* **9,** 70 (1967).
38. M. Iguchi, H. Nakaniski, and M. Hasegawa, *J. Poly Sci.,* A1, **6,** 1055 (1968).
39. M. Hasegawa, *Japan Chemical Quarterly,* **1,** 45 (1968).
40. G. R. DeMare, M. C. Fontaine, and P. Goldfinger, *J. Org. Chem.,* **33,** 2528 (1968).
41. C. O. Kunz, P. C. Long, and A. N. Wright, *Polymer Eng. and Sci.,* **12,** 209 (1972).
42. A. N. Wright, *Polymer Eng. and Sci.,* **11,** 416 (1971).
43. C. O. Kunz and A. N. Wright, *J. Chem. Soc, Faraday Trans.,* I, **68,** 140 (1972).
44. V. J. Mimeault and A. N. Wright, in *Reactivity of Solids,* ed. J. W. Mitchell, R. C. DeVries, R. W. Boberts, and P. Cannon, p. 543, Wiley, New York (1969).
45. A. N. Wright, in General Electric Publication GP-67-0382.
46. A. N. Wright, *Nature,* **215,** 935 (1967).
47. D. H. Maylotte and A. N. Wright, *Faraday Discussion (Chemical Soc.),* **58,** 292 (1974).
48. A. Christopher, A. K. Fritzsche, and A. N. Wright, *Photochem. of Macromolecules,* ed. by R. E. Reinisch, p. 117 Plenum Press, New York (1970)
49. M. Kryszewski, *Pure Appl. Chem.,* **31,** 21 (1972); M. Kryszewski, A. Galeski, W. Jablonski, and S. Sapieka, *Polymer,* **15,** 211 (1974).
50. V. A. Belyi, and D. A. Rodchenko, *Lakokrasoch. Mater, Ikh Premin.,* 31 **(1971)**
51. A. N. Wright, Postprints of Soc. Plastics Engineers Regional Technical Conference on 'Photopolymers—Principles, Processes, and Materials', Ellenville, N.Y., 1967, p. 110.
52. A. N. Wright, unpublished results.
53. R. D. Doepker, *J. Phys. Chem.,* **72,** 4037 (1968).
54. F. M. Bukanaeva, *Zh. Prikl. Khim.,* **43,** 395 (1970).
55. A. N. Wright and R. C. Merrill, U.S. Patent, 3 743 532 (1973).
56. A. N. Wright, U.S. Patent 3 635 750 (1972).
57. Y. Segui, A. Bui, and H. Carchano, *Thin Solid Films,* **22,** S15 (1974).
58. G. B. Kistiakowsky and T. A. Walter, *J. Phys. Chem.,* **72,** 3952 (1968).
59. P. P. Budenstein, P. I. Hayes, J. L. Smith, and W. B. Smith, *J. Vac. Sci. Technol.,* **6,** 289 (1968).
60. W. Vollmann, and H. U. Poll, *Thin Solid Films,* **26,** 201 (1975).
61. M. Rost, H.-J. Erler, H. Geigengack, O. Fiedler, and C. Weissmantel, *Thin Solid Films,* **20,** S15 (1974).
62. J. M. Tibbett, M. Shen, and A. T. Bell, *Thin Solid Films,* **29,** L43, (1975); H. R. Anderson, Jr., F. M. Fowkes, and F. H. Hielscher, *J. Poly Sci., Polymer Phys. Ed.,* **14,** 879 (1976).
63. A. N. Wright, U.S. Patent 3 522 076 (1970).
64. E. V. Wilkus and A. N. Wright, *J. Poly Sci.,* A-1, **9,** 2071 (1971).
65. M. Shimizu, H. D. Gesser, and M. Fujimoto, *Can. J. Chem.,* **47,** 1375 (1969).
66. L. J. Rigby, *Trans. Faraday Soc.,* **65,** 2421 (1969).
67. D. L. Percy and M. W. Roberts, *Chem. Comm.* 147(1972).
68. Y.-H. Pao and P. M. Rentzepis, *Appl Phys. Letters,* 93 (1965).
69. D. H. Maylotte, U.S. Patent 3 679 461 (1972).
70. M. S. Toy, *J. Poly Sci.,* C, **34,** 273 (1971).

184

71. M. M. Millard, *J. Appl. Poly. Sci.*, **18**, 3219 (1974).
72. A. N. Wright and R. C. Merrill, U.S. Patent 3 521 339 (1970); A. N. Wright, U.S. Patent 3 522 226 (1970).
73. R. R. Hentz and M. Burton, *J. Am Chem. Soc.*, **73**, 532 (1951).
74. A. N. Wright and W. F. Mathewson, Jr., U.S. Patent 3 713 874 (1973).
75. T. F. Thomas and H. J. Rodriguez, *J. Am. Chem. Soc.*, **93**, 5918 (1971).
76. H. J. Rodriguez, J.-C. Chang and T. F. Thomas, *J. Am. Chem. Soc.*, **98**, 2027 (1976).
77. A. N. Wright and W. F. Mathewson, Jr., U.S. Patent, 3 619 259 (1971).
78. C. H. Bamford, J. F. Bingham, and H. Block, *Trans. Faraday Soc.*, **66**, 2612 (1970).
79. K. Hayakawa, K Kawase, and H. Yamakita, *J. Poly Sci.*, A-1, **8**, 1227 (1970).
80. K. Hayakawa, K. Kawase, and H. Yamakita, *J. Poly Sci.*, **12**, 2603 (1974).
81. C. O. Kunz, and A. N. Wright, unpublished results.
82. R. N. Smith, D. A. Young, E. N. Smith and C. C. Carter, *Inorg. Chem.*, **2**, 829 (1963).
83. A. R. Blake and K. E. Hodgson, *J. Chem. Soc.*, (A) 254 (**1966**).
84. R. N. Smith, R. A. Smith, and D. A. Young, *Inorg. Chem.*, **5**, 145 (1966).
85. D. G. Williamson, and K. D. Bayes, *J. Am. Chem. Soc.*, **89**, 3390 (1967).
86. A. Forchioni and C. Willis, *J. Phys. Chem.*, **72**, 3105 (1968).
87. A. Bukowski, and S. Porejko, *J. Poly Sci.*, A-1, **8**, 2491 (1970).
88. A. N. Wright, U.S. Patent 3 677 800 (1972).
89. A. N. Wright, and H.-D. Becker, U.S. Patent 3 625 745 (1971).
90. A. N. Wright and E. D Foster, submitted to *J. Radiation Research.*
91. P. L. Kronick and M. E. Schafer, U.S. Govt. Res. Develop. Rep., **70**, 42 (1970).

9

The Application of Plasmas to the Synthesis and Surface Modification of Polymers

D. T. Clark, A. Dilks and D. Shuttleworth

University of Durham

1. INTRODUCTION

The past few years have witnessed a growing awareness of the great potential of the field which might loosely be denoted by the term 'plasma chemistry' in areas of polymer chemistry and physics. The wide-ranging capability in respect of both *in-situ* polymer synthesis and in the surface modification and degradation of polymers is already apparent and it is clear that these topics should receive some attention in any critical review of polymer surfaces.

The glow-discharge synthesis of polymers has been a particularly active area of research, especially in industrial laboratories where the particular advantages of producing pore-free, uniform films of superior physical, chemical, electrical, and mechanical properties were appreciated at an early stage. In this short review we briefly consider the preparation, characterization, and properties of glow-discharge synthesized films.

The second major area of interest in the application of plasma techniques to polymers is in the surface modification of polymers. This may conveniently be divided into three sections: namely plasma polymerization at surfaces, surface modifications effected by direct and radiative energy transfer, and finally the use of plasmas for selectively etching or removing organic polymeric phases.

The predominant emphasis in this work relates to organic-based systems and this reflects the overall balance of work published to date. It is undoubtedly the case, however, that some of the most important potential applications are in the field of inorganic systems. The background provided here, however, should provide the uninitiated with some feel for the sorts of applications which should be possible.

In the first category of surface grafting there are many potential applications, but two which immediately spring to mind and have received considerable attention to date relate to the modification of the surfaces of fibres for improving wettability[1] characteristics and for flame retardancy purposes.[2] In the second category the surface modification effected by direct and radiative energy transfer from plasmas excited in inert gases has been used to improve adhesive bonding,[3] wettability characteristics,[4] and printability of polymers. Indeed, the possibility of selectively modifying surface properties whilst retaining desirable bulk properties has much to commend it and the technique therefore has many potential applications. The use of plasmas, particularly those excited in oxygen, for selectively etching or removing organic polymeric phases is also an important area of application, and indeed such techniques are routinely used to thin samples for direct investigation by electron microscopy.[5]

2. FUNDAMENTAL ASPECTS OF PLASMAS

(i) Definition

A plasma may be defined as a gaseous state consisting of atoms, molecules, ions, metastables, and excited states of these, and electrons such that the concentration of positively and negatively charged species is roughly the same. The characteristic feature of plasmas of interest in the work outlined here is that the Boltzmann temperature of the ions and molecules is roughly ambient whilst that of the electrons is some two orders of magnitude greater. The plasmas of primary interest in chemistry, therefore, are those which might be termed 'cool', as opposed to those of more interest to physicists which in the same parlance might be termed 'hot'.

Theoretically, a plasma, which has colloquially been referred to as a 'fourth state of matter', may be characterized in terms of the average electron temperature and the charge density within the system. For simple systems (e.g. inert-gas plasmas) the

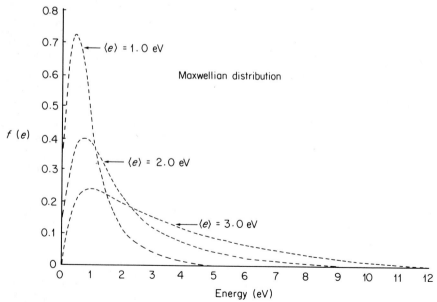

Figure 1. Schematic Maxwellian energy distribution function for electrons in an inductively coupled plasma.

solution of the relevant equations, which need not be of direct concern to us here, leads to a Maxwellian distribution of energies (Figure 1). For 'real systems' which are considerably more complex the form of the distribution has been analysed experimentally by probe measurements[6] and direct electron sampling.[7] The distribution in both cases corresponds closely to that predicted theoretically for the simpler systems. As will become apparent in section (5ii) of this article plasmas are in general copious sources of electromagnetic radiation particularly in the UV and vacuum UV, and indeed plasmas are often used as sources in these regions. In addition, the relatively smaller output in the visible region gives rise to characteristic colours for plasmas excited in a given system and hence the apellation 'glow discharge'.

Having briefly defined what we mean by the term 'plasma' we now consider the most commonly employed experimental techniques for their generation.

(ii) Plasma Techniques

In broad outline there are three distinct aspects which are of interest: namely the source of electrical power to sustain the plasma, the coupling mechanism, and what may loosely be termed the 'plasma environment'. This is illustrated schematically in Figure 2 and the combination selected for a given investigation is dependent on a number of factors such as cost, ease of construction, and convenience. Whilst most of the early work involved AC and DC electroded discharges the greater flexibility and closer control over operating parameters has of recent years shifted the emphasis towards the investigation of inductively coupled RF and microwave plasmas, and the predominant emphasis in this work will be in this area. Recent reviews provide a good background to much of the older work.[8-10]

Turning now to sample handling there are two situations of common interest. Firstly the production of polymer films by plasma polymerization of appropriate monomers, either on an inert substrate or by direct interaction with, for example, a polymer (as in grafting, etc.), or the use of plasmas excited in inert gases, for example as a means of selective energy transfer (both direct and radiative) to the surface of a polymer to effect appropriate modifications. The two basic

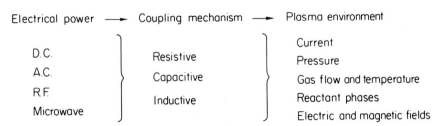

Figure 2. Schematic illustration of the elements of a glow-discharge experiment.

instrumental features which circumscribe the requirements are a means of introducing the 'monomer' into an appropriate reactor in which the plasma is excited, and a means of introducing substrates to interact with the plasma. The most usual arrangement involves straightforward flow systems with conventional vacuum-line techniques. In the case of electroded discharges the polymer is deposited on an electrode and must subsequently be removed for investigation, which is invariably somewhat inconvenient. With suitable vacuum-line techniques and provision for heating or cooling sample reservoirs it is possible to introduce 'monomers' encompassing a wide range of volatilities into the plasma reactor. The great flexibility of electrodeless designs makes it a straightforward matter to interface to instrumentation employed in characterization of polymers, or surface-modified polymers produced by plasma techniques, without removing samples into the atmsophere with all the attendant difficulties which this involves.

Electrode discharges can employ AC, DC, or RF power and are generally resistively coupled. This arrangement allows the possibility of using both static and dynamic flow systems using bell-jar-type arrangements or cylindrical reactors, respectively. Electrodeless RF discharges may be either capacitively or inductively coupled and are well suited to flow systems, with copper coils being wound around the external surface of a typically cylindrical reactor. Electrodeless microwave discharges may also be excited inductively, using tuned cavities which can be in a demountable form to attach conveniently to a cylindrical reactor.

The operating parameters of a glow-discharge experiment are those of power input and operating pressure. Powers may range from 0.1 W to a few kilowatts, although typical experiments involve powers of a few tens of watts or less. Low-frequency and DC discharges are generally characterized in terms of the voltage and current supplied to the electrodes. Typical operating voltages being in the range $\sim$10–100 V and $\sim$1 A at pressures of $\sim$1 Torr. For RF and microwave plasmas the situation is somewhat less straightforward and the requisite instrumentation to measure the power in the plasma is relatively expensive. Most of the early work was predominantly concerned with DC and AC electroded discharges, with the emphasis being on the broad outline of chemical processes occurring, with rather less attention being paid to the operating parameters of the plasma.

Low power levels ($<$1.0 W) are generally difficult to sustain, but when using RF power this can be accomplished by pulsing the power input. Thus, with switching in the microsecond range it becomes possible to obtain stable plasmas operating at low average power loadings.

Operating pressures can range from $\sim$0.01 to $\sim$1.0 Torr with R.F. and microwave discharges, and up to 1 atm for DC discharges.

When working at high power levels ($>$50 W), some external cooling of the reactor zone may be necessary, particularly in the case of DC electroded discharges at high pressures and, to a lesser extent, microwave cavities.

(iii) Reactive species in plasmas

The reactive species in a plasma resulting from ionization, fragmentation, and excitation processes arising from collisions involving electrons accelerated by the

Energies associated with a glow discharge

eV

Electrons	0 - 20
Ions	0 - 2
Metastables	0 - 20
U.V. /visible	3 - 40

Bond energies:

C—H	4.3	C=O	8.0
C—N	2.9	C—C	3.4
C—Cl	3.4	C=C	6.1
C—F	4.4	C≡C	8.4

Figure 3. Energies available in a glow discharge and some typical bond energies.

electric field, include ions, metastables, neutral molecules, and free radicals in ground and excited states. Together with the electrons and electromagnetic radiation previously noted, therefore, a typical plasma constitutes a relatively complex entity overall (Figure 3) and not surprisingly it is only for the simplest systems that any great attempt has been made to fully characterize a given system under standard operating conditions.

In general terms, therefore, although under a given set of conditions a plasma excited in a given manner in a given monomer will produce a well-characterized series of products, at this stage of development, it is only possible to suggest broad outlines of the likely mechanistic schemes involved.

With a wide variety of ions, radicals, and excited states available from excitation of plasmas, even in simple monomers, it is clear that a variety of mechanisms are available for polymerization.[11] At relatively high pressures the interaction between reactive intermediates generated in the plasma and uncharged monomer can lead to the production of relatively high molecular weight materials in homogeneous polymerization processes the material then settling as a fine powder at the bottom of the reactor.[12-14] Since there is a concentration profile of ions, radicals, and neutral species extending through the glow region and into the region immediately outside of the reactor, the material produced in one region of the plasma can be different from that produced in another.[15,16] Polymerization also occurs at substrate/plasma interfaces (heterogeneous polymerization),[17-19] and indeed glow-discharge polymerization provides, under appropriate conditions, a most convenient means of producing thin uniform films.[9]

As will become apparent in the ensuing discussion the reactive intermediates involved are such that extensive molecular rearrangements are involved and, in general, the films produced are extensively cross-linked.

Although not of direct relevance to this article it should be noted that under appropriate conditions it is possible to reduce bimolecular and surface processes to a minimum such that little high molecular weight material is produced. It then becomes possible to utilize plasma chemistry for synthetic purposes in, for example, effecting molecular rearrangements.[20] In economic terms it is clear that this is an area of considerable potential which as yet has received little exploitation.

The basic simplicity of inert-gas plasmas has allowed a much more complete elaboration of the relative importance of direct and radiative energy transfer processes and a brief discussion of recent work in this area is presented in section (5).

(iv) Advantages and disadvantages of the glow-discharge technique

The principal advantages and disadvantages of the glow-lischarge technique are set out in Table 1.

The glow-discharge synthesis of polymer films, or modification of polymer surfaces, is a process of such flexibility that either batch or continuous operations can be employed, whichever is the most desirable. This may prove to be a facility of great importance, especially when considering the integration of the technique into industrial processing or treatments. Furthermore, the 'clean' nature of vacuum treatments has obvious advantages when considering the problems of solution work, etc.

The glow-discharge technique, consisting of essentially one step, often proves to be cost effective both in terms of power consumption and labour requirements. This contrasts with conventional solution techniques which often employ several steps and require solvents of high purity. Such synthetic procedures are often carried out at temperatures greater than ambient.

Radio-frequency and microwave generators are relatively inexpensive *ca* $2000, and vacuum systems are readily constructed. Pumping requirements need not be elaborate since modest operating pressures are normally employed (~0.1–0.01 Torr).

Plasma techniques facilitate great control over operating conditions in terms of

Table 1

Advantages

1. Applicable to batch or continuous processing.
2. Low initial capital outlay.
3. Suitably applied to a wide range of systems.
4. Close control over experimental conditions.

Disadvantages

1. Cannot produce films to a specific formula.
2. Thick films are brittle and discoloured.

pressure, flow rate, and 'on-off' power control. The introduction of 'monomers'* over a wide range of volatilities is easily accomplished by heating or cooling sample reservoirs. The only constraint on 'monomers' or modifying gases is that they should not decompose under the operating conditions of pressure and temperature. Even so, a wide range of materials can be investigated by glow-discharge methods. For *in situ* polymerizations it is possible to coat the sample with the appropriate monomer prior to introduction into the plasma reactor.

Turning now to the disadvantages set out in Table 1, it can be seen that a previously unprepared film cannot be synthesized to a given formula, and indeed it is not possible to modify or synthesize a 'new' polymer surface and say, *a priori*, what the resultant composition will be. Having first completed the experiment, however, the glow-discharge technique may be relied upon to give compositions and surface properties to a high degree of reproducibility.†

When preparing thick films by plasma techniques, the products tend to be brittle and discoloured. Should this be an undesirable property it is readily circumvented by forming (grafting) the film on to a polymeric surface with the required bulk properties.

3. POLYMER CHARACTERIZATION TECHNIQUES

Having examined the principles and the techniques involved in the application of gas plasmas to polymers we now briefly consider the analytical techniques which have been employed to study these materials.

Some of these techniques are displayed in Table 2, which while being by no means comprehensive does encompass those which have provided the most insight

Table 2

A.	Bulk properties	
	1.	Microanalysis
	2.	Electron spin resonance spectroscopy
	3.	Nuclear magnetic resonance
	4.	Dielectric properties
	5.	Differential scanning calorimetry and thermal gravimetric analysis
	6.	Infrared spectroscopy
B.	Surface properties	
	1.	Contact angle
	2.	Microscopic studies
	3.	Reflectance IR
	4.	ESCA

*The term 'monomer' is used here as the generic name for the low molecular-weight starting material used in these polymerization studies. Owing to the complex nature of the processes taking place in a gas plasma, it is often not possible readily to identify the true precursor to polymerization.
†The possibility of preparing 'tailor-made' films by varying the composition of the starting gas mixture has been investigated by several authors.[21,22]

into the structure (in both the chemical and physical sense) of these materials. The table is divided in two sections, namely bulk and surface properties, and in a chapter such as this it is appropriate that surface characterization should be considered in somewhat more detail in this discussion. However, it is worthwhile at this stage briefly to summarize the results obtained by bulk studies, the results, of course, having relevance only to plasma-polymerized material.

Plasma-polymerized films have been shown to be of a reproducible nature, both in terms of composition and properties, and to be extensively cross-linked and that this cross-linked matrix 'traps' a number of more volatile products of the plasma reaction; the polymers when freshly prepared contain a large number of unpaired spins and these trapped radicals convey a reactive nature upon the polymer films inasmuch that they react with oxygen in the atmosphere. A much more detailed and complete discussion of the bulk properties of plasma-polymerized materials can be found in the relevant literature.[9,10,23-25] Section B of Table 2 gives a few of the surface studies employed in the study of plasma polymerized and modified films. The remainder of this article will be devoted to these studies, and in particular to the role ESCA has played in the interrogation of plasma-treated surfaces.

Contact-angle determinations have been used extensively to investigate the polarity of these surfaces,[26,27] and the properties have been shown to reflect those of their conventional counterparts. For example, plasma-prepared fluorocarbon films are of low surface free energy, as are the surfaces of regular fluorocarbon polymers. Oxygen-containing surfaces exhibit wettabilities appropriate to high-energy surfaces, just as do those of appropriate oxygen-containing polymers.

The gross morphological features of plasma polymerized and modified surfaces have been investigated using electon microscopy. Generally the technique has been employed to investigate the homogeneity of the films.

Reflectance IR studies have produced results in agreement with those of transmission IR and the technique has provided further information on the structure of these systems and their reactivity (cf. Chapter 15).

The application of ESCA to the study of structure and bonding in polymer surfaces will be described more fully in a later chapter (Chapter 16). At this point it is sufficient to merely set out some of the information levels available from ESCA studies which may be of use in the study of plasma polymerized and modified surfaces; these are indicated in Table 3.

All of these features have been used in the study of these systems and appropriate examples are provided in subsequent sections.

Table 3

1.	Elemental analysis
2.	Functional group analysis
3.	Shake-up studies to investigate unsaturation
4.	Angular studies
5.	Kinetic studies

In summary, the most important spectroscopic techniques for the investigation of structure and bonding of plasma-polymerized materials are IR and ESCA, ESR usefully provides information about radical sites, whilst contact-angle measurements provide complementary data to ESCA with regard to the outermost sample surface.

4. PLASMA POLYMERIZATION

(i) Introduction

Plasma polymerization by a variety of means has been an active area of research over the past 10 years with the predominant emphasis being on the investigation of organic systems. Despite the fact that in excess of 200 monomers,[28] (ranging from aromatic and polyaromatic systems to chloro- and fluorocarbons and siloxanes) have been studied, the relatively poor characterization of the products and non-systematic investigation of the various operating parameters implies that it is only in the past few years that any real progress in a semiquantitative sense has been made.

The major points of interest in plasma polymerization may be elaborated as follows:

(a) Under what conditions may a plasma excited in a given monomer produce a polymeric film and how does the rate of deposition depend on the operating parameters?
(b) How does the structure of a plasma-polymerized film depend on the operating parameters?
(c) Under a given set of operating conditions how reproducible are the results?
(d) How does the structure of a plasma-polymerized film depend on the nature of the precursor and how does this vary as a function of polymer formed in different regions of a plasma reactor?

Such questions are circumscribed by having available techniques for characterizing the samples and for monitoring the rate of deposition of films. In the earliest investigations, particularly relating to electroded systems, the available tools, namely microanalysis and a reasonably sensitive balance, allowed the broad features of such questions to be established for appropriate systems. However, the comparatively large amount of sample required obviated the investigation of either the early stages of the plasma polymerization or of plasma polymerization in appropriate 'monomers' which proceed at a slow rate.

The advent of ESCA as a spectroscopic tool has, however, transformed this situation, and in the same experiment it is possible directly to monitor rates of deposition and obtain information on structure and bonding in the polymer films.

As pointed out earlier, the systems polymerized under plasma conditions are numerous and it is hoped to exemplify some of the more important surface features of these polymers by recourse to the work which has been carried out in this laboratory. In recent years the emphasis in plasma polymerization has swung from purely hydrocarbon systems to fluorocarbons, presumably in an attempt to produce polymers of similar surface properties to regular fluoropolymers by the much simpler techique of glow discharge. In addition to this the number of investigations carried out by means of ESCA has greatly increased.

(ii) Glow-discharge polymers

The polymers formed from glow-discharge reactions tend to exhibit surface structures quite different from that of the starting 'monomer' owing to fragmentation of the starting materials. An illustrative example of this is displayed in Figure 4 which shows the ESCA spectra of the plasma polymer formed from vinylidene fluoride monomer. As is evidenced by the spectra the polymer contains a number of carbon environments, in contrast to that of polyvinylidene fluoride (Figure 5) which contains only CF_2 and CH_2 environments. The broad chemical environment of the polymer surface is easily determined by following a recognized deconvolution procedure. This analysis, also shown in Figure 4, suggests five components of binding energy: 293.6 ± 0.2; 291.1 ± 0.2; 288.3 ± 0.1; 286.8 ± 0.2, and 285.8 ± 0.1. By comparison with model systems these may be assigned to $\underline{C}F_3$ structural features; $\underline{C}F_2$ features attached to carbons not bearing fluorine substituents (cf. polyvinylidene fluoride); carbons with more than one β fluorine (secondary shift owing to fluorine 0.7 eV);[29] and carbons bonded directly to carbon and hydrogen but with a fluorine substituent in a β position. By examining the ESCA spectra of polymer films produced by glow discharge in vinylidene fluoride over a range of powers, pressures, and thicknesses, it has been shown that the films are of reproducible stoichiometry and structure for a fixed position in the

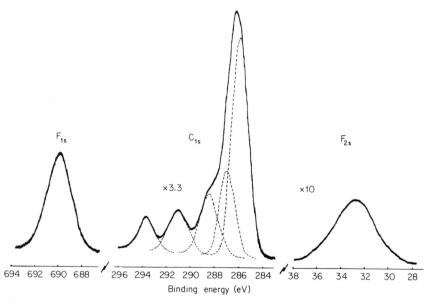

Figure 4. F_{1s}, C_{1s}, and F_{2s} spectra of plasma-polymerized vinylidene fluoride.

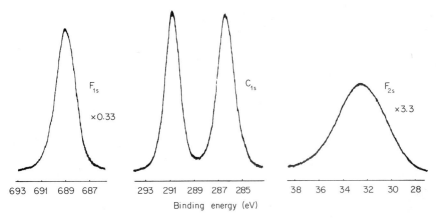

Figure 5. F_{1s}, C_{1s}, and F_{2s}, spectra of polyvinylidene fluoride powder.

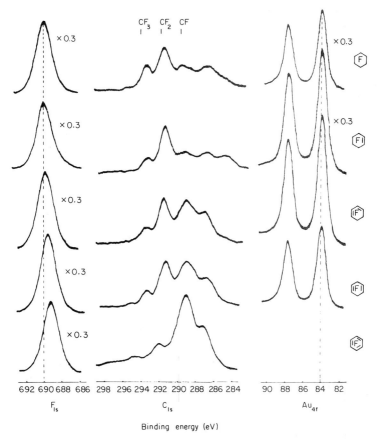

Binding energy (eV)

Figure 6. F_{1s}, C_{1s}, and Au_{4f} spectra obtained from the polymerization (by glow discharge) of a series of related 'monomers' (1.0 W, 0.1 Torr, 1.0 min), on gold substrates.

reactor. Studies which have been made at much higher power loadings and at lower pressures indicate that the structure of the polymer is somewhat (but not grossly) dependent on the location of the substrate on which polymer build-up is being monitored with respect to the RF coils in inductively coupled plasmas.[15]

It can be inferred from the wide distribution of carbon environments in the polymer that the resultant film will be considerably cross-linked. This result has previously been inferred from IR studies and solubility measurements.

Although we have mentioned the departures of polymer structures from 'monomer' structures it is possible to draw certain comparisons. For example Figure 6 shows the ESCA spectra obtained from the deposition of plasma polymer on to gold foil from a range of related 'monomers' under the same conditions of power and pressure and for identical times, namely, 1.0 W, 0.1 Torr, and 1.0 minute (inductively coupled RF plasma). These monomers were perfluorobenzene, perfluorocyclohexa-1,4-diene, perfluorocyclohexa-1,3-diene, perfluorocyclohexene, and perfluorocyclohexane. As evidenced from the spectra the resultant polymer displays a relationship between the fluorine content of the 'monomer' and that of

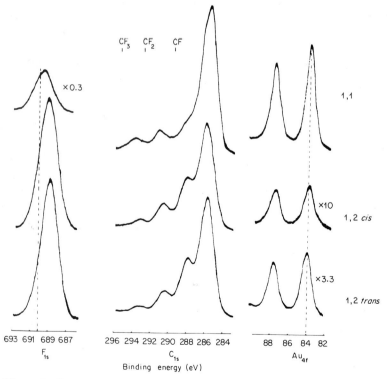

Figure 7. F_{1s}, C_{1s}, and Au_{4f} spectra obtained from the polymerization (by glow discharge) of three difluoroethylene isomers (1.0 W, 0.2 Torr, 30 sec), on gold substrates.

the product polymer. Further trends may be discerned, inasmuch that the CF_3 content of the polymer appears to rise as the degree of unsaturation falls in the starting material.

The spectra displayed in Figure 7 are those obtained from polymers formed from 1,1-difluoroethylene, 1,2-*trans* difluoroethylene and 1,2-*cis* difluoroethylene. It is apparent that the polymer formed from the 1,2 derivatives differ from those formed from the 1,1 derivative while the '1,2' polymers are similar to each other.

The plasma polymers which have been prepared in this laboratory have been shown to be vertically homogeneous on the ESCA scale. This, however, may not be the case in all instances. Vertical homogeneity may be determined by recourse to angular-dependent studies, and investigation of relative peak intensities for levels of different escape-depth dependence a topic which is discussed in greater detail in Chapter 16.

Further information relating to the degree of unsaturation within a plasma polymer may be drawn from the results of IR work. Unsaturation is shown to be present in the polymers formed from perfluorobenzene by means of ESCA, as the spectra shows in figure 8. The deconvolution here gives six peaks; the small peak at high binding energy, however, is at too high a binding energy to correspond to a primary photoionization peak and has been assigned to a shake-up peak (originating in $\pi \to \pi^*$ transitions accompanying core ionizations) on the $\underline{C}-F$ structural features. Typical energy separations of shake-up and primary peaks being *ca.* 6.5 eV.

Kinetic studies of glow-discharge polymerizations are typically carried out by weighing substrates after a given deposition time. It is possible directly to monitor the initial build-up of polymer film by use of ESCA substrate-overlayer technique. Using this technique it has proved possible to deposit films of known thickness by glow discharge and thereby measure mean free paths of electrons as a function of kinetic energy.[30]

Kinetic information is readily obtained from the spectra presented in Figures 6, 7 by monitoring the attenuation of the substrate signal. Taking first Figure 6, it can be seen that in going from perfluorobenzene to perfluorohexane the substrate signal intensity increases. Since discharges were carried out for equal lengths of time we may infer that polymerization efficiencies under these conditions are in the order

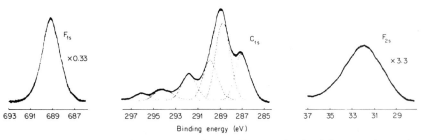

Figure 8. F_{1s}, C_{1s}, and F_{2s} spectra of the polymer obtained from a plasma in perfluorobenzene

perfluorobenzene > perfluorocyclohexa-1,4-diene ~ perfluorocyclohexa-1,3-diene > perfluorocyclohexene ~ perfluorocyclohexane.

Similarly, the difluoroethylene polymers exhibit polymerization efficiencies in the order: 1,1 < 1,2 *trans*< 1,2 *cis*.

(iii) Surface modification by plasma polymerization

Graft polymerization at surfaces by plasma techniques is a field of surface modification which has shown great potential. It is a field in which ESCA is particularly applicable. The treatment of wool with a variety of both volatile[31] and non-volatile[32] monomers has resulted in considerable success in shrink, soil, and flameproofing of the fibre whilst not altering the bulk properties of the material. The advent of ESCA has allowed for the direct study of structure and bonding of both the substrate and grafted materials. It is important to distinguish between processes in which the polymer is merely deposited into the surface of the fabric, and actual chemical bonding of monomers to the substrate which greatly extends the life of the coating: these distinct situations have been observed for the example of polypropylene and wool exposed to plasmas excited in hexafluoroethane and tetrafluoroethylene.[31] The spectrum of the polypropylene sample after exposure to the fluorocarbon discharge shows five signals in the C_{1s} region corresponding to $\underline{C}F_3$, $\underline{C}F_2$, $\underline{C}=O$, [$\underline{C}F$ and $\underline{C}-O$] and $\underline{C}H_2$ structural features. A similar spectrum is exhibited by the wool sample after a similar treatment, although the overall ratio of $\underline{C}H_2$ to C attached to fluorine is greater due to a lower incorporation of fluorocarbon species. The distinguishing feature between these two experiments, however, is that several of the signals due to carbon attached to fluorine on the wool substrate are ~1 eV lower in binding energy than the corresponding signals of the thicker fluoropolymer film on the polypropylene. This is strongly indicative that the particular structural features are directly bonded to the wool substrate, since the lowering of the binding energy is due to the lack of β fluorine substituent effects[29] as are observed in the fluorocarbon polymer.

An extension of this work has been in the grafting of non-volatile species on to the fabric by exposing impregnated samples to RF glow discharges.[32] Although the percentage incorporation of monomers can be determined crudely by conventional techniques, the use of ESCA has enabled accurate determinations to be made at the outermost surface of the material and, moreover, has detected degradation of the monomers themselves (i.e. depletion of bromine from tetrabromophthalic anhydride and tribromomethanilic acid monomers).

5. PLASMA MODIFICATION OF POLYMER SURFACES

(i) Introduction

The modification of polymer surfaces by electrical discharges (RF, microwave, corona) excited in a variety of gases as a technique to improve their surface properties has been the subject of extensive research in both industrial and academic laboratories. The major virtues of these techniques are that they involve clean reactions which take only seconds to achieve the required results and are therefore ideally suited to flow systems, and whilst producing profound changes in the surface properties of the polymer, the properties of the bulk material, for which it was originally chosen, remain essentially unchanged.[4] The thickness of the modified layer may be as much as several micrometres, depending on the parameters of the plasma (pressure, power, gas, flow rate, treatment time, etc.); however, the surface characteristics of the material are determined solely by the composition of the outermost few monolayers. The choice of ESCA for the study of such systems, therefore, owing to its unrivalled surface sensitivity, is mandatory. In this section we devote our attention to inert gas plasmas and oxygen plasmas, representing the two broad categories of such modifications. The first category involves only energy transfer from the plasma, whilst the second also includes the incorporation of the sustaining gas into the final product.

(ii) Modification of polymer surfaces by inert gas plasmas

The unique capability of ESCA to sample the outermost few tens of ångströms offers a technique to delineate the relative roles of direct-energy transfer from active species in the plasma and radiative energy transfer from the vacuum UV component of the electromagnetic radiation emitted from the plasma.[7,33] Using an inert gas as the sustaining medium, the reaction initiated in a saturated polymer via either energy transfer process is thought to involve a cross-linking mechanism, in which excited states of the polymer chain undergo homolytic bond cleavage, producing a wide variety of free radicals and unsaturated centres, which by radical additions produce cross-links between adjacent polymer chains. At high power densities, ablation of small volatile fragments from the polymer surface cannot be discounted.[34] However, at the relatively low power loadings employed in the investigations described below, ablation is expected to be minimal.

As a simple prototype for more complicated systems the polymer chosen for

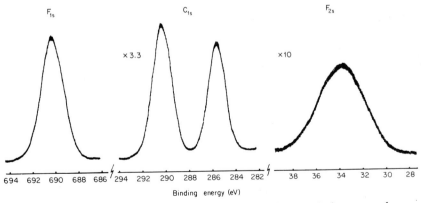

Figure 9. Core-level spectra of an ethylene-tetrafluoroethylene copolymer (52% TFE).

investigation is an ethylene-tetrafluoroethylene copolymer containing 52% tetra-fluoroethylene (Figure 9). A previous ESCA study of this copolymer has revealed its structure to be largely alternating.[35] The shift in binding energy for the C_{1s} levels is sufficiently large that the signals arising from photoemission from the tetrafluoroethylene components are well resolved with respect to the ethylene components (shift ~4.7 eV). It is therefore relatively easy to monitor changes in structure arising from plasma treatment by monitoring the components of the C_{1s} levels and also the F_{1s} levels. The F_{1s} and F_{2s} levels span a large range in mean free path of the photoemitted electrons, and the monitoring of the relative intensities of these levels therefore provides a convenient means of establishing the homogeneity or otherwise of the surface regions.[7]

The choice of an electrodeless inductively coupled RF plasma enables close consideration of the variables which are likely to be of importance and also allows considerable flexibility in terms of reactor design and configurations for introducing and removing samples, particularly for kinetic studies.

As a starting point it is worth while to consider the energy available from such plasmas, and Table 4 summarizes the main features for the inert gases employed in this work. Although the positive ions and neutral species in the plasma have near-ambient kinetic energy, the ions and metastables have sufficient energy to cause ionization of the polymer through ion neutralization and Penning ionization processes, respectively, with the exception of the 9.92 eV metastable state of krypton which is lower than the ionization potential of a typical alkane. The mean free paths for the ions and metastables *a priori* might be expected to be strongly dependent of the energy (kinetic or electronic) and size of the species in question. At ambient kinetic energy their range is of the order of a few monolayers in polymers.[33,36]

The electromagnetic radiation associated with inert-gas plasmas is predomin-

Table 4

Inert gas	Ionization potential		Metastable states		Resonance lines	
	Designation	Energy	Designation	Energy	Designation	Energy
He	$^2S_{1/2}$	24.586	^1S	20.615	HeI	21.217
			^3S	19.818	HeII	40.811
Ne	$^2P_{3/2}$	21.564	$^3P_0$	16.795	NeI	16.671
	$^2P_{1/2}$	21.661	$^3P_2$	16.619		16.848
					NeII	26.813
						26.910
Ar	$^2P_{3/2}$	15.759	$^3P_0$	11.723	ArI	11.623
	$^2P_{1/2}$	15.937	$^3P_2$	11.548		11.823
					ArII	13.302
						13.479
Kr	$^2P_{3/2}$	13.999	$^3P_0$	10.562	KrI	10.032
	$^2P_{1/2}$	14.665	$^3P_2$	9.915		10.643
					KrII	12.858
						13.514

antly in the vacuum UV. In the pressue range used for RF discharges the output is mainly in the form of the line spectra of the gases. Clearly, the attenuation coefficients for the electromagnetic radiation will be dominated by photoionization processes. At slightly longer wavelengths components of the Rydberg transitions (which converge on the ionization limits) also have substantial cross-sections, and since for the higher members the orbitals involved are so diffuse the concept of locally excited (e.g. $\sigma \rightarrow \sigma^*$) transition loses its meaning, and effectively such excited states in their chemical behaviour will have a close similarity to the molecular ions. Ultraviolet radiation has been demonstrated to have long-range effects in polymeric materials, and in appropriate cases modifications have been evident to a depth of several micrometres.[4]

Whilst electrons play a dominant role in the plasma itself, in the interaction with polymers it seems likely that their role will be secondary. The mean free paths of electrons near zero kinetic energy is of the order of hundreds of ångströms; this being the case, direct-energy transfer in the surface region is likely to be relatively small and dominated by phonon excitation.

A valence ionized polymer chain produced in the condensed media can undergo a number of transformations which we now consider for the prototype case of the ethylene-tetrafluoroethylene copolymer (Figure 10). The available valence ionized states correspond to removal of F_{2p} lone pairs or electrons dominantly of C–C or C–H bonding characteristics. Such ions could undergo unimolecular reactions such as carbon–carbon or carbon–hydrogen bond cleavage; however, in such a rigid matrix the cage effects would be expected to dominate, more particularly for the former reaction. The more likely process would seem to be ion neutralization by means of the electron flux through the sample leading to electronically excited systems having sufficient localized internal energy for homolytic C–F, C–C, and C–H bond cleavage. The greater mobility of hydrogen and fluorine atoms almost certainly dominates the following reaction sequences.

Elaboration of the processes involved in the interaction of inert-gas glow discharges with polymers may be achieved by means of ESCA investigations.[7,33]

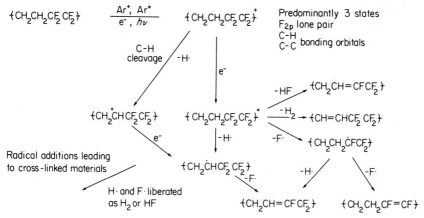

Figure 10. Cross-linking reaction scheme.

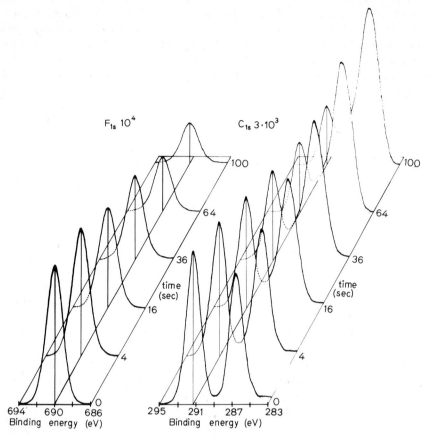

F_{1s} 10^4 C_{1s} $3 \cdot 10^3$

Figure 11. Core-level spectra of the copolymer after increasing times of exposure to an argon plasma at 0.1 Torr and 0.1 W.

Figure 11 shows the F_{1s} and C_{1s} levels for a sample exposed to an argon plasma at 0.1 Torr and 0.1 W. The spectra were recorded as a function of time of exposure to the plasma. The loss of fluorine and CF_2-type structural features is readily apparent from the decrease in intensity of the F_{1s} signal and the CF_2 component of the C_{1s} levels. The concomitant appearance of signals at intermediate binding energy in the C_{1s} spectrum, owing to the CF and CH structural feature, provides strong evidence supporting the reaction scheme of Figure 10. The increase in intensity of the overall total C_{1s} spectrum is consistent with a cross-linking process, since the net effect of cross-linking is to reduce interchain separation (i.e. the surface essentially shrinks) and the number of carbon atoms per unit volume in the surface thereby increases. The surface nature of the modification can be amply demonstrated by the angular dependence of the relative signal intensities (cf. Chapter 16).[33]

Early studies of the cross-linking of polymers by inert-gas plasmas employing less surface-sensitive techniques (in particular the measurement of gelation masses) were unable to delineate the relative roles of direct and radiative energy transfer processes.[4] The results obtained pertained predominantly to the radiative energy-transfer processes because of their long-range effects. More recently[7,33] ESCA has allowed this distinction, which is of considerable importance since the properties of the polymer are determined predominantly by the modification of the outermost surface layers, and which was thought likely to be dominated by direct energy transfer processes because of their short-range nature.

The kinetic model developed for analysis of the ESCA data is given by the equation,

$$\frac{I^t}{I_0^{TOT}} = [1 - \exp(-d/\lambda)] \, \exp(-K_s t) + \exp(-d/\lambda) \cdot \exp(-K_b t)$$

where I_0^{TOT} is the total integrated intensity of the ESCA signal for a given structural feature, for the initial system; I^t is the intensity at time, t; d the depth to which direct-energy transfer processes are important; λ the mean free path of the photoemitted electrons; K_s a pseudo rate constant encoding all rate processes in the surface layer, d; and K_b a pseudo rate constant encoding all rate processes in the subsurface and bulk of the polymer. The derivation of this equation may be found elsewhere.[7,33] Figure 12 shows a logarithmic plot of I^t/I_0^{TOT} versus time of exposure for the CF_2 data corresponding to the spectra in Figure 11. The distinct curvature is as might be expected for a two-component system.

The analysis in terms of the two-component system is straightforward, since as t becomes large the dominant contribution is from the component of the small exponent. The plots of $\ln(I^t/I_0^{TOT})$ versus t therefore approach linearity with the absolute value of the slope corresponding to the composite rate constant for the

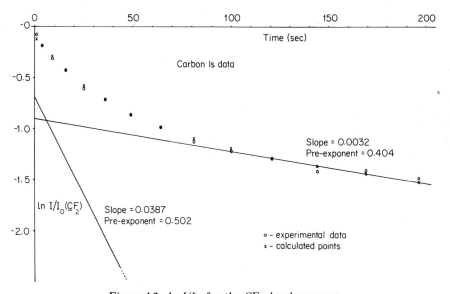

Figure 12. $\ln I/I_0$ for the CF_2 levels versus t.

slower reaction K_b assigned to the radiative energy transfer processes in the subsurface and bulk of the sample. The intercept of this line, extrapolated to $t = 0$, gives information on the depth of importance of the direct energy-transfer processes involving the ions and metastables, since it may be assigned the value of $-d/\lambda$. Replotting the differences of this extrapolated line from the experimental data at low t yields a second straight line of absolute slope equivalent to K_s, the composite rate constant for processes in the surface layer of thickness d, involving both direct and radiative energy transfer. The kinetic model has been found to be an extremely good approximation over a wide range of power loadings, pressures, flow rates, and sustaining gases.[33,36] Here, K_s is typically an order of magnitude or so greater than K_b. Figure 13 shows K_s values for the different inert gases at power loadings of 0.4, 5.0, and 10.0 W and average K_b values from the same data for the CF_2 and F_{1s} levels (K_s and K_b are plotted as a function of ionization potential merely for convenience – metastable energy, atomic radius, etc. may equally well have been chosen). The trends in the surface rate constant are clearly evident with apparently closely linear correlations. The trends for the radiative process in the bulk, however, are not so clear, but are eminently reasonable since overlap terms are involved for the absorption of UV radiation. The values of K_s derived from the F_{1s} data (not shown in Figure 13) are within the experimental error of the related value derived from the CF_2 data, suggesting that the fast reaction in the surface layer induced predominantly by direct energy transfer involves conversion of CF_2 structural features to CF-type environments. This is not the case, however, for K_b, for which the values obtained from F_{1s} information are generally ~1.3 times greater than those derived from CF_2 data, which clearly indicates that the CF sites

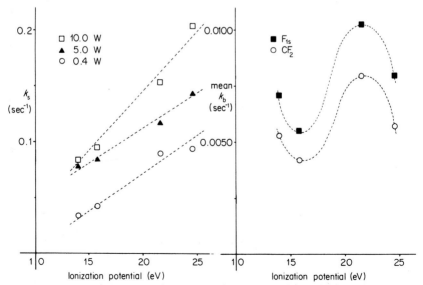

Figure 13. K_s (at power loadings of 0.4, 5.0 and 10.0 W) and mean K_b, versus ionization potential of the sustaining gas, the pressure being ~0.2 Torr.

may undergo further reaction to yield carbon not directly attached to fluorine, by a radiative process.

The depth d, to which direct energy transfer processes are important (derived from the pre-exponents in equation (1)), is also a function of the physical properties of the sustaining gas, ranging from ~1 monolayer for krypton to ~3 monolayers for helium (Figure 14) and is also dependent to a lesser extent on the power loading,[36] as might be expected since the temperature is ambient and the kinetic energy of the ions and metastables will increase with the power.

The dependence of d, K_s, and K_b on the operating parameters of the plasma may indirectly provide diagnostic information on the plasma itself.

The relative reactivities of particular structural features in the polymer, for example CH_2 versus CF_2 in the copolymer under discussion here, may, in principle, for appropriate systems be directly obtained from the ESCA spectrum, i.e. by monitoring the relative intensities of signals due to CF and CH carbon environments. However, the region of interest in the spectrum is unresolved (cf. Figure 11). The use of difference spectra, however, largely overcomes this problem as can be seen from Figure 15. The difference spectra are generated by subtracting the

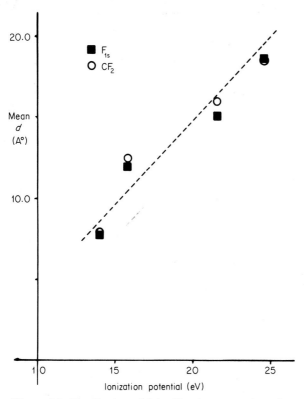

Figure 14. Depth to which direct energy transfer processes are important, d versus ionization potential of the sustaining gas.

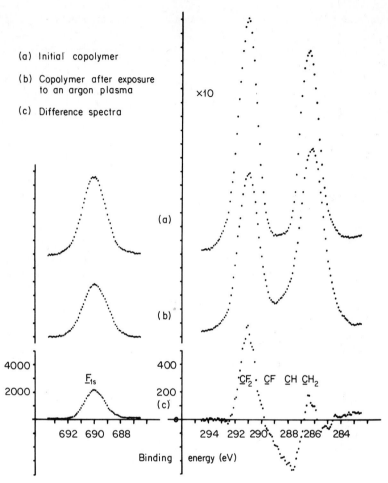

(a) Initial copolymer

(b) Copolymer after exposure to an argon plasma

(c) Difference spectra

Figure 15. (a) F_{1s} and C_{1s} spectra of initial copolymer; (b) F_{1s} and C_{1s} spectra of a sample treated for a short time in an argon plasma; (c) difference spectra, i.e. (a)–(b).

spectra of the treated sample from that of the starting material in a digital fashion.[37] The decrease in F_{1s}, CF_2, and CH_2 is readily apparent from Figure 15(c), and from the distinct asymmetry of the region of intermediate binding energy in the C_{1s} spectrum it is clear that both CF and CH features are produced and that more CH is produced than CF, reflecting a greater reactivity of the CH_2 sites. This conclusion has been supported by comparison with theoretical calculations on model systems.[36]

(iii) Modification of polymer surfaces by oxygen plasmas

An area of considerable importance is the surface oxidation of polymers, in terms of both improvement of surface properties by increasing their surface free

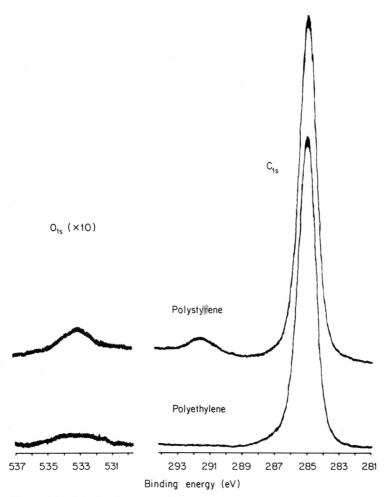

Figure 16. Core-level spectra for high-density polyethylene and poly-styrene films.

energy or wettability and in the understanding of oxidative degradation processes in general. The oxidation of high-density polyethylene and polystyrene provides an excellent example of the great applicability of ESCA coupled to glow-discharge techniques in this area. The core-level spectra of the starting materials (Figure 16) shows a small degree of oxidation when the O_{1s} region is magnified by a factor of 10. The samples are clearly distinguishable owing to the shake-up satellite observed for polystyrene, arising from $\pi \to \pi^*$ excitations accompanying core ionization.

Figure 17 shows the spectra after 16 sec exposure to the plasma at 100 mW. Both samples exhibit a broad, intense oxygen signal and a great deal of structure in the C_{1s} spectrum owing to a variety of carbon oxygen environments. The assignment of the components of the C_{1s} spectra is straightforward with a background knowledge of related systems: $\sim$286.6 eV, carbon singly bonded to

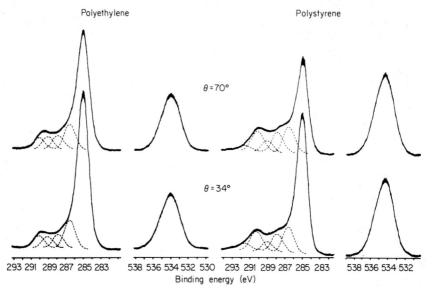

Figure 17. C_{1s} and O_{1s} spectra (recorded at $35°$ and $70°$ electron take-off angles) of samples of polyethylene and polystyrene exposed to an oxygen plasma for 16 sec (0.2 Torr and 0.1 W).

oxygen; ~287.9 eV, carbon double bonded to oxygen/carbon singly bonded to two oxygen atoms; ~289.1 eV carboxy-type carbon; ~290.4 eV, carbon associated with three oxygen atoms (i.e. carbonate); and shake-up satellite at ~291.6 eV.

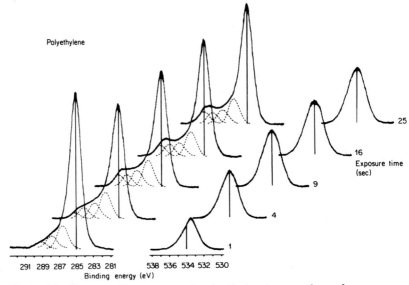

Figure 18. O_{1s} and C_{1s} spectra of polyethylene versus time of exposure to an oxygen plasma (0.2 Torr and 0.1 W).

The greater susceptibility of polystyrene to the oxidation is evidenced by a larger oxygen signal, and even at this extremely low power loading the reaction is rapid and essentially complete after a few tens of seconds as far as ESCA is concerned. The enhancement of surface features at grazing take-off angle demonstrates the surface nature of the modification, and furthermore the higher absolute intensity of the O_{1s} signals at grazing angle is indicative of a reaction contained within the outermost monolayer or so.[33]

At this low power loading the reaction may be readily followed as a function of time. This is demonstrated by Figure 18 for polyethylene. This opens the way to kinetic studies, although the full potential has not been exploited at the time of writing this review. Clearly, in the very initial stages a greater proportion of singly bonded oxygen is produced, evidenced both from the C_{1s} spectrum and a distinct shift in the centroid of the O_{1s} spectrum.

Information concerning the unsaturated centres in polystyrene can be derived by monitoring the shake-up intensity (Figure 19). The shake-up intensity is greatly reduced with the first second of reaction due to loss of unsaturation and substituent effects in the surface regions. As the reaction proceeds the intensity tends towards a value consistent with the shake-up and CH signals arising only from unmodified polystyrene of the bulk.[38] Clearly the extent of oxidation in the outermost monolayer or so becomes very high. The development of the various carbon—oxygen features can be straightforwardly monitored, as in Figure 20. For the two samples the concentration of carbon singly bonded to oxygen and carboxy-type carbon become closely similar, although their rate of production is faster in polystyrene. The greater incorporation of oxygen in polystyrene is clearly

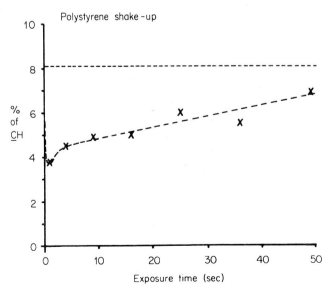

Figure 19. Shake-up intensity of polystyrene relative to the CH signal versus time of exposure to an oxygen plasma (0.2 Torr and 0.1 W).

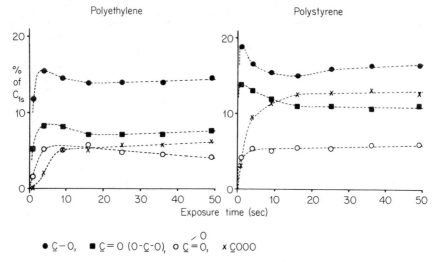

Figure 20. Intensity of various carbon–oxygen signals of polyethylene and polystyrene relative to the total C_{1s} signal versus time of exposure to an oxygen plasma (0.2 Torr and 0.1 W). 15, 2321 (1977).

in the form of ketone/methylenediether and carbonate environments. This degree of information has not been possible – particularly for the early stages of reaction and the outermost surface of the sample – by any technique apart from ESCA. The investigation of the interaction of polymers with glow discharges excited in oxygen by means of ESCA is in its preliminary stages at the time of writing this review. However, the potential is clear and a great deal more information should be forthcoming in the very near future.[39]

6. CONCLUSION

The procedures and illustrative examples outlined here emphasize the great potential of glow-discharge techniques as a means of both the synthesis, and the surface modification of polymers. This account briefly touches upon the applications which are currently being pursued, and serves to illustrate the unique capabilities of glow-discharge techniques, particularly when coupled with the latest spectroscopic tools for surface characterization.

REFERENCES

1. M. M. Millard, K. S. Lee, and A. E. Pavlath, *Text. Res. J.*, 42, 307 (1972).
2. D. M. Soignet, R. J. Berni, and R. R. Benerito, *J. Appl. Polym. Sci.*, 20, 2483 (1976).
3. H. Schonhorn, Chapter 10, this book.
4. M. Hudis, in *Techniques and Applications of Plasma Chemistry*, eds. J. R. Hollahan and A. T. Bell, Ch. 3, Wiley, New York, 1974.
5. R. S. Thomas, in *Techniques and Applications of Plasma Chemistry*, eds. J. R. Hollahan and A. T. Bell, Ch. 8, Wiley, New York, 1974.

6. E. O. Johnson and L. Malter, *Phys. Rev.*, **80** (1), 58 (1950).
7. D. T. Clark and A. Dilks, A.C.S. Centennial Meeting, New York, April 1976, *International Symposium on Advances in Characterization of Polymer Surfaces*, ed. L. H. Lee, Vol. 2, p. 101, Academic Press, 1976.
8. *Techniques and Applications of Plasma Chemistry*, eds. J. R. Hollahan and A. T. Bell, Wiley, New York, 1974.
9. A. N. Mearns, *Thin Solid Films*, **3**, 201 (1969).
10. V. M. Kolotyrkin, A. B. Gil'man, and A. K. Tsapuk, *Russ. Chem. Phys.*, **36** (8), 579 (1967).
11. Cf. Chapters 16 and 8.
12. H. Yasuda and E. Lamaze, *J. Appl. Polym. Sci.*, **16**, 597 (1972).
13. L. F. Thompson and K. G. Mayhan, *J. Appl. Polym. Sci.*, **16**, 2317 (1972).
14. L. F. Thompson and K. G. Mayhan, *J. Appl. Polym. Sci.*, **16**, 2291 (1972).
15. D. W. Rice and D. F. O'Kane, *J. Electrochem. Soc.*, **123** (9), 1308 (1976).
16. H. Yasuda, M. O. Burngarner, H. C. Marsh, and N. Morosoff, *J. Polym. Sci. Polym. Chem. Ed.*, **14**, 195 (1976).
17. T. Williams and M. W. Hayes, *Nature*, **216**, 614 (1967).
18. A. M. Wrobel, M. Kryszewski, and M. Gazicki, *Polymer*, **17**, 673 (1976).
19. A. R. Westwood, *Eur. Polym. J.*, **7**, 363 (1971).
20. H. Suhr, *J. Pure and Appl. Chem.*, **39** (3), 395 (1974).
21. J. R. Hollahan and R. P. McKeever, *Advan. Chem. Ser.*, **80**, 272 (1969)
22. J. P. Redmond and A. F. Pitas, NASA Contract Rep. NASA-CR-94310 (1968).
23. J. P. Wightman and N. J. Johnson, *Advan. Chem. Ser.*, Washington, **80**, 322 (1969).
24. D. D. Neiswender, *Advan. Chem. Ser.*, **80**, 338 (1969).
25. A. Bradley and J. P. Hammes, *J. Electrochem. Soc.*, **110** (1), 15 (1963)
26. B. D. Washo, *J. Macromol. Sci. – Chem.*, **A10** (3), 559 (1976).
27. P. J. Dynes and D. H. Kaeble, *J. macromol. Sci. – Chem.*, **A10** (3), 535 (1967), and references therein.
28. M. Millard, in *Techniques and Applications of Plasma Chemistry*, eds. J. Hollahan and A. T. Bell, Ch. 5, Wiley, New York, 1974.
29. D. T. Clark and W. J. Feast, *J. Macromol. Sci. Revs. in Macromol. Chem.*, **C12**, 191 (1975).
30. D. T. Clark and D. Shuttleworth, *J. Polym. Sci., Poly. Chem. Edn.*, in press.
31. M. M. Millard and A. E. Pavlath, *J. Macromol. Sci. – Chem.*, **A10** (3), 579.
32. A. Pavlath and K. S. Lee, *J. Macromol. Sci. – Chem.*, **A10** (3), 619 (1976).
33. D. T. Clark and A. Dilks, *J. Polym. Sci., Poly. Chem. Edn.*, **15**, 2321 (1977).
34. E. Mathias and G. H. Miller, *J. Phys. Chem.*, **71** 2671 (1967).
35. D. T. Clark, W. J. Feast, I. Ritchie, W. K. R. Musgrave, M. Modena, and M. Ragazzini, *J. Polym. Sci., Poly. Chem. Edn.*, **12**, 1049 (1974).
36. D. T. Clark and A. Dilks, *J. Polym. Sci., Poly. Chem. Edn.*, in press.
37. D. T. Clark and A. Dilks, *J. Elect. Spec.*, **11**, 225 (1977).
38. D. T. Clark and A. Dilks, *J. Polym. Sci., Poly. Chem. Edn.*, **15**, 15 (1977).
39. D. T. Clark and A. Dilks, *J. Polym. Sci., Polym. Chem. Edn.* (in press). Cf. D. T. Clarke, A. Dilks, and H. R. Thomas, 'The Application of ESCA to Polymer Degradation', Chap. 4 in *Polymer Degradation*, Ed. N. Grassie, Applied Science Publishers, London, 1977.

10

Surface Modification of Polymers for Adhesive Bonding

Harold Schonhorn

Bell Telephone Laboratories, New Jersey

I. INTRODUCTION

Although adhesive-bonding technology has a rather long and varied history, the science of adhesion is relatively new. In fact, one may argue that the technology has far outstripped the science. Often, opposing viewpoints have been formulated to account for the large body of experimental data. In principle, the merit of any theory is to account for the existing body of data and, hopefully, to provide insight to guide the scientist to perform new experiments, leading to further understanding. As one examines the science of adhesion, it soon becomes apparent that a wide variety of disciplines in both physics and chemistry is needed and that no one discipline is sufficient to enable the investigator to understand, in detail, the operations of a particular system. Surface chemistry, surface physics, polymer physics, rheology, fracture mechanics, etc. are but a few of the important fields of scientific endeavour with which the investigator should be familiar before beginning an excursion into the science of adhesion.

The purpose of this paper is to review surface modifications of polymers for adhesive bonding. It is therefore appropriate that we analyse briefly the various theories of adhesion and the nature of polymer surfaces before and after suitable modification. Further complicating the issue is that in many instances, no surface treatment of the polymer is required prior to adhesive bonding. Therefore, it is clear that a detailed knowledge of the surface structure and composition of the polymer should provide clues to resolve this problem.

2. NATURE OF POLYMER SURFACES

The nature of the polymeric surface is involved in the phenomena of adhesion, adhesive joint strength, wettability, and spreading of a polymer in contact with another phase. At present, there is a considerable body of experimental information on the wetting of polymers, polymer melt surface tension, interfacial tension between different pairs of polymer melts, and the modification of polymer surfaces by additives and chemical reactions.

Special problems associated with polymers arise from their molecular weight heterogeneity and the existance of many classes of polymer chain defects due to compositional, optical, geometrical, and head-to-tail isomerism, as well as in-advertent branching. An immediate consequence of this architectural irregularity of the backbone chains of polymers is the possibility of several distinct modes of solidification from the melt. We shall explore, in section 5ii, what profound influence the mode of crystallization has on generating particular surface properties for semicrystalline polymers.

Since we are concerned with the behaviour of a treated polymer in an adhesive joint, we are necessarily concerned with the mechanical properties of polymers. Bulk properties of polymers exhibit time-dependent mechanical properties, including a remarkable degree of viscoelasticity. Dynamic mechanical experiments yield both an apparent elastic modulus (in tension or shear) E, and the mechanical damping tan δ, where δ is the loss angle. At a convenient frequency of, say 100 Hz, a typical linear amorphous polymer, in the glassy region below T_g has a modulus of about 10^{10} dynes/cm^2. As the temperature is increased the next region of mechanical response is the transition region, including T_g, which can be taken to be in first approximation the point of inflection of the modulus curve or the maximum in the damping curve. In this region the modulus drops by a factor of 1000. The third region is called the 'rubbery plateau' where the modulus remains roughly constant. As the temperature is further increased this is followed by a region of elastic or rubbery flow, and finally liquid flow with very little elastic recovery. In these regions the damping curve again begins to increase and the modulus may drop below 10^5 dynes/cm^2. In a cross-linked (amorphous) polymer the last two regions are absent; since there is no flow, the cross-links prevent the chains from slipping past each other. The presence of crystallinity increases the modulus significantly above T_g, with a sharp drop in the modulus and increase in mechanical damping curve at the melting point T_M. Thus, in crystalline polymers the demarcation point between the elastomeric (rubber-like) 'solid' and rubbery 'melt' can be taken to be given by T_M, whilst for amorphous solids it lies in the vicinity of the end of the rubbery plateau region.

The fact that the demarcation between the 'melt' and the 'solid' is far from sharp reinforces a need for a detailed study of the surface and interfacial tensions of polymer melts as related to adhesion phenomena. The complicated dynamical mechanical response of polymers to applied stresses preclude the reliable estimation of surface tensions or surface free energies of polymers by such methods as mechanical necking experiments which have been applied to other classes of solids. Ultimately, it is the many internal degrees of freedom implied by the macro-molecular nature of polymers which are responsible for this extensive and varied time dependence of their mechanical and thermal properties.

Synthetic organic polymers are basically van der Waals or molecular solids in terms of their local surface properties, even though they may remain solid at much higher temperatures than other members of that class by virtue of their large molecular weight or the presence of cross-linking. The fact that even the bulk polymer is non-uniform in composition, molecular weight, and structure leads to many new ways in which the interfacial region can differ from the bulk. One must deal with 'contamination' from within as well as without by surface-active substances. With some notable exceptions, most polymers which we consider are hydrophobic and polar sites in such a hydrophobic matrix present special problems. Last, but not least, polymer surfaces are usually not pretreated extensively, unless they are being prepared for adhesive bonding.

The surface tension of liquids or interfacial tension of liquid pairs can be reliably measured experimentally. This is true also of the viscous polymer melts. On the other hand, *there exists currently no direct, reproach-free method of measuring the surface tension or specific surface free energy of a macroscopic surface of a polymeric solid.* The surface free energies, which are discussed and possibly even measured, of a crystalline nucleus of a polymer within an amorphous matrix have little relation to the surface tensions or specific surface free energies of a macroscopic polymer surface in contact with vapour or a simple non-polar liquid.

Polymers introduce further problems since the definition of a polymeric substance allows for possibilities of isomerism and molecular-weight heterogeneity. Thus, the surface tension of an isotactic polymer melt should (and can) differ from that of the syndiotactic polymer melt. Even in the absence of isomerism a heterodisperse polymer melt will not possess precisely the same molecular-weight distribution in the immediate vicinity of a phase boundary as in the bulk (e.g. because of a difference of configurational entropy in the interface region of shorter chains as compared to longer chains). Furthermore, polymer solids can be quenched or otherwise prepared in metastable states which can exhibit hysteresis in surface as well as bulk properties. In short, the large size, the extensive range of mechanical viscoelastic relaxation times, and the molecular inhomogeneity of polymer molecules aggravate characteristic difficulties in applying continuum, pheno-menological theory (hydrodynamic or thermodynamic) to the description of experimental surface phenomena.

3. THEORIES OF ADHESION

A proper exposition of the surface properties of polymers with respect to adhesion and adhesive joint strength should begin by defining these terms. *Adhesion,* as used in this article, refers only to the attractive forces exerted between a solid surface and a second phase (either liquid or solid). Adhesion is concerned with the phenomenon of making an adhesive joint (i.e. wettability, relative surface energetics of both phases, and kinetics of wetting).[1,2] These are purely surface considerations. *Adhesive joint strength* is the breaking strength of a bonded assembly.[3] Once an adhesive joint has been formed, interfacial forces are no longer of primary concern, since interfacial separation probably never occurs under ordinary failure conditions, except at points where molecular contact did not exist prior to breaking the adhesive joint. What is of prime importance is the mechanical response of the composite to an applied stress. Probably a more realistic description of the breaking strength of an adhesive joint would be based on mechanical deformation theory of adhesive joint strength.

There is, however, one important instance where interfacial forces play a dominant role and that is with respect to the permanence of a bonded structure. As we shall describe in a later section, the thermodynamic work of adhesion is a useful criterion for determining the composite lifetime when subjected to hostile environments. Chemisorption rather than physical adsorption should be the primary driving force in preparing more permanently bonded structures.

(i) The electrical theory

Voyutskii published a review paper[4] in which he discussed the merits and shortcomings of the then existing adsorption theory and the diffusion and electrostatic theories of adhesion. This paper[4] will be used as the basis for analysis. Voyutskii describes in an orderly fashion Derjaguin's criticisms of the adsorption theory. Derjaguin states:[5]

(a) 'That the work of peeling an adhesive film may be as great as 10^4-10^6 erg/-cm^2, whereas the work required to overcome the molecular forces is only 10^2-10^3 erg/cm^2. It is claimed that this shows the actual work of adhesion to be orders of magnitude higher than could be expected to result from molecular forces.'

Peeling experiments gives no direct measure of interfacial molecular forces, and that the overwhelming proportion of work is expended in deforming the material prior to the break[3]

(b) 'That the work of adhesion depends on the rate of separation of the adhesive film, whereas the work expended on overcoming the molecular forces should not depend on the rate at which the molecules separate.'

Again, interfacial forces are not being measured; rather, the rheological properties of the adhesive film are being measured and there is, in general, a dependence on rate.

(c) 'That the adsorption theory cannot explain the adhesion between non-polar high polymers nor the adhesion between such non-polar polymers as polyisobutylene, natural rubber, and gutta-percha, and a number of substrates.'

Clearly, the intermolecular forces involved in cohesion and adhesion are of the same kind; that the existence of universal dispersion forces means that all the materials exert an attractive force (adhesion) for all other materials. In essence, the matter of polarity or non-polarity may be of minor significance in practical adhesive joining.

These three items, in addition to the observation that rupture can lead to electrification of the rupture surfaces produced and sometimes even to electrical discharge and electron emission during the process of separation, led Derjaguin to propose the electrical theory of adhesion.[5] This theory treats the adhesive—adherend system as a capacitor which is charged owing to the contact of two different substances. Separation of the parts of the capacitor, as during breaking of the joint, leads to separation of charge and to development of a potential difference which increases until a discharge occurs. Adhesion is presumed because of the existence of the electrical double layer. Voyutskii says:[4] 'Although the electrical theory is an advance on the adsorption theory, a number of factors limit its application to the mutual adhesion of high polymers.' He then proceeds to enumerate five distinct general systems to which Derjaguin's theory is not applicable.

The electrical phenomena, on whose existence the theory is based, are phenomena which manifest themselves only when the adhesive–adherend system is broken, and there is not, *a priori* reason to believe that phenomena resulting from the breaking of an adhesive–adherend system have any connection whatsoever with the phenomenon of adhesion which is involved in the making of an adhesive–adherend joint. Furthermore, there seems no reason to believe that the two electrically charged surfaces obtained when rupture of an adhesive–adherend system is made to occur are identically the same two electrically neutral surfaces which were placed in contact with each other initially to form the system.

(ii) The diffusion theory

The diffusion theory of Voyutskii,[6–8] which has been applied previously only to the adhesion of polymeric materials to themselves and to other polymers, states that adhesion is due to mutual diffusion of surface layers of polymer molecules, each into the other, to form an interwoven network. That such mutual diffusion of certain combinations of polymeric systems does occur cannot be denied; that it occurs prior to wetting of one polymer by the other is seriously open to question, since this would imply that diffusion can occur prior to or without contact. The notion that diffusion is the prime cause of adhesion is, in our view, incorrect; that it is the result of adhesion is a far more reasonable point of view.

Voyutskii's arguments for the diffusion theory of adhesion are all based on measurements of the breaking strength of adhesive joints. He measures these strengths as a function of time of contact between polymers, temperature, polymer type, molecular weight, viscosity, etc. and points out that the functional dependence of strength on some of these parameters is similar to that expected for a diffusion process; therefore, adhesion is a result of diffusion. This concept carries with it the consequence that there does not exist an interface where the two polymer surfaces join, but a region of variable composition.

Voyutskii also says:[4] 'It is obvious that the diffusion of molecules of one high polymer into another is nothing less than solution. The importance of mutual solubility in the adhesion of high polymers was first indicated by the present author, who suggested that the adhesion of high polymers to one another should be used as a criterion of their compatibility.' Subsequently, Derjaguin, who also emphasized the connection between compatibility and adhesion, suggested the use of compatibility to assess adhesion.

The importance of the mutual solubility of components for adhesion, this being mainly determined by the relationship between the polarities of the high polymers, is in complete accordance with the empirical law of de Bruyne,[9] according to which strong adhesion is possible only when both high polymers are either polar or non-polar and hindered when one is polar and the other non-polar.

The de Bruyne rule[9] has been shown, in part, to be incorrect.[10]

The diffusion theory cannot explain the joining of polymeric materials to metals, glass, or other hard solids, since it is difficult to understand how adhesion to these materials can result from diffusion of the polymer into such materials.

It would be instructive to take each of Voyutskii's points in support of his diffusion theory of adhesion, and to show how they are much more reasonably treated, in a consistent manner, in terms of adhesion as a surface phenomenon.

(iii) The adsorption theory

Although this chapter is concerned primarily with the formation of an interface between surface modified polymer adherends, the same guidelines are applicable when considering metals and metal oxides. In the following sections we shall consider a possible mode for the generation of weak boundary layers in polymers, and methods designed to eliminate mechanically weak surface layers in polymers.

Formation of Interface (Wettability)

Two materials probably adhere, at least initially, because of van der Waals' attractive forces acting between the atoms in the two surfaces. Interfacial strengths based on van der Waals' forces alone, far exceed the real strengths of one or other of the adhering materials. This means that interfacial separation probably never occurs to any sensible extent when mechanical forces are used to separate a pair of materials which have achieved complete interfacial contact (probably a highly unlikely situation), or a number of separate regions of interfacial contact. As we shall describe later (section 6), environmental effects may violate this concept. It follows, then, that breaking the joint mechanically, in general, tells nothing directly about interfacial forces.

Van der Waals' forces are operative over very small distances. Hence, in order that materials adhere, the atoms in the two surfaces must be brought close enough together for these forces to become operative. If we had a piece of A (solid) and B (solid) and each had an absolutely clean, smooth (on an atomic scale), planar surface, and if these surfaces were brought together in a perfect vacuum, all attempts to separate them would result in failure in either A or B (Figure 1(a)). But real surfaces differ from these ideal surfaces in that they are rough and contaminated, and both of these imperfections contribute to a greatly decreased real area of contact between the surfaces of A and B (Figure 1(b)). In general, however, where they have achieved contact — that is, where they have been brought close enough for van der Waals' forces to become operative — they have adhered, and when they are separated mechanically, a little of A remains on B, and B on A,

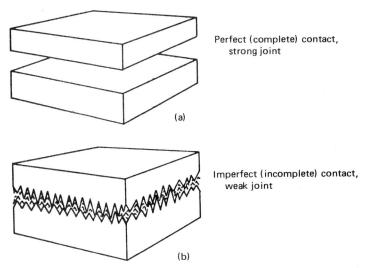

Perfect (complete) contact, strong joint

(a)

Imperfect (incomplete) contact, weak joint

(b)

Figure 1. (a) Ideal surfaces: clean, atomically smooth, planar; (b) real surfaces: dirty, rough, non-planar.

depending on the geometry in the neighbourhood of each area of contact and the cohesive strength of each. The general assumption based on visual examination is that either the solids did not adhere, or the failure was in adhesion. The first statement, however, is incorrect, because surely some areas of A and B achieved interfacial contact, and the second is incorrect because where the surfaces were not in contact there was no adhesion and caution must be exercised when discussing the failure of something that did not exist.

This point of view leads to the conclusion that to get A and B to make a stronger joint we need to increase their real area of contact. This means that one or both of the materials must be made to conform better to the surface roughness of the other. This implies, in a practical sense, that one of the materials should be fluid is a necessary, but may not be a sufficient, condition, for whilst a high-viscosity fluid makes a sizeable contact angle with the solid, its tendency to create a large interfacial area of contact may be relatively poor. The result is that it may do a great deal of bridging, fail to displace adsorbed air, and achieve little penetration into the surface roughness of the solid. Resultant stress concentrations due to poor wetting become important. However, if the fluid member spontaneously spreads on the solid, the interfacial area of contact increases, because the fluid can now flow more completely into the micro or submicro pores and crevices in the surface of the solid and can displace gas pockets and other contamination. In addition, the zero contact angle tends to minimize stress concentrations. The effect of creating a spontaneous spreading situation, then, is twofold, i.e. the real area of contact is increased and stres concentration is minimized.[10]

Specifically, only van der Waals' forces have been mentioned in connection with the preceding treatment. This is not to be construed as meaning that other molecular forces may be excluded from participation in adhesion. In the initial process involving the establishment of interfacial contact, all the molecular forces involved in wetting phenomena can be considered to be important. Chemisorption is not excluded, but if it is to occur, molecular contact must have already been established i.e. van der Waals forces' must already be operative. Therefore, any such chemical reaction as does occur, occurs after adhesion has taken place.

(iv) Rheological theory, weak boundary layers

The effect of wettability of epoxy adhesives on thermoplastics and wetting of cured epoxy adhesives by molten thermoplastics have been demonstrated.[10-13] Although the wettability of both chlorotrifluoroethylene homopolymer and polyethylene are similar, strong joints can be made to the former but not to the latter at temperatures well below their respective melting temperature. Polyethylene and probably many other melt-crystallized polymers seen to have associated with them a weak boundary layer which precludes the formation of strong adhesive joints, even though extensive interfacial contact occurs between the polymer film and the epoxy adhesive.[14,15]

4. CHARACTERIZATION AND STRUCTURE OF POLYMER SURFACES

To characterize an interface fully, one is interested in the values of various physical and chemical properties of the matter composing it, first, as a function of the depth into the interface and, second, as a function of two directions normal to the depth. This requires combinations of many analytical tools and techniques. Each experimental technique provides data on a specific property (or properties) which is an average over a characteristic depth of sampling its range. In the case of polymer surfaces one is interested in: (1) the elemental and molecular compositions; (2) the distribution in molecular weight, cross-link density extent of branching and chain isomers; (3) variation in density and percent of crystallinity; (4) changes in morphology or extent of ordering; (5) the nature and extent of preferential orientation of the molecular segments; (6) the distribution of regions of varying surface free energy (surface energetics), patches of high polarity, etc.; (7) the superficial surface area (particularly of finely divided polymer samples); and (8) the microtopography of the surface.

Some techniques, for example LEED or Auger spectroscopy, have not yet been successfully applied to polymer surfaces for various reasons. Thus, the high-energy electron beams used to excite Auger emission may cause radiation damage to polymer surfaces. This has also limited the application of electron and ion microprobe techniques. The nature of the nuclei composing organic polymers prevents the use of Mössbauer spectroscopy, except in certain polymers such as the polysiloxanes. Newer techniques appear very promising, but have not actually been developed for applications to polymer surfaces; examples are secondary ion mass spectroscopy or ion neutralization spectroscopy. Certain techniques such as ellipsometry [16,17] have been very successfully applied to the study of monolayers of adsorbed polymers; they have not been applied directly to a bulk polymer surface. Finally, we must not forget that in studying say a polymer—metal interface, the presence of metal or metal oxide can be detected by a variety of techniques (X-ray fluorescence or electron microprobe (particularly as attachments to a scanning electron microscope), atomic absorption spectroscopy, etc. Barring effects due to voids or non-polymeric impurities, etc. such studies provide some insight into the distribution of the polymer in the interface by the argument that regions containing high concentrations of metal or metal oxide are largely devoid of polymer. Some useful techniques for the characterization of adherend surfaces (to polymers) have been recently reviewed by Hamilton.[18]

A well known technique to determine surface composition and some information on orientation and/or degree of crystallinity is attenuated total reflectance (ATR) infrared spectroscopy.[16] This technique is often augmented by comparison of its results with results of standard and modified infrared transmission studies on bulk films (*ca.* 1 mil or more in thickness). Following Harrick[19] in ATR the penetration of the incident beam into the surface region of a polymer film adhering to say a KRS-5 crystal is approximately one-fifth of the wavelength or about 2 μm for radiation of frequency 720 cm^{-1}. It thus has a range of about 1 μm and measures a relatively thick surface layer. Good contact is required between the sample and the spectrometer prism, which may be a source of difficulty with rough or inflexible films.

An example of the application of this technique is Luongo and Schonhorn's[20] infrared study of the surface region of polyethylene (PE) nucleated on gold or aluminium (high-energy surfaces) and polytetrafluoroethylene (PTFE, a low-energy surface). The chemical composition of the surface region and the bulk appears spectroscopically to have the same chemical nature, but there are characteristic differences in the intensities of certain infrared bands which are known from studies of bulk samples[21] to be suitable for following relative changes in crystallinity.

The Henke-design X-ray tube[22] generates long-wavelength or 'soft' X-rays and thereby extends the usefulness of X-ray fluorescence analysis down to atomic number 5, by what is termed *soft X-ray spectroscopy*. The effective depth of analysis is about 1 μm into the surface.[23,24] It is a valuable technique in its own right as well as a useful complement to the infrared ATR technique.

X-ray photoelectron spectroscopy (ESCA) developed by Siegbahn and co-workers allows chemical structure analysis even in polymers in a range 1000 times less deep than either ATR or soft X-ray spectroscopy. Basically, this is because photoelectrons ejected from atoms deep in the sample are either self-absorbed by the sample or lose part of their kinetic energy and contribute only to the spectral background. The effective escape depth varies with the electron kinetic energy and surface structure, but is of the order of a few tens of ångströms.[23-25] Besides providing a semiquantitative elemental analysis, ESCA spectra exhibit small shifts (chemical shifts) owing to the environment of an atom in a molecule. Thus, for example in the carbon spectrum from poly(vinyl fluoride) there is a doublet corresponding to the carbon atoms in the CHF and CH$_2$ groups. The electron density is less around the fluorine-substituted carbon than around the carbon with only hydrogen substituents, and the binding energy is correspondingly increased. Clark and co-workers[26] have exploited the ESCA technique with particular reference to the surface treatment of hydrocarbon and fluorocarbon polymers and the locus of failure of adhesive joints.

Certain spectroscopic techniques can only be applied to the much higher area internal surfaces attained at the interface between polymer and dispersed, small filler particles. A variety of nuclear magnetic resonance (NMR) studies have been carried out on filled polymers; usually the filler employed was carbon black.[27] Proton spin resonance studies support the view that a portion of polymer in the vicinity of the filler is immobilized.

Lipatov and Fabuliak[28] had earlier reported large shifts with temperature in the minima in the NMR signal of poly(methyl methacrylate) and polystyrene filled with silica. Somewhat smaller effects were observed with these polymers when filled with finely dispersed PTFE. These authors also interpret their results as due to a decrease of segmental motion, mainly determined by a decrease of the number of conformations in the interface rather than due to significant changes of interaction energy at the surface.

Recently, direct metallization of fluoropolymers has been explored as a surface treatment for adhesive bonding. ESCA examination of such surfaces revealed extensive defluorination and generation of intermediate organometallic complexes.[29]

Among the most extensively studied polymeric surfaces have been those of fluoropolymers, both control surfaces and those subjected to special surface treatments. These have been studied by combinations of infrared spectroscopy (ATR and transmission), ESCA, soft X-ray spectroscopy, contact angle hysteresis and electron and optical microscopy.[23,25,30-33] The polymers most studied are untreated commercial polytetrafluoroethylene (Teflon TFE) and tetrafluoro-ethylene-hexafluoropropylene copolymer (Teflon FEP) films and the same films subjected to a variety of surface etchings intended to increase the adhesiveness of these to other polymers, for example polyurethane, epoxy- or phenolic-resin adhesives. These include treatment with (1) sodium in naphthalene/tetrahydrofuran, (2) sodium in liquid ammonia, or (3) molten potassium acetate at 325°C. The colour of the fluorocarbon surface is changed dark brown to black by these treatments. Microscopic measurements indicate that the colour change is localized within a surface region of a few tenths of a micrometre.[32] The advancing contact angle of water (25°C) on PTFE changes from 108°C for the unmodified film to 62–66° for the etched films[32] Similarly, Teflon FEP exhibits a decrease of the advancing water contact angle from 109° to 52°.[23] The infrared spectra of the original PTFE (either of a composite layer built up from seven 5 μm films or the differential spectrum) is greatly changed in the region of 1600 cm^{-1} as a result of the three etching treatments. The presence of the intense absorption band at 1600–1700 cm^{-1} has been ascribed to the presence of conjugated double bonds and/or the valence vibrations of C $=$ O groups.[32] Surface films subjected to Na in naphthalene/tetrahydrofuran or to potassium acetate treatment exhibit spectra indicating the presence of the hydrophilic OH and CO groups as well as unsaturation. Films treated with Na in anhydrous ammonia exhibit indications of surface NH groups as well.[32] These results are consistent with the work of Dwight and Riggs[23] Collins *et al.*,[25] and Clark and co-workers[26], using among other techniques ESCA and soft X-ray spectroscopy. The presence of nitrogen in the sodium/ammonia etched films of both Teflon TFE and FEP is confirmed. The soft X-ray data indicate that the depth of etch after 60 seconds of exposure is about 0.3 μm in PTFE and 0.07 μm in FEP.

In summary, a number of general observations can be made:

(1) Current surface characterization techniques appear to fall broadly into two categories – those whose effective range is measured in ångströms and those whose effective range is about 1000 times larger. Clearly, the wetting behaviour of a surface such as the polyfluorocarbons above follows the ESCA characterization of a *surface layer* measured in ångströms. On the other hand, significant effects on the elastic properties of a thin crystalline film of PE can arise owing to a modified degree of crystallinity in a surface *region* whose extent is measured in micrometres or tens of micrometres.

(2) Both have to be separately investigated – the thin surface layer may have a different composition or structure from the thicker surface region, and both may have properties that differ from the bulk polymer.[23,34]

(3) The absence of properties that appear to respond to an intermediate range, say of several hundred ångströms, is clearly in part a consequence of the lack of suitable probes within that range. This intermediate region may possess unusual properties since the average diameter (radius of gyration) of a polymer coil has this dimension.

5. SURFACE MODIFICATION OF POLYMERS FOR ADHESIVE BONDING

A large amount of work aimed at chemically modifying polymer surfaces has been reported. In an extensive review,[35] Angier classifies these as surface grafting by irradiation, chemical surface grafting, and chemical treatment of surfaces. The first two categories involve the production of a surface layer of a second polymer, grafted to the base polymer. There may be more or less interpenetration or copolymerization, depending largely on the permeability of the base polymer to the monomer of the second. Initiation of the polymerization may be by high-energy radiation (such as γ-rays), lower-energy radiation (ultraviolet) or the usual chemical initiating species such as peroxides (formed *in situ* by oxidation of the base polymer surface) or ozone.

Among the chemical treatments summarized by Angier are oxidation, especially of polyethylene, to enhance its wettability and adhesion. Acid and alkali treatments, halogenation, sulphonation (concentrated or fuming sulphuric acid or chlorosulphonic acid), and treatment with chlorosilanes have been applied to a number of different polymers. Fluorinated polymers may be modified by treatment with alkali metals dissolved in liquid ammonia.

More recently, the surface layer of polyethylene has been cross-linked by reaction with gaseous fluorine.[36] Olsen and Osteraas[37,38] have reported a variety of modifications of polyethylene surfaces by reactive sulphur compounds, carbenes, and nitrenes, as well as difluorocarbene ($: CF_2$) modification of several other polymers.[39] Amino groups have been attached to several polymers with the aid of radio-frequency plasmas;[40] this, and several other chemical modification techniques, has been used to permit binding of the blood anticoagulant, heparin, to plastic surfaces.[41] Nitrogen-containing functional groups can also be incorporated in polycarbonates by treatment with diamines or polyamines, such as triethylene-tetramine.[42]

In the following sections we describe in detail two procedures for modifying polymer surfaces because they are important from a scientific as well as practical point of view.

(i) Surface cross-linking (CASING technique)

Conventional techniques for treating polyethylene surfaces in order to obtain strong adhesive joints, such as surface oxidation by corona discharge[43] or flame treatment,[44] are commonly believed to be effective because they create wettable polar surfaces on which the adhesive may spread spontaneously and thus provide extensive interfacial contact. However, as pointed out previously, extensive interfacial contact is a necessary, but not a sufficient, condition for forming strong joints. It is here suggested that the primary function of surface-oxidation techniques is to remove the weak boundary layer. In fact, if surface oxidation alone occurred without removal of the weak boundary layer, only weak adhesive joints would be obtained.

The low mechanical strength of the weak boundary layer, which prevents the formation of strong adhesive joints, can be increased rapidly and dramatically by allowing electronically excited species of rare gases to impinge upon the surface of a large number of polymers. As these metastable and ionic gases come into contact with polyethylene, for example, they cause abstraction of hydrogen atoms. The polymer radicals formed by this process interact to form cross-links and unsaturated groups without appreciable scission of the polymer chain. The cohesive strength of the surface region is increased markedly by the formation of a dense gel matrix, and the wettability of the surface is relatively unaffected. This surface-treatment technique is called CASING (Cross-linking by Activated Species of Inert Gases).[14,15] CASING allows us to form strong adhesive joints to a variety of polymers with conventional adhesives.

Contact time of activated gas with the polymer film of as little as 1 second under relatively mild conditions resulted in greatly improved adhesive joint strength for an epoxy adhesive on polyethylene. Longer contact times were required for polymers such as PTFE. Helium, argon, krypton, neon, and xenon, and even hydrogen and nitrogen, were all effective cross-linking agents, although nitrogen markedly changed the wettability of the surface.

A tenfold or greater increase in lap—shear joint strength was produced by bombardment with activated research-grade helium, although no change in wettability of the polymer was observed. Infrared examination of polyethylene film which has been held at a pressure of 0.05 mm for several hours, then bombarded for 1 hour with activated research-grade helium at 1 mm pressure, and finally kept at a pressure of 0.05 mm for 16 hours at room temperature to permit dissipation of radicals formed during bombardment.[45] showed only the formation of *trans*-ethylenic unsaturation at the surface by attenuated total reflectance techniques. Transmission spectra of treated and untreated films were identical, with no peak attributable to carbonyl or hydroxyl groups, indicating that unsaturation, as was the case for cross-linking, occurs only at or near the surface of the polymer during bombardment. Apparently the improvement in adhesive joints strength by CASING is primarily due to increasing the mechanical strength of the polymer in the surface region through formation of a densely cross-linked matrix.

It has been shown that by treating the surface of materials like PE with activated rare gases, an increase in wettability is not necessary in order to make a strong adhesive joint. In fact, if the surface treatment is carried out in elemental fluorine[36] (no excitation), a strong joint can be made, even though the surface wettability has been decreased, provided the weakness in the surface region is removed.

(ii) Heterogenous nucleation of polymer melts on surfaces

The question arises why the melting of a polymer on to a high-energy surface (i.e. metal, metal oxide) generates a strong joint, provided the polymer has wet the substrate, whilst the free polymer film prepared by conventional techniques requires a surface treatment prior to joining. Apparently, the substrate has a profound effect on the ultimate mechanical properties in the interfacial region of the polymer.

Surface studies on crystallizable polymers (e.g. PE) have ignored, in general, the nature of the nucleating phase (i.e. vapour, solid, or liquid) used to generate the solid polymer. There has been a neglect concerning the details of formation of the polymer melt–nucleating phase interface which, on solidification by cooling, results in a polymer solid–nucleating phase interface.

Extensive heterogeneous nucleation of polyethylene melts on high-energy surfaces results in generation of transcrystallinity in the interfacial region $[(S-L) \rightarrow (S-S)]$ (see Figure 2). It has been observed that there is a variation in the extent of supercooling which may depend upon surface energy and interatomic spacing in the substrate.[46] Effective nucleating agents allow for only small supercooling. Others have concluded that stresses set up at the interface during cooling from the melt are important in determining the subsequent morphology.[46] Vapour phases are apparently ineffective nucleating agents.

When polymers are solidified in contact with a vapour phase, nucleation is precluded at the liquid–vapour interface and is apparently initiated in the bulk. Sufficient supercooling has not occurred at the liquid–vapour interface to nucleate the interfacial region before nucleation occurs in the bulk. Apparently, this is the reason for the lack of a well-defined transcrystalline region when PE is nucleated in

Figure 2. Transcrystallinity in the interfacial region for polyethylene melt cooled in contact with a high-energy surface.

contact with a vapour phase. As crystallization proceeds in the bulk, polymer molecules which cannot be accommodated into the crystal lattice during crystallization are rejected to the interface.

Employing high-energy surfaces for the nucleation of a polymer melt is effective only if sufficient time is allowed for the polymer melt to achieve extensive and intimate contact with the substrate. This is a kinetic requirement. If sufficient time has not been allowed, considerable interfacial voids result and nucleation generally occurs in the bulk. If sufficient time is allowed for spreading to occur, a situation results in which interfacial voids are precluded and nucleation occurs predominately at the S–L interface. The mere presence of a high-energy surface in itself does not ensure that extensive and intimate contact occurs, and a highly nucleated surface region results upon solidification of the polymer melt.

At the high-energy, solid-polymer melt interface a region of substantial mechanical strength is generated by extensive wetting and subsequent nucleation and crystallization of the polymer. During crystallization from the melt, species contributing to the generation of weak boundary layers are rejected from the interface into the bulk. Apparently, at the surface of the metal oxide, numerous crystallization nuclei are formed and spherulites which grow from the nuclei now can propagate in only one principal direction, since growth in the lateral directions is inhibited by neighbouring spherulites. In this way, only very narrowly divergent spherulite sectors develop, which give an overall appearance of a rod-like build-up.

Since strong adhesive joints can be formed by melting on to a high-energy surface, we can inquire whether the surface generated at the high-energy solid-polymer melt interface is amenable to conventional adhesive bonding when the metal is removed. This is indeed the case. It is important to remove the metal by dissolution rather than by peeling which disrupts the surface region of interest. This can be seen by examining the bondability of both the polymer and foil surfaces after peeling the foil from the polymer. In both cases the joint strengths are low. When the foil is peeled, cohesive failure occurs in the polymer, exposing two new surfaces which are not amenable to adhesive bonding.

From the above analysis we can conclude that the weakness in the surface region of many polymers, particularly PE, is not an intrinsic property of the polymer, but is dependent on the manner in which the surface region is formed from the melt. If care is taken in the preparation of the polymer sheet (i.e. nucleated in contact with a high-energy substrate) to prevent any mechanical work upon removal of the polymer from the substrate, then it is possible to prepare this polymer for adhesive bonding without cross-linking the surface region. Solvent extraction (e.g. xylene) of the transcrystalline region showed no evidence of a gel fraction.

6. ENVIRONMENTAL ASPECTS OF ADHESIVE JOINT STRENGTH

Although the final adhesive joint strengths may be extremely high, they are subject, in many cases, to the viscissitudes of environmental factors. In particular, simultaneous exposure to excessive humidity (or bulk water), temperature and stress leads to premature failure of the composites at stress levels well below those attained for dry adhesive joints. Whilst a variety of procedures have been adopted to obviate this premature failure, among them the use of silane coupling agents, it is not clear from the recent literature that this effect has been explained adequately. In this section we shall present a simplified version of the interface and the principal factors involved in environmental failure. These considerations lead to a prognosis for exposure to hostile environments, which has utility in not only adhesive bonding but also in coatings technology.

A proper exposition of environmental factors and their influence on the performance of adhesive joints is best viewed by considering some recent work in adhesive bonding. Collating these views yields a simplified approach which accounts for the large body of the diverse experimental evidence.

Andrews and Kinloch[47] have described an adhesive failure energy θ, which is based on the earlier efforts of Gent et al.[48] The adhesive failure energy is comprised of two major components, namely, (1) the energy to propagate a crack through the unit area of interface in the absence of viscoelastic energy losses, i.e. an 'intrinsic' adhesive failure energy, θ_0, and (2) the energy Ψ, dissipated viscoelastically within the adhesive in the propagating crack, again reference to the unit area of interface. θ_0 should be rate and temperature independent since its value depends upon the nature of the bonding at the interface. The energy Ψ, dissipated viscoelastically, is by definition, rate-temperature dependent.

Further, Andrews and Kinloch[47] show that (see Chapter 3)

$$\theta_0 = \theta \tau_0 / \tau$$

where τ is the cohesive failure energy. Methods are described for obtaining θ, τ, and τ_0 for a variety of rates and temperatures to permit a computation of θ_0. For cases where only secondary forces are operative it is shown[48] that $\theta_0 \sim W_A$, where W_A is the thermodynamic work of adhesion. When covalent bonding occurs across the interface, $\theta_0 > W_A$.[49] Furthermore, θ_0, was found to be independent of test geometry.

Johnson et al.[50] considered an alternative approach based on earlier findings of Roberts[51] and Kendall[52] who used, in adhesion studies, smooth rubber spheres and glass spheres, respectively.

In the absence of surface forces, the contact radius a_0 of the two spheres is given by the generalized Hertz equation[50] (see Chapter 2)

$$a_0^3 = R P_0 / k \tag{1}$$

where k is related to elastic constants of the two spheres and $R = (R_1 R_2)/R_1 + R_2$, where R_1 and R_2 are the radii of the spheres pressed together under a load P_0. Johnson et al. noted that at low loads the contact areas between these bodies were considerably larger than those predicted by the Hertz theory[53] and tended towards a finite value as the load was reduced to zero. Once again, for dry contact, the joint strength was high. However, suspension of the joined rubber spheres in a solution of sodium dodecyl sulphate resulted in agreement with the Hertz theory; the spheres separated. Johnson et al.[50] considered that the total energy of the system U_T is comprised of three main terms, the stored elastic energy U_E, the mechanical energy is the applied load U_M, and the surface energy U_S. Their analysis leads to the following equilibrium situation which includes surface effects

$$a^3 = \frac{R}{K} \{P + 3\gamma\pi R + \sqrt{[6\gamma\pi RP + (3\gamma\pi R)^2]}\} \tag{2}$$

When the interfacial tension is zero, equation (2) reverts to equation (1). At zero applied load the contact area is finite and given by

$$a^3 = R(6\gamma\pi R)/K \tag{3}$$

Johnson et al. show that separation of the spheres will occur when

$$P = -\tfrac{3}{2}\gamma\pi R \tag{4}$$

which is independent of the elastic modulus. It is inherent in the approach of Johnson et al. that no interfacial bonding other than secondary valence forces are operative. It would appear that there is a strong similarity between θ_0 and P for the situation where $\theta_0 \sim W_A$.[47] The analysis of Johnson et al. implies a Griffith approach to the propagation of a brittle crack.[54] A clear analysis of the effect of the surfactant is somewhat lacking in the approach of Johnson et al.

A more detailed look at environmental factors pertaining to adhesive joint strengths was undertaken by Owens[55] and Gent and Schultz[48]. Owens[55] demonstrated that in the presence of a particular liquid phase, if the thermo-dynamic work of adhesion of a composite A/B is near zero ($W_{AB}^L \sim 0$), the adhesive joint would be unstable and, delaminate in a reasonably short period of time. In the absence of a liquid phase, the work of adhesion is large (corresponding to large θ_0 and P) and $W_{AB} > 0$. Under these conditions, delamination may be retarded for an indefinite interval. Excluding chemisorption,[56] (i.e. where $\theta_0 \gg W_{AB}$) and interdiffusion,[57] it appears that at least two criteria are necessary to induce premature failure in the presence of stress and a fluid phase.[55]

(a) The liquid must be immiscible with the polymer or other members of the composite structure, since miscibility of the liquid with one or more members of the composite would sufficiently plasticize that member to modify the modulus and possibly preclude failure.
(b) The liquid or fluid phase must fully interact (i.e. in a surface-chemical sense) with either one or both of the members of the composite. For example, in a composite consisting of PE and another member, to facilitate environmental failures, the fluid phase should react principally via dispersion forces ($\gamma_{LV} \sim \gamma_{LV}^d$).

A simple theory may be evolved which considers the features of the foregoing analysis.[58] The work W per unit length required to separate an adhesive joint A–B may be represented as

$$W = W_S + W_P + W_E + W_T \tag{5}$$

where W_S is the reversible work to form new surfaces, W_P the work expended in plastic flow, W_E the loss in stored elastic energy, and W_T is associated with the work to desorb chemisorbed sites and to break tie molecules.[59] Employing a Griffith approach we may represent the work required to form an elliptical crack of length $2a$ as

$$W = 2aW_{AB} + W_P^A(a) + W_P^B(a) - \frac{\pi a^2 \sigma_0^2}{E_e} + \omega_T a \tag{6}$$

where W_{AB} is the thermodynamic work of adhesion, $W_P(a)$ a function of a, σ_0 the applied stress, E_e the effective Young's modulus, and ω is associated with the work to desorb sites and fracture primary valence bonds. At a critical stress σ_f and critical crack length $2a_s$, beyond which crack propagation proceeds spontaneously,

$$\frac{\partial W}{\partial a} \bigg|_{\substack{\sigma_0 = \sigma_f = 0 \\ a = a_s}}$$

and

$$2W_{AB} + W'^{(A)}(a) + W'^{(B)}(a) + \omega_T'(a) - \frac{2\pi a_s \sigma_f^2}{E_e} = 0 \tag{7}$$

In the presence of an immiscible liquid phase the above argument leads to

$$2W_{AB}^L + W_p'^{(A)}(a) + W_p'^{(B)}(a) + W_T'(a) - \frac{2\pi a_L (\sigma_f^L)^2}{E_e^L} = 0 \tag{8}$$

If the liquid is immiscible with either member of the composite A—B then $E_e = E_c^L$ and $a_S = a_L$. Combining equations (7) and (8) yields

$$\frac{(\sigma_f^L)^2}{(\sigma_f)^2} = \frac{2W_{AB}^L + W_p'^{(A)}(a) + W_p'^{(B)}(a) + \omega_T'(a)}{2W_{AB} + W_p'^{(A)}(a) + W_p'^{(B)}(a) + \omega_T'(a)} \tag{9}$$

Since

$$W_{AB}^L = W_{AB} + (2\gamma_{LV} - W_{AL} - W_{BL}) \tag{10}$$

We obtain, by assuming that W_{AB}, $W_p'^{(A)}(a)$, $W_p'^{(B)}(a)$ and $\omega_T'(a)$ are independent of the liquid phase, the following:

$$(\sigma_f^L/\sigma_f)^2 = 1 - 2k_{AB}(W_{AB} - W_{AB}^L) \tag{11}$$

If we ignore the work expended in plastic flow and desorbing chemisorbed sites equation (9) yields

$$(\sigma_f^L/\sigma_f)^2 = W_{AB}^L/W_{AB} \tag{12}$$

which was discussed earlier.[58] Following the arguments of Owens,[55] one can obtain W_{AB}^L and W_{AB} by considering the various interfacial tensions.

To obviate the influence of the liquid phase, several choices appear plausible:

(a) To modify the interfacial tension so that $W_{AB}^L \sim W_{AB}$,
(b) Induce chemisorption and covalent bonding across the interface to minimize the work of adhesion contribution. The joint strength will then be governed by terms consisting of W_P and ω_T. This is apparently what Andrews and Kinloch[47] find when $\theta_0 \gg W_{AB}$.

It would appear that the action of adhesion modifiers or promoters may serve the important function of modifying the interfacial tension to maintain effectively a similar value for W_{AB} and W_{AB}^L. Coupling agents which invariably lower the surface energy of the substrate appear to protect that surface from an incursion of humidity or a water environment. The work of Owens[55] illustrates quite pointedly that surface modification of a polymer surface may improve the dry strength of the composite, but may contribute to the poor environmental behaviour of the surface treated adherend if the modified surface interacts more favourably (in a surface chemical sense) with the liquid phase. The results of Johnson et al. for the spontaneous delamination of rubber spheres when exposure to a dilute aqueous solution of sodium dodecylsulphate agree favourably with the observation of Owens. In Table 1 are shown the data for γ_{LV}^d and γ_{LV}^p values for a variety of surface-active agents.[55] These data clearly show that sodium dodecyl sulphate behaves more like a non-polar liquid and consequently would interact more favourably with the rubber surface. If γ_S of the rubber is 35 erg/cm^2 [50] and γ_S^d is assumed to be 25 erg/cm^2 it can be shown easily how the results of Johnson et al. revert to the classical Hertz theory. Since $W_{coh} = 70$ erg/cm^2 we must compute W_{coh}^L for the rubber—sodium dodecyl sulphate solution system. As shown by Johnson et al.

$$P = -(3/2)\gamma\gamma\pi R$$

Table 1

Properties of solutions

	Concn (%)	γ_{IF}^{*}	γ_{LV}^{d}	γ_{LV}^{P}	γ_{LV} (erg/cm^2)
Water			21.8	51.0	72.8
Sodium n-octyl sulphate	3.5	14.8	26.6	14.8	41.4
Sodium n-decyl sulphate	1.0	11.5	28.0	11.5	39.5
Sodium n-dodecyl sulphate	0.5	8.2	29.0	8.2	37.2
Sodium n-tetradecyl sulphate	0.2	8.6	28.1	8.6	36.7
Sodium n-hexadecyl sulphate	0.05	7.2	28.8	7.2	36.0
Sodium diisobutyl sulphosuccinate	1.0	28.0	16.6	26.7	43.3
Sodium diisoamyl sulphosuccinate	1.0	2.0	23.7	1.9	25.6
Sodium di(ethylhexyl) sulphosuccinate	1.0		1.2	24.7	1.1	25.8

*Interfacial tension of solution versus n-hexadecane, γ_{LV} = 27.4 erg/cm^2.

for the rubber joint. Introduction of a liquid phase modifies the above equation to

$$P = -(3/2)\gamma_{SL}\pi R \qquad (13)$$

where γ in equation 9 is actually γ_{SV} and in equation 13, γ_{SL}. For the adherence of similar materials $W_{coh} = 2\gamma_{SV}$. In the presence of a liquid phase $W_{coh}^{L} = 2\gamma_{SL}$. To compute γ_{SL} we use the modified Fowkes expression[60]

$$\gamma_{SL} = \gamma_{SV} + \gamma_{LV} - 2[(\gamma_{LV}^{d}\gamma_{SV}^{d})^{\frac{1}{2}} + (\gamma_{LV}^{P}\gamma_{SV}^{P})^{\frac{1}{2}}] \qquad (14)$$

and the data in Table 1. Upon substitution of the appropriate values, in equation 13 we find $W_{coh}^{L} \simeq 0$, indicating a tendency for the composite to separate spontaneously. It may have been fortuitous for Johnson *et al.* to select SDS as a surfactant for, if others had been used where the polar interaction is in excess of the dispersion interaction, W_{coh}^{L} would remain large and positive. In fact if water were used, $W_{coh} \simeq W_{coh}^{L}$.

It is evident from the data and the arguments in this chapter that the interfacial tensions at the metal oxide—polymer interface and the interaction with a humid environment may be responsible for premature delamination of the composite under low levels of stress. Modification of the interfacial tensions by judicious use of particular silicone coupling agents, which not only chemisorb on to the metal oxide but may also modify the polymer surface by cross-linking, appears to be a valid explanation for the observed wet strength of the silicone-treated composite structure. Clearly, this approach may be extended to other combinations of adherends and guide one in a proper selection of surface treatments. Although the initial joint strengths of surface-treated adherends ($\sigma_0 \gg W_A$) may exceed those where only interfacial forces are operative ($\sigma_0 \sim W_A$), in the presence of an adverse environment, as shown by Owens,[55] these strong joints may deteriorate rapidly ($\sigma_0^{L} \ll W_A$). What appears to be a troublesome notion — that high-energy surfaces when surface treated by adhesion promotors to yield low-energy surfaces (i.e. typically with glass, coupling agents lower γ_c to low levels) often yield more

232

permanent joints (higher wet strength) — appears to be resolvable in the light of the arguments presented in this report. Although the same degree of wetting is achieved for both treated and non-treated substrates, it is clear that the interfacial structure, for the same extent of interfacial contact area, is of prime importance.

REFERENCES

1. E. R. Houwink and G. Salomon, eds., *Adhesion and Adhesives*, Vol. 1, Elsevier, New York, 1965.
2. G. Salomon, in Houwink and Salomon[1], pp. 1–28.
3. *Ibid.*, pp. 29–52.
4. S. S. Voyutskii, *Adhesives Age*, **5**, 30 (1962).
5. B. V. Derjaguin and N. A. Krotova, *Dokl. Akad. Nauk. SSSR*, **61**, 849 (1948).
6. S. S. Voyutskii and Yu. L. Margonlin, *Usp. Khim.*, **18**, 449 (1949); *Rubber Chem. Technol*, **30**, 531 (1957)
7. S. S. Voyutskii and V. M. Zamazii, *Kolloidn. Zh.*, **15**, 407 (1953); *Rubber Chem. Technol*, **30**, 544 (1957).
8. S. S. Voyutskii and B. V. Shtarkh, *Kolloidn. Zh.*, **16**, 3 (1954); *Rubber Chem. Technol.*, **30**, 548 (1957).
9. N. A. de Bruyne, *Aircraft Engr.*, **18**, 53 (Nov. 1939).
10. L. H. Sharpe and H. Schonhorn, *Advan. Chem. Ser.*, **43**, 189 (1964).
11. H. Schonhorn and L. H. Sharpe, *J. Polymer Sci.* **B2**, 719 (1964).
12. *Ibid.*, **A3**, 3087 (1965).
13. L. H. Sharpe, H. Schonhorn, and C. J. Lynch, *Intern. Sci. Technol.*, **26**, 1964 (April).
14. R. H. Hansen and H. Schonhorn, *J. Polymer Sci.*, **B4**, 203 (1966).
15. H. Schonhorn and R. H. Hansen, *J. Appl. Polymer Sci.*, **11**, 1461 (1967).
16. M. Rosoff, *Physical Methods in Macromolecular Chemistry*, ed. B. Carroll, M. Dekker, New York, 1969, p. 1.
17. R. Peyser, D. J. Tutas, and R. Stromberg, *J. Polymer Sci.*, A-1, **5**, 651 (1967).
18. W. C. Hamilton, *Appl Polymer Symposium*, No. 19, 105 (1972) and reference cited therein.
19. N. J. Harrick, *J. Phys. Chem.*, **64**, 110 (1964).
20. J. P. Luongo and H. Schonhorn, J. Polymer Sci., A-2, **6**, 1649 (1968).
21. S. Krimm, C. Y. Liang, and G. B. B. M. Sutherland, *J. Chem. Phys.*, **25**, 549 (1956).
22. B. L. Henke, *Advances in X-ray Analysis*, **13**, 1 (1970).
23. D. W. Dwight and W. M. Riggs, *J. Colloid and Interface Sci.*, in press.
24. D. M. Hercules, *Anal. Chem.*, **44** (5), 106 (1972); W. M. Riggs and R. P. Fedchenko, *Amer. Lab.*, **4** (11), 65 (1972).
25. C. G. S. Collins, A. C. Lowe and D. Nicholas, *Europ. Poly. J.*, **9**, 1173 (1973).
26. D. T. Clark and W. J. Feast, *J. Macromol. Sci.-Revs. Macromol. Chem.*, **C12** (2), 191–286 (1975).
27. G. Kraus, *Adv. in Polymer Sci.*, **8**, 155 (1971).
28. Y. S. Lipatov and F. G. Fabuliak, *Vysokomolekul Soedin*, **10**, 1605 (1968); *Academia Nauk, SSR*, **10A**, 1605 (1968).
29. R. F. Roberts and H. Schonhorn, *ACS Polymer Preprints*, **16** (2), 146 (1975).
30. H. Schonhorn and F. W. Ryan, *J. Adhesion* **1**, 43 (1969).
31. K. Hara and H. Schonhorn, *J. Adhesion*, **2**, 100 (1970).
32. F. K. Borisova, G. A. Galkin, A. V. Kiselev, A. Y. Korolev, and V. I. Lygin, *Kolloidnyi Zhurnal.*, **27**, 320 (1965).
33. D. H. Kaelble and E. H. Cirlin, *J. Polymer Sci.*, A-2, **9**, 363 (1971).

34. D. R. Fitchmun and S. Newman, *J. Polymer Sci.*, A-2, **8**, 1545 (1970).
35. D. J. Angier, in *Chemical Reactions of Polymers*, ed. E. M. Fettes (*High Polymers*, Vol. 19), Interscience, New York, 1964, p. 1009.
36. H. Schonhorn and R. H. Hansen, *J. Appl. Polymer Sci.*, **12**, 1231 (1968).
37. D. A. Olsen and A. J. Osteraas, *J. Polymer Sci.*, A-1, **7**, 1913, 1921, 1927 (1969).
38. A. J. Osteraas and D. A. Olsen, *J. Appl. Polymer Sci.*, **13**, 1537 (1969).
39. D. A. Olsen and A. J. Osteraas, *J. Appl. Polymer Sci.*, **13**, 1523 (1969).
40. J. R. Hollahan, B. B. Stafford, R. D. Falb, and S. T. Payne, *J. Appl. Polymer Sci.*, **13**, 807 (1969).
41. R. I. Leininger, R. D. Falb, and G. A. Grode, *Ann. N.Y. Acad. Sci.*, **146**, 11 (1968).
42. J. R. Caldwell and W. J. Jackson, Jr., *J. Polymer Sci.*, C, **24**, 15 (1968).
43. H. E. Wechsberg and J. B. Webber, *Mod. Plastics*, **36**, 101 (1959).
44. J. A. Boxler, S. P. Foster, and E. E. Lewis, in *Abstracts 132nd Meeting Amer. Chem. Soc., New York, Sept., 1957*, Vol. 17, No. 2, p. 58.
45. L. A. Wall and R. B. Ingalls, *J. Polymer Sci.*, **62**, 56 (1962); *J. Chem. Phys.*, **35**, 370 (1961).
46. A. Sharples, *Polymer Crystallization*, St. Martin's Press, New York, 1966, pp. 21–22.
47. E. H. Andrews and A. J. Kinloch, *Proc. R. Soc. (Lond.)*, **A332**, 385 (1973); E. H. Andrews and A. J. Kinloch, *ibid.*, 401 (1973).
48. A. N. Gent and J. Schultz, Paper presented at the 162nd Meeting, American Chemical Society, Organic Coatings and Plastics Chemistry, September 1971; *Preprints*, **31**, 113 (1971); A. N. Gent and J. Schultz, *J. Adhesion*, **3**, 281 (1972).
49. G. E. Koldunovich, V. G. Epshtein, and A. A. Chekanova, *Sov. Rubb. Technol.*, **29**, 22 (1970); E. B. Trostyanskaya, G. S. Golovkin, and G. V. Komarov, *Sov. Rubb. Technol.*, **25**, 13 (1966).
50. K. L. Johnson, K. Kendall and A. D. Roberts, *Proc. R. Soc. (Lond.)*, **A324**, 301 (1971).
51. A. D. Roberts, Ph.D. dissertation, Cambridge University, Cambridge, 1968.
52. K. Kendall, Ph.D. dissertation, Cambridge University, Cambridge, 1969.
53. H. Hertz, *Miscellaneous Papers*, Macmillan, London, p. 146
54. E. H. Andrews, *Fracture in Polymers*, American Elsevier, New York, 1968.
55. D. K. Owens, *J. Appl. Polym. Sci.*, **14**, 1725 (1970).
56. D. D. Eley, *Adhesion and Adhesives, Fundamentals and Practices*, Wiley, New York, 1954.
57. S. S. Voyutskii, *Autohesion and Adhesion of High Polymers*, Wiley–Interscience, New York, 1963.
58. H. Schonhorn and H. L. Frisch, *J. Polym. Sci., Polym. Physics Ed.*, **11**, 1005 (1973).
59. H. D. Keith, F. J. Padden, and R. G. Vadimsky, *Science*, **150**, 1026 (1965); *J. Polym. Sci.*, **4**, 267 (1966).
60. F. M. Fowkes, in *Surfaces and Interfaces, I. Chemical and Physical Characteristics*, ed. J. J. Burke, N. L. Reed, and W. Weiss, Syracuse University Press, Syracuse, New York, 1967, p. 197.

11

Modification of Polymer Surfaces by Silicone Technology

P. J. Clark

Dow Corning Ltd, Barry, Glamorgan

1. INTRODUCTION

It is often necessary, for a variety of reasons, for the surface of an article fabricated from a polymeric material to be modified. Silicones, in one form or another, have been used to perform some of these modifications. Synthetic fibres can be lubricated with silicone fluids to reduce their coefficients of friction and aid subsequent manufacturing processes, while finished textiles can be treated with silicone polymers to impart water repellancy. Silane coupling agents are used to improve the adhesion of polymeric resins to glass fibres, whilst low-viscosity silicone fluids aid the release of plastics and rubbers from moulds. In most cases the silicone is a separate, sometimes mobile, layer on the surface of the polymer, and as such is prone to removal by solvent treatment or abrasion.

The technology described in this chapter deals with the modification of polymer surfaces by silicone polymers which are incorporated either at the polymerization stage of the organic polymers or during the forming process, whether that be extrusion, injection moulding, or any of the less common means of manufacturing products from organic polymers. Amongst the silicones which have been used are both reactive and non-reactive polymers and these have been evaluated in both thermoplastic and cross-linkable organic polymers.

2. NON-REACTIVE ADDITIVES

(i) Addition to thermoplastics

Initially, it was thought necessary[1,2] to give the silicone some partial compatibility with the polymer by using a block copolymer consisting of polydimethylsiloxane and the organic resin to be modified. For this purpose, block copolymers of polydimethylsiloxane with polystyrene, polyethylene, and poly-isoprene were prepared and incorporated in the relevant polymers. Examination of a compression-moulded disc of polystyrene containing 1% of a block copolymer of polydimethylsiloxane and polystyrene showed a reduction in the critical surface tension of wetting, σ_c, from 36.5 mN m^{-1} for the unmodified polymer to 24.5 mN m^{-1} for the modified version. Even after severe abrasion of the surface the value of σ_c remained unchanged, owing to the even dispersion of the block copolymer throughout the polystyrene matrix.

To manufacture a different copolymer for each organic polymer to be modified is impracticable. Some are relatively easy to make, for example those with polystyrene, whereas others with condensation polymers such as polyesters are rather more difficult to prepare. It was decided, therefore, to examine the behaviour of polydimethylsiloxane homopolymers in a range of organic polymers. This work showed that, provided the molecular weight of the polydimethylsiloxane is sufficiently large, corresponding to a viscosity of at least 12 500 centistokes (cS) at 25°C, then although the silicone is incompatible with organic polymers it can be made to form a stable dispersion in the polymer matrix. Moreover, the silicone does not migrate to the surface through the polymer matrix, even at temperatures above the glass transition temperature.[3] Figure 1 shows the effect of different amounts of polydimethylsiloxane, designated Dow Corning® 200 fluid, 30 000 cS, on the coefficient of friction of polystyrene.

These values were obtained from a machine developed by Dow Corning called the LFW 6 tester, in which the raised lip of a moulded disc of polymer is rotated against a steel surface. The instrumentation allows calculation of the coefficient of friction at a range of velocities and loadings, PV limits, and wear rates. An organic lubricant, carnauba wax, has little effect on the coefficient of friction. However, 0.5% polydimethylsiloxane reduces the coefficient of friction considerably, especially at low velocity. More important is the observation that 2% silicone addition not only results in a very low coefficient of friction but also maintains that low value for a wide range of velocities. A similar effect is seen when the coefficient of friction is plotted against pressure.

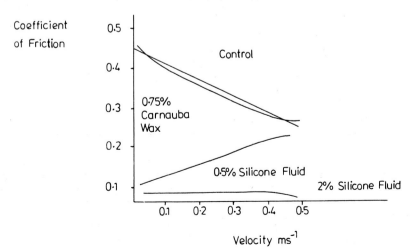

Figure 1. Coefficient of friction of polystyrene.

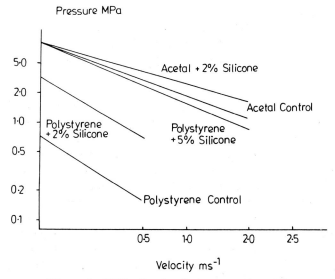

Figure 2. PV limits of polystyrene and acetal.

The effect of adding silicone fluid has just as dramatic an effect on PV limits (Figure 2). An addition of silicone fluid has a much greater effect on the PV limits of polystyrene, a polymer with a fairly high control coefficient of friction, than on those of acetal, a polymer well known for its low coefficient of friction and performance as an engineering material.

Table 1 Wear rate of polystyrene

Weight % silicone	Wear rate[a] (mm hour^{-1})
0	10.54
0.5	4.06
1.0	0.64
2.0	0.13
5.0	0.03

[a]0.51 m s^{-1} and 1 MPa.

Table 2 Physical properties of polystyrene

Conc. of silicone fluid	Tensile stress at yield (MPa)	Tensile stress at break (MPa)	Elongation at break (%)	Tensile modulus (GPa)	Deflection temperature (K)	Vicat softening point (K)
Control	38.9	38.9	1.5	2.9	374	379
0.5%	39.0	38.9	1.5	2.9	373	379
1%	38.8	36.1	2.0	2.9	373	379
2%	34.9	28.7	4.0	2.8	373	379

The important point to emerge is that polystyrene, a relatively inexpensive plastic, can be made to perform as well as an expensive material such as acetal in most of its applications by the addition of only 2% of polydimethylsiloxane. This is further shown in Table 1, showing the effect of polydimethylsiloxane concentration on wear rates; the optimum level of silicone is between 1% and 2%.

The presence of the additive has little effect upon other physical properties. Table 2 shows tensile and thermal properties of general-purpose polystyrene. Very

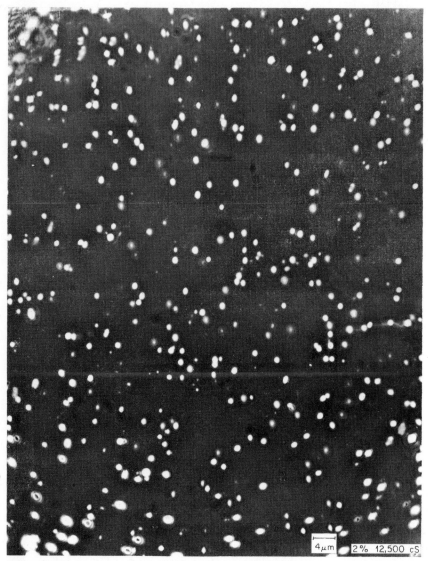

Figure 3. Dispersion of droplets of polydimethylsiloxane (approximately 1–5 μm in diameter) in polystyrene.

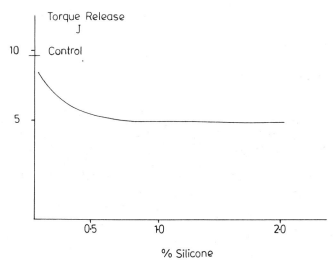

Figure 4. Torque release of polystyrene.

little change is observed in either tensile stress or elongation at break and, unlike most organic additives, the silicone has no effect upon deflection temperature or Vicat softening point. This is to be expected when it is remembered that polydimethylsiloxane is incompatible with polystyrene and is present as a dispersion of droplets approximately 1–5 μm in diameter (see Figure 3).

The addition of polydimethylsiloxane to thermoplastics also has a beneficial effect on release properties. Figure 4 is a plot of release torque for a bottle screw cap, moulded on a transfer press, against percentage silicone. The optimum level of silicone addition is between 0.5% and 1%, when the most difficult release problems are solved without use of external release sprays. This is important because silicone, when used as an internal release agent, does not deleteriously affect subsequent finishing operations on the mouldings. Thermoplastics can therefore be painted, plated, welded, etc. in exactly the same manner as unmodified materials.

(ii) Addition to cross-linkable polymers

So far we have discussed the addition of high-viscosity silicone fluids to thermoplastics. Similar effects are observed when polydimethylsiloxane is added to cross-linkable organic resins which are in the form of latices. In this case, however, silicone emulsions are used as the means of incorporation. Cross-linkable organic resin latices are used as treatments for textiles and as adhesives, both in general usage and as binders for fibres in the production of non-woven textiles. Table 3 shows the effect of addition of a silicone fluid emulsion, Dow Corning® 1111, on the abrasion resistance of a carboxylated styrene–butadiene resin latex. A modified

Table 3 Abrasion resistance of carboxylated styrene—
butadiene resin

	Cycles to form a wear scar[a]	
	Dow 893	Dow 221
Control	< 100	< 20
2% Silicone (DC 1111)	> 300	> 300

[a]Evaluated on a modified crockmeter.

crockmeter was used to obtain the results from cured, cast films of the resin. An increase in abrasion resistance of at least a factor of three can easily be achieved with this type of system. This leads to improved performance of non-woven textiles and coating systems based on these organic latices.

High-viscosity silicone fluids are difficult to add to liquid organic resins such as polyurethane prepolymers. Because of the lack of compatibility of poly-dimethylsiloxane fluids with liquid organic resins two separate phases are present in the matrix. However, unlike the thermoplastic system in which the silicone remains dispersed in a rigid matrix, the organic resin remains sufficiently liquid for long enough prior to cross-linking to allow the silicone droplets to coalesce and form a separate layer from the organic resin. This problem can be overcome by use of a special emulsifier which aids the formation of a stable dispersion of the silicone fluid in a liquid resin. A combination of emulsifier and fluid exist in the form of Dow Corning® Q2-3107 fluid.

Table 4 shows the effect of adding Q2-3107 on the coefficient of friction obtained from evaluation of a cast film on a modified crockmeter. In this test a steel ball reciprocates across the surface of a flat test piece which is mounted on a floating bed connected to a load cell. The longitudinal force transmitted to the test piece is continuously recorded and may be used to calculate the coefficient of friction. The test is usually run until the surface wears through creating a wear scar, or for a minimum of 120 minutes after a steady state is achieved.

The control sample indicates that, in a very short running time, the coefficient of friction increases to the extent that the surface begins to break down and form a wear scar. The incorporation of Q2-3107 (2%) allows the film to withstand the test

Table 4 Coefficient of friction[a] of polyurethane

	Time (minutes)	Coefficient of friction
Control[b]	0	0.38
	10	0.88
	15	1.1 wear scar
2% Q2-3107	0	0.10
	180	0.10 no wear scar

[a]Evaluated on modified crockmeter.
[b]Cyanaprene A-9, Moca Cure.

Table 5 Physical properties of polyurethane

	Tensile stress at break (MPa)	Elongation at break (%)	Tensile stress (MPa) at			Tear strength (mg m^{-1})
			100%	200%	300%	
Control	46.4	590	6.3	8.2	11.6	8.5
2% Q2-3107	39.9	550	6.7	8.6	12.1	10.0

conditions more than 10 times longer than the unmodified film with no noticeable change in appearance or coefficient of friction, which itself is only a quarter of the initial value of the unmodified resin. This addition of silicone fluid has little effect upon tensile physical properties, as can be seen from Table 5. Surface adhesion of a polyurethane, Mobay Multrathene® F-242 with a 1,4 butane diol cure, cast on a steel plate, was determined using a Kiel peel tester. In this test half-inch strips are pulled at 180° to the surface at a velocity of 0.5 cm sec^{-1}. The average adhesion value for the control sample was 0.9 kg cm^{-1} and for the sample containing 2% Q2-3107 it was 0.8 kg cm^{-1}. The difference between these two values is within the limits of accuracy of the machine and so it can be said that there is essentially no difference in adhesion.

The ability of the liquid polyurethane containing Q2-3107 to wet a surface and spread is much improved. The surface tension of Adiprene® L-100 is reduced from 44 mN m^{-1} to 29 mN m^{-1} for the addition of 2% of the silicone. This results in more uniform thin coatings and much improved degassing of cast articles and films. There are obvious applications for this type of additive to improve the abrasion resistance of conveyor belting and chute linings and to reduce the wear rates and improve lubrication of gears, bushes, and shaft seals made from polyurethanes. Surface coatings may also be made more abrasion resistant.

3. REACTIVE ADDITIVES

All of the silicones discussed so far can be classed as non-reactive additives. It is possible, when dealing with cross-linkable organic resins, to improve the permanence of the additives by using a silicone capable of reacting with the resin during the curing process. This is of most importance when the organic resin is likely to come into contact with solvents during its use as a fabricated part. A typical example of this is the textile resin which may be used as a top finish or as a binder for non-woven textiles. The use of such textiles may involve constant contact with solvents when used, for example, as a filtration medium, or intermittent contact when, in the form of a garment, dry-cleaning processes are involved.

The organic polymers of interest for surface modification are listed in Table 6. With the exception of the polyesters and epoxies these can all be cured by reaction with carbinols such that the carbinol forms a cross-link point. Polyesters and polyurethanes can be prepared by condensation and addition reactions, re-

Table 6 Cross-linkable organic polymers

Polymer	Reactive group(s)	Reactive repeat group
Polyurethane	$-NCO$	$-NH-\underset{\underset{O}{\parallel}}{C}O-$
Epoxy	$-CH-CH_2$ $\diagdown O \diagup$	
Polyester	$-CH=CH_2$	$-\underset{\underset{O}{\parallel}}{C}O-$
Acrylic	$-COOH$ $-NHCH_2OH$	
Styrene–butadiene	$-COOH$	
Formaldehyde-based resins	$-NHCH_2OH$ $-CH_2OH$	

spectively, incorporating the carbinol functional polymer as an integral part of the linear polymer. Epoxies can be cured by reaction with amines or acid anhydrides.

For these reasons the reactive silicone additives were chosen to have either carbinol or amino functionality, and these are illustrated in Table 7. All of these

Table 7 Reactive silicone polymers

		Average OH equal weight
Q4-3667	$HOCH_2$ ————————— CH_2OH	1200
Q4-3557		400
DC 1248		2000

			Approx. weight % $-NHCH_2CH_2NH_2$
DC 531	Reactive group	$-\overset{\mid}{\underset{\mid}{Si}}(CH_2)_3NHCH_2CH_2NH_2$	1.4
DC 536	Reactive group	$-\overset{\mid}{\underset{\mid}{Si}}(CH_2)_3NHCH_2CH_2NH_2$	4.6

products can be incorporated at the curing stage and, depending upon their functionality, may or may not affect the cross-link density of the system. All, however, will have the effect of flexibilizing to some degree the resin in which they are incorporated.

It is also possible to incorporate them into prepolymers. Q4-3667 can be reacted at the initial polymerization stage of, for example a polyurethane, replacing some of the organic diol, to form linear prepolymers which may be isocyanate terminated. The other products can be added at the same stage, although as their functionality increases it becomes difficult to prepare useful prepolymers unless they are used from solution or dispersion in an organic solvent.

Although these additives may be used at fairly low levels of addition, they still show a marked influence on the surface properties of polymers in which they are incorporated. Table 8 shows the effect of two levels of addition of silicone polycarbinol copolymers on the water-contact angle of two polyurethane coating polymers, one based on a Dow Chemical polyether, the other on a Hooker Chemical polyester. At a 1% addition level a marked increase in the contact angle is observed, indicating an increase in hydrophobicity to a level nearly that of pure polydimethylsiloxane. This value is not altered by increasing the level of silicone copolymer present. It would be expected that DC 1248 and Q4-3557 should increase the hydrophobicity of an organic resin, both copolymers being based on polypropylene oxide and thus being water insoluble. The similar effect of Q4-3667, a water-soluble copolymer based on polyethylene oxide suggests that when the

Table 8 Water-contact angle of modified polyurethanes

	Water-contact angle Polyether[a] base	Polyester[b] base
Control	$72°$	$67°$
1% DC 1248	$97°$	—
4% DC 1248	$96°$	$96°$
4% Q4-3557	—	$95°$
4% Q4-3667	—	$93°$

[a] Dow Chemical Voranol P-2000 and CP-260.
[b] Hooker Chemical S-100-90.

Table 9 Coefficient of friction of modified Polyurethanes

	Coefficient of friction Polyether[a] base	Polyester[b] base
Control	1.20	1.25
1% DC 1248	0.70	0.83
4% DC 1248	0.70	0.53
4% Q4-3667	—	0.64

[a] Dow chemical Voranol P-2000 and CP-260
[b] Hooker Chemical S-100-90.

Table 10 Permeability of modified polyurethanes to water vapour

	Permeability (mg cm^{-2} mm^{-1} hour^{-1})	
	Polyether[a] base	Polyester[b] base
Control	0.80	17.6
1% DC 1248	0.83	—
4% DC 1248	1.00	7.5
4% Q4-3557	—	14.6
4% Q4-3667	—	25.3

[a]Dow Chemical Voranol P-2000 and CP-260
[b]Hooker Chemical S-100-90

copolymer is chemically bound into the urethane polymer only the silicone portion is presented to the polymer—air interface. Little difference is observed between polyurethanes based on the polyether or the polyester.

These additives affect the dynamic coefficient of friction (Table 9). Similarly, reduction in coefficient of friction achieved using reactive silicone additives is not as great as that achieved with the non-reactive silicone, Q2-3107, but is significant. Unlike non-reactive silicones, however, the reactive additives, chemically bound into the matrix, can modify bulk properties of the polyurethane. Table 10 shows their effect on the permeability of cast polyurethane films; the polyether-based polymer is a top coating finish deposited on leather.

Differences in the permeability are observed when DC 1248 is incorporated, depending upon whether the polyurethane is polyether or polyester based. In the former case, the permeability is increased 25%, whereas in the latter it is reduced 57% for a 4% addition of silicone. There is also a large difference of effect when comparing one silicone additive with another, from a 57% reduction to a 44% increase in permeability in the same polyurethane. These figures suggest that the permeability may depend to a large extent on the degree of compatibility of the

Table 11 Tensile properties of modified polyurethane[a]

	Level of addition (%)	Control	DC 1248	Q4-3557	Q4-3667
Tensile stress at break (MPa)	1	31.2	34.8	37.4	30.6
	2		25.8	34.4	26.9
	4		28.8	39.9	33.5
Elongation at break (%)	1	930	1000	1020	980
	2		1020	1060	960
	4		990	1120	1030

[a]Hooker Chemical S-100-90.

Table 12 Hydrolytic stability of modified polyurethane[a]

	4% silicone after 1 hour water boil			
	Tensile stress at break	% change over dry control	Elongation at break (%)	% change over dry control
Control	13.6	−56%	710	−24%
DC 1248	22.3	−29%	940	+1%
Q4-3557	20.9	−33%	1070	+15%
Q4-3667	32.6	+4%	1100	+18%

[a]Hooker Chemical S-100-90.

silicone copolymer with the polyurethane and the functionality of the reactive additive.

Table 11 shows the effect of reactive silicone additives on the tensile properties of a polyester-based polyurethane. The tensile stress at break shows both increases and reductions, depending upon the level and type of silicone additive. Elongation at break, however, shows an improvement in all cases, 4% of Q4-3557 giving a 20% increase.

Polyurethanes generally have poor hydrolytic stability, most important for applications in which the polymer is used at high surface area to weight ratios, such as paints and textile coatings. Table 12 shows that, after a 1 hour boil in water, the tensile strength of a polyester-based polyurethane is reduced by 56% and the elongation at break by 24%. Addition of 4% of a reactive silicone improves the hydrolytic stability dramatically, in each case giving either no change or an increase in the elongation at break. The best result is achieved with Q4-3667 which gives a 4% increase in tensile strength and an 18% increase in elongation at break. Modification with these additives could therefore have a very beneficial effect on

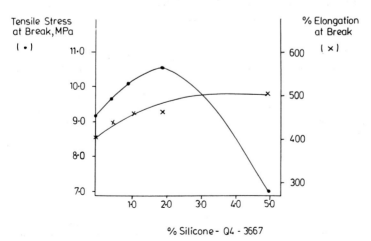

Figure 5. Tensile properties of modified acrylic (Rohm & Haas Rhoplex HA-12.

Table 13 Water contact angle and permeability of modified acrylic[a]

	Contact angle	Permeability ($mg\ cm^{-2}\ mm^{-1}\ hour^{-1}$)
Control	76°	46
1% Q4-3667	71°	60
5% Q4-3667	12°	62

[a]Rhom & Hass Rhoplex® HA-8.

some applications of polyurethanes by increasing their useful life under wet or humid conditions.

The reactive silicone polycarbinol, Q4-3667, can also be incorporated in acrylic resins to modify surface and bulk properties. The acrylic resins which have been studied are latices of reactive acrylics used as binders in the non-woven textile industry. The effect of adding Q4-3667 to one such resin, Rhoplex® HA-8 of Rohm & Hass, is shown in Table 13, where the water-contact angle and permeability of a cast film are given. Unlike polyurethanes, the water-contact angle is virtually unchanged for the addition of 1% and drops to 12° when 5% of the silicone is present. This is probably due to the lower reactivity of Q4-3667 with the acrylic resin than with the polyurethane, where it is incorporated in the prepolymer. This would lead to concentration of the silicone copolymer at the surface of the cast film and as Q4-3667 is water soluble would give rise to the low water-contact angle. The permanence of this effect is not yet known. Table 13 shows that, at both 1% and 5% addition levels, the permeability of the acrylic is increased by 35%. The combination of increased hydrophilicity and improved permeability would enhance the comfort of fabrics coated with such resins.

An interesting and useful phenomenon is demonstrated by Figure 5, showing the variation of both tensile strength and elongation at break of Rhoplex® HA-12 acrylic resin at varying levels of addition of Q4-3667. At an addition level of approximately 2% both properties are improved, the tensile strength by 13% and the elongation at break by 25%. In the manufacture of non-woven textiles this should allow less resin binder to be used to achieve the same tensile strength, thus reducing manufacturing costs. An added advantage is that lower resin content should give a softer fabric with better handle.

Most of the work carried out with reactive silicones has been directed towards the modification and improvement of resins used in the textile and surface coating industries. However, the potential for these additives covers the whole range of cross-linkable organic resins and applies to nearly every application where the nature of the surface is important.

4. SUMMARY

Non-reactive silicones can be used in both thermoplastic and thermosetting organic resins for applications where reduced coefficient of friction, improved abrasion resistance, and lower wear rates are desired. These include gear wheels, cams, bearings, and abrasion resistance coatings in both mechanical engineering and textile industries. At the same time, these additives improve the wetting, levelling and degassing of liquid resins, and in some cases the melt flow of thermoplastics. Mould release is improved to the degree that external release agents are unnecessary, while there is no effect on paintability, welding, vacuum plating or any other finishing process. Because the levels of silicone required to achieve these results are so low, little effect is seen upon either the tensile or thermal properties of the polymers.

The reactive silicones perform in a similar manner to the non-reactive ones, the major difference is that they are completely substantive in the presence of solvents. They give reduced coefficients of friction, better abrasion resistance, improved wetting and spreadability, and at the same time allowing the modification of bulk properties such as permeability and mechanical performance. It is possible to adjust the hydrophobic/hydrophilic balance of a resin surface to some degree and the hydrolytic stability of polyurethanes can be much improved. When used to modify textile resins, the handle of textiles coated with such materials can be improved and the tendency of polymers such as polyurethanes to block can be either reduced or eliminated completely.

These additives, therefore, offer a wide scope for anyone wishing to improve the performance of an organic polymer, especially if that improvement requires modifications to the surface.

REFERENCES

1. M. J. Owen and T. C. Kendrick, *Macromolecules,* 3 (4), 458–461 (1970).
2. T. C. Kendrick and M. J. Owen, British Patent 1 257 304.
3. M. P. L. Hill, P. L. Millard, and M. J. Owen, in L. H. Lee (Ed.), Plenum Press, New York. *Advances in Polymer Friction and Wear,* A.C.S. Symposium, Los Angeles, 1974, p. 469.

12

Epitropic Fibres

G. A. Gamlen

I.C.I. Fibres Ltd. and University of Salford

1. INTRODUCTION

The special physical property of a fibre, whether staple or continuous filament, is that it has great length and very small thickness. It follows that its surface area to volume ratio is extraordinarily high and therefore that the character of the fibre reflects to a large extent the chemical and physical properties of the surface. Normally, the surface is identical in composition with the bulk polymer and so there are no major discontinuities in properties from the outside to the inside of the fibre. However, because it is mainly the surface properties which are 'seen' by the world outside the fibre it is only necessary to modify the surface skin to make a major change to the fibre properties − and 'skin-deep' in this case means 0.5−2 μm. Moreover, if the surface can be changed without affecting the bulk polymer properties then the fibre will retain all its useful properties such as tenacity, modulus, extension, etc. and can continue to be woven or knitted into fabrics on existing conventional equipment or shaped and incorporated into the usual industrial products.

2. THE DEVELOPMENT OF EPITROPIC FIBRES

Such modification at will would be particularly suitable for synthetic fibres because, in general, they are essentially passive in character, that is to say that they are odourless, tasteless, colourless, hydrophobic, non-conducting, etc. and so they form an ideal carrier on which to graft the additional desirable properties.

A technique to achieve this modification by fusing high-melting powders into the surface has recently been invented by I.C.I. Fibres and the resulting products are known as 'epitropic' fibres. The definition of an epitropic fibre is 'a fibre whose surface contains partially or wholly embedded particles which modify one or more of the properties' (Figure 1).

The invention was made in the course of research to provide electrostatic protection for nylon carpets, even at low humidities. A number of ways of making antistatic carpet fibres have been explored in the past, including incorporating humectants into the polymer or very fine conducting metal fibres into the yarn

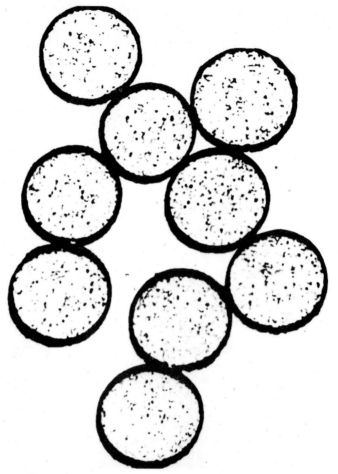

Figure 1. Cross-section of conducting epitropic fibre.

bundle. The humectants, however, are only effective at high humidities as they work by increasing the moisture regain. Nowadays, many buildings are centrally heated to rather higher temperatures than hitherto, and the relative humidity in winter often falls to a very low level and renders the humectant ineffective. On the other hand, the addition of very fine stainless steel fibres to yarn does give an antistatic fibre which is independent of humidity and which is effective when only 0.1–0.5% of these fibres are present in the carpet. However, there are problems associated with processing such yarns in normal equipment, and the price of the stainless steel fibre is high. Of the alternative conducting materials, carbon black is much cheaper than stainless steel and so attempts have been made to achieve electrically conductive polymers by loading them with special grades of carbon. Unfortunately, so much carbon black has to be incorporated (Boyd and Bulgin[1] found that viscose rayon needed 29% by weight of carbon black) that the resulting polymer has poor fibre-forming properties.

It was at this point that our Research Department hit on the idea of modifying the surface only. There are a number of ways in which this can be done, but all were rejected because of known deficiencies, for example although resins can be used to bond particles to the fibre surface[2] it was found that continued flexing of such fibres can cause the resin surface to crack away. Similarly, a combination of heat and a plasticizer can be used to soften the surface so that the particles then adhere.[3] This method is very difficult to control and adequate, safe plasticizers are not available for all synthetic fibres. Other techniques which have been proposed include coating solution spun fibres with particles while still in the gel state, and coating the spinning threadline of a melt-spun polymer before it has completely solidified.

However, in the case of epitropic fibres an entirely new technique for impregnating the surface with particles was invented which depends on the use of drawn bicomponent fibres ('heterofils'). Any bicomponent fibre can be used, such as the core/sheath or side/side type, but it is rather convenient to use a core/sheath heterofil in which the sheath has a lower softening point than the core. The sheath can then be softened by an accurately controlled heat treatment and the particles embedded in it while the core remains unaffected. The particles are then an integral part of the fibre surface (Figure 2), and the continuous nature of the path can be demonstrated by using the fibre as a wire in a lamp circuit.

The same technique can be used for a wide variety of particles as long as the particles have a higher melting or decomposition point than the sheath. There is an apparent restriction on the particle size which can be used, larger particles being more difficult to embed. Very good results are usually obtained with particles less than 5 μm in size.

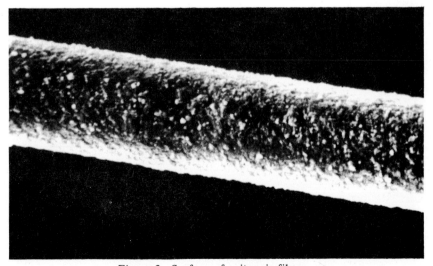

Figure 2. Surface of epitropic fibre.

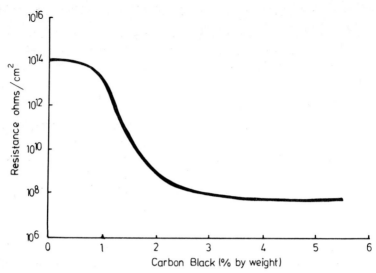

Figure 3. Resistance of 3.0 DTex polyester fibre with embedded carbon black (% by weight).

In this way, it is possible to metallize or colour the surface, to alter the surface tension, the surface friction, and the handle of fabrics made from such fibres.

The first epitropic fibres were electrically conducting and as they have already been described elsewhere,[4] will be dealt with very briefly.

3. ELECTRICALLY CONDUCTING EPITROPIC FIBRES

Carbon black has been embedded in the surface of both polyester and nylon sheath/core bicomponent fibres. The relationship between the weight of embedded carbon and the resistance is shown in Figure 3. As can be seen, the resistance levels off once about 2.5% by weight of carbon has been embedded and has a value of about 10^8 ohm/cm (which is independent of humidity) compared to a value of 10^{14} ohms/cm at 40% relative humidity for the untreated polyester fibre.

Some typical properties of a nylon monofilament and polyester staple are shown in Table 1.

Table 1

	Nylon epitropic monofilament	Polyester epitropic staple
Decitex	20	3
Tenacity (gm/dtex)	2.5	3.5
Extension at break (%)	40	40
Specific gravity	1.14	1.30
Electrical resistance (ohms/cm)	4×10^6	1×10^8

4. SPECIFIC EXAMPLES OF THE USE OF CONDUCTING EPITROPIC FIBRES

There are a number of areas where antistatic protection is required, such as electronic assembly lines and computer rooms. To illustrate the variety of end-uses possible, three examples will be considered.

(i) Protective clothing

Incorporation of, say, 2% epitropic fibre into 100% polyester reduces the surface resistivity from 5×10^{13} ohms/cm^2 to 3×10^7 ohms/cm^2 (Figure 4) and suggests that an interconnecting network of epitropic fibres exists within the matrix of polyester fibres. At this level substantial antistatic protection has been achieved, particularly as such fabrics show satisfactorily short charge-decay times.

(ii) Air filters

Dusts frequently carry an electrostatic charge, and if filtered out on a non-conducting fabric the charge can accumulate and ultimately may discharge, igniting the dust or any inflammable gases in the vicinity. The use of epitropic fibre in such a filter, provided it is earthed, meets the required specification for use in hazardous environments such as coal mines.

(iii) Carpets

There is a nuisance element rather than a hazard about experiencing a shock on walking over a carpet. Surprisingly, the charge may accumulate to over 5 kV. The lower the humidity the greater will be the voltage generated; below 30% RH

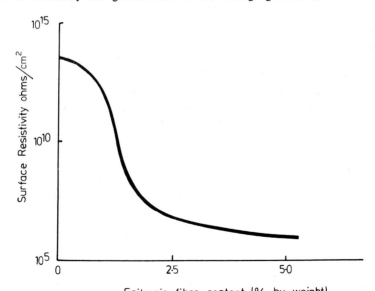

Figure 4. Polyester woven fabric (4 oz/sq. yd).

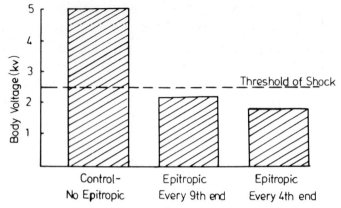

Figure 5. Body voltage developed walking over nylon carpets with and without epitropic monofilament at 21°C and 25% RH, wearing leather-soled shoes.

(relative humidity) both natural and synthetic fibres will give static problems in carpets. The threshold value of sensitivity for static discharge has been found to be about 2.5 kV; below this value static dissipation is unnoticed. It has been found that by introducing an epitropic monofilament into carpet yarn at manufacture it is possible to make bulked, continuous-filament nylon carpets which are antistatic. It is unnecessary to have epitropic monofilament in every end; depending on the carpet weight and construction one filament in every fourth to tenth end is sufficient to reduce body voltage to below 2 kV at 25% RH, which means that only 0.1–0.2% epitropic monofilament is required (Figure 5).

5. GENERAL APPLICABILITY OF EPITROPIC FIBRES

Once the general principle has been grasped it is very easy to think of many ways in which fibres can usefully be modified; some examples are given in Table 2. All these modifications have been explored; many have interesting specialist applications. Depending on the grade of carborundum used it is possible to make a saw with an extremely fine blade — just a hair's breadth in fact — which can saw through metals and alloys such as brass, or polymers such as polystyrene, and follow with ease a complicated path because the cutting edge extends right round the fibre and the saw does not have to be turned.

Table 2

Particle type	Effect
Pigment	Fibre colouration
Metal powder	Metallized fibre surface
Magnetic metal oxides	Magnetizable fibres
Abrasive metal oxides	Polishing and cutting fibres
Treated silica	Hydrophobic fibres

6. APPLICATION TO POLLUTION CONTROL

The main emphasis of this chapter is on the hydrophobic fibres which are produced by embedding treated water-repellent silica particles in the fibres. Such fibres when knitted or woven into a fabric then possess the merit of repelling water while retaining the ability to allow gases to pass through the cloth; fabrics with this property are known as 'ventile'. The waterproofing effect is due to the very fine layer of air which is trapped by the particle on the surface of the fibres and so operates in much the same way as a duck's feather. (Water/air contact angles are in excess of $160°$.)

Fabrics made from these materials have obvious applications for sporting gear and tents, as they can support a head of some 35 cm of water without losing their ventile characteristics. Although these fabrics are impermeable to water they are completely permeable to non-aqueous solvents such as paraffin and, in fact, it is possible to separate a mixture of paraffin and water by passing it through a filter made of hydrophobic epitropic fibres. However, if the head which the fabric can support is exceeded it is possible to drive the water through the fabric. This has an interesting and important consequence if the water contains fine particles of oil, as passage through the fabric causes the oil droplets to coalesce into larger drops which can then float up to the surface and be skimmed off. The applications for this effect in pollution control are obvious and legion; about half of all manufacturing industry produces an oil-contaminated effluent.

In these days when concern for the environment is very much to the forefront of everyone's mind, the problem of how to deal with aqueous effluents which are contaminated with small quantities of oil is a pressing one as it exists on such a large scale. Certain industries such as steel and engineering use large quantities of oil as an integral part of their processes, and the transportation of oil round the world by sea, road, and rail inevitably presents problems of how to deal with an oily effluent when the tanks are washed out. A large sea-going tanker may discharge about 90 000 tonnes of water per year probably containing up to 90 tonnes of oil. Fortunately, the disposing oily waste by dumping, whether on land or sea, is now less and less practised for both economic and environmental reasons. The rapid rise in the price of oil since 1973 has made it well worth recovering, and much more stringent regulations to protect the environment are being introduced and enforced. The proposed standards for effluent dumped at sea require that is shall not contain more than 100 p.p.m. (parts per million) of oil if discharged further than 12 miles from the coastline and not more than 15 p.p.m. if discharged within 12 miles.

To achieve these standards requires the removal of the small oil droplets that are present, and this is usually beyond the capability of the gravity separators hitherto employed. These separators are designed to allow sufficient residence time for the less dense oil droplets to float to the top of the water. It is then an easy matter to remove the oil layer by separating phases, for example by skimming it off. The design of such separators has steadily improved from the initial American Petroleum Institute (A.P.I.) design through to sophisticated tilted-plate arrangements. However, gravity separation becomes proportionately less effective as the

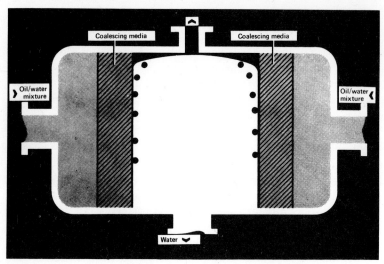

Figure 6. The Flofoil process (schematic).

size of the oil droplet reduces and the conventional designs become very inefficient when the size falls below 60 μm. As dispersions containing droplets below this size are readily created by pumping or turbulence, a new mechanism for separating this oil was wanted and the need has now been filled by the use of special cartridges incorporating hydrophobic epitropic fibres.

The basic process is shown in Figure 6. The oil—water mixture is passed through porous media containing the epitropic fibre, capture of oil droplets occurs with coalescence, and large oil droplets are formed on the exit side. These detach from the media and rise to the surface of the purified water for subsequent removal by skimming or pumping.

The three characteristics of the coalescing media are therefore oil capture, oil migration, and oil release, and each can be investigated and optimized independently by modifications to the composite cartridge. For example, with oil in water dispersions, where the oil droplets are below the 10 μm level, electrostatic charges can influence the migration of oil droplets to a surface. With the majority of oil in water dispersions, the droplets are stabilized by negative charges. As most common surfaces such as silica and glass also have a negative surface or zeta potential, these materials are not ideal for causing coalescence. The development by I.C.I. of saffil fibres based on alumina changed this position as these fibres do have

Table 3

	Approx. surface area (m^2/g)
Alumina fibres	150
glass fibres	0.5
Nominal diameter 3 μm	

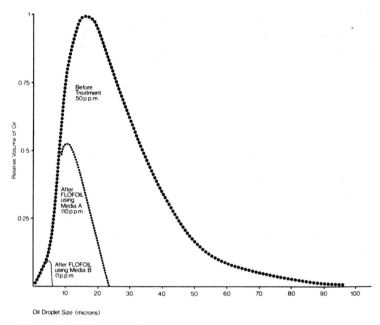

Figure 7. Efficiency of Flofoil in treating effluent from an A.P.I. separator.

advantageous positive zeta potentials which attract small droplets to their surface, thereby allowing coalescence. Such fibres have the further advantage of an extremely high surface area (Table 3) which is an important feature in the design of an efficient coalescer.

This flexibility in design, which allows the appropriate coalescing medium to be selected for a particular effluent, ensures that the required reduction in level of oil is achieved in the treated effluent. The choice depends primarily on the oil droplet size distribution in the incoming effluent; Figure 7 shows oil droplet size variation for effluent containing 50 p.p.m. of oil from an A.P.I. separator. Most of the oil content is associated with droplets in the 20 μm size. If a final oil content of 10 p.p.m. is required then a medium can be chosen which removes all the droplets above 23 μm but none below 5 μm. For a final oil content less than 1 p.p.m., then, the medium selected removes all droplets greater than 5 μm.

By combining this system, Flofoil, with other processes most of the desired solutions can be obtained, e.g. Figure 8 illustrates in schematic form how refinery effluent can be treated to reach a final standard of:

Total ether extractable oil content (by IR analysis)	5 p.p.m.
BOD	20 p.p.m.
Suspended solids	30 p.p.m.

In conclusion, this review has shown how the epitropic technique for modifying the surfaces of fibres can be used to achieve a wide range of novel effects. Some of

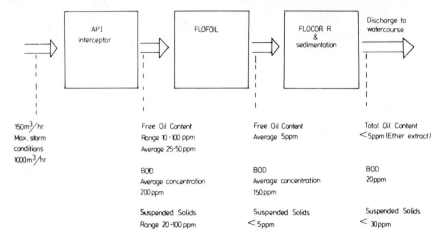

Figure 8. Solution to effluent problem at refinery producing 100 000 barrels/-day.

these are of academic interest only; some are already being used commercially and other useful applications are being designed.

ACKNOWLEDGEMENTS

Acknowledgements are due to many but mostly to Mr. J. E. McIntyre's Section (particularly V. S. Ellis, A. J. East, B. Walker, and M. Woods) at I.C.I. Fibres and Mr. T. Roxby's Pollution Control Systems Group (particularly D. B. Chambers) at I.C.I., Hyde.

REFERENCES

1. J. Boyd and D. Bulgin, *J. Text. Inst.*, 66, **48**, (1957).
2. Sumitomo Electric U.K., 943 836; Nippon U.S. 3 032 855.
3. British Celanese U.K., 341 057; Eastman Kodak 1 094 986; Dr. Plate, U.K., 1 146 918.
4. V. S. Ellis, *Textile Manufacturer*, July 1974.

13

On the Origin of the Amorphous Component in Polymer Single Crystals and the Nature of the Fold Surface

John D. Hoffman and G. Thomas Davis

Institute for Materials Research, National Bureau of Standards, Washington, D.C. 20234

1. INTRODUCTION

The objective of this chapter is to present a possible resolution to the problem of 'regular' folds versus an 'amorphous' layer as a model for the surface of a chain-folded polymer single crystal. There is a considerable body of experimental evidence (e.g. density and heat of fusion data) that points to the existence of an 'amorphous' layer that is in some way associated with the fold surface in single crystals. This has been widely interpreted as leading inexorably to the conclusion that most of the folds are either long and loose, or that non-adjacent re-entry with intervening loops occurs on a large scale, or both (Figure 1(a)). On the other hand, there also exists an equally credible body of experimental evidence, based mainly on crystallographic, spectroscopic, and morphological observations, that suggests that certain polymer single crystals exhibit rather regular folding with mostly adjacent re-entry (Figure 1(b)). From this it is frequently concluded that there is little place for the amorphous layer suggested by the density and heat of fusion data. Thus, acceptance of the aforementioned conclusions leads to a paradox, since 'regular' folding and the presence of an 'amorphous' component have been hitherto regarded as mutually exclusive or contradictory concepts. In the present work we shall suggest that the amorphous layer can in fact exist in most (but perhaps not all) cases, and that when it occurs that it is to a considerable extent the result of polymer molecules that are *physically adsorbed* on a fairly regularly folded surface (Figure 1(c)). The adsorbed layer is expected to resemble a two-dimensional random coil (or a highly flattened three-dimensional one) with 'loops' or traverses between the points of attachment on the fold surface, and may be detachable under appropriate circumstances. If this suggestion is correct, the paradox is essentially resolved, since then neither the evidence for an amorphous layer nor that for rather regular folding with mostly adjacent re-entry need be rejected.

260

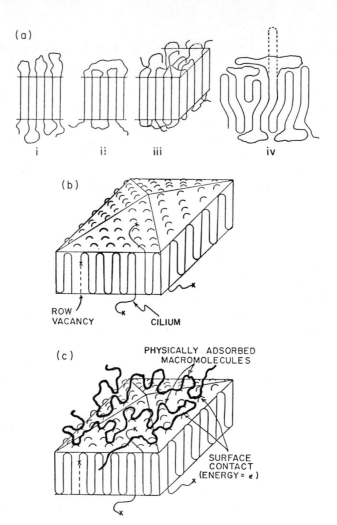

Figure 1. Models for polymer single crystals (schematic).
(a) Models with loose loops where amorphous phase is
part of the fold zone: (i) adjacent re-entry with loose
loops in growth plane (see Zachmann[11]); (ii) nonadjacent
reentry with loops in growth plane (see Frank[13]); (iii)
nonadjacent reentry commonly known as the 'switch-
board' model (see Flory[14]); (iv) mostly adjacent re-entry
with some loose folds and some 'buried' folds (see Keller
et al.[12]). (b) Model with regular folds and adjacent
re-entry forming a smooth or slightly rough fold surface
with no amorphous phase. (c) Proposed model with
mostly adjacent re-entry and smooth or slightly rough
fold surface and same crystal defects as in (b), but with an
amorphous phase that consists mostly of polymer molecules that are physically absorbed on sites on the fold
surface.

2. EXPERIMENTAL EVIDENCE RELATING TO THE ADSORBED LAYER THEORY

Lauritzen and Hoffman were originally led to the suggestion of an adsorbed layer on polymer crystals by analysis of growth rate and lamellar thickness data.[1] Parameters resulting from the best fit of the data could be interpreted in terms of a model where a three-dimensional random coil in the subcooled liquid or solution state was adsorbed on the lateral surface as a quasi two-dimensional coil, the segments of which then migrated across that surface to the site of crystallization. This led to the subsequent prediction that a broadly similar adsorption effect probably took place on the fold surface.[2]

Before proceeding with the main discussion, we briefly outline the experimental evidence leading to the apparently contradictory conclusions noted above regarding the fold surface. In the interest of brevity, we concentrate on polyethylene, and even this case is not treated exhaustively.

The density of polyethylene single crystals formed from solution has been studied extensively. Despite some early data showing near-perfect single crystal densities,[3] the main body of results[4-6] in the literature leave little doubt on two points. One is that the density is well below that of a 'perfect' crystal predicted from unit cell dimensions ($\rho_{25°C} = 1.00$ g/cm^3) and the other is that the density increases as the crystals become thicker.[7-10] Several models for a non-crystalline phase associated in some manner with the fold surface have been proposed to account for this density deficiency. Zachmann[11] has considered loose loops between adjacent crystal sequences, whereas Keller *et al.*[12] have attributed it to a combination of loose loops and 'buried' folds. A surface region composed of long cilia and loose loops resulting from non-adjacent re-entry within the growth plane was suggested by Frank[13] and a more or less random re-entry of molecules into the crystal has been advanced by Flory.[14] (The presence of a density defect, however, does not necessarily prove the existence of extensive non-adjacent re-entry or a large number of loose loops.) We note that the density of the single crystals of a given overall thickness has been reported to depend upon the solvent from which the crystal was grown,[9] for example, crystals from *n*-hexadecane were consistently about 0.004 g/cm^3 more dense than those of equal thickness grown from tetralin and decalin. This implies that the nature of the amorphous surface layer is dependent on the solvent. It has also been suggested that the density of the crystals may increase slightly with time when stored in the presence of solvent.[4]

Studies of the heat of fusion of polyethylene single crystals have also been made which point to the conclusion that the observed heat of fusion for single crystals is always substantially below the theoretical value for a crystal of infinite thickness ($\Delta h_f \cong 293$ J/g).[15-17] A small part of the reduced heat of fusion can be explained by the heat stored in the folds[2] (*ca.* 15 J/g), but the deficiency is still so large as to clearly suggest the existence of a non-crystalline component that is in some way associated with the fold surface of the crystal. More recently, low angle X-ray intensity measurements on single-crystal mats of polyethylene have been interpreted in terms of a two-phase model.[18-19] Also, the low-angle X-ray spacing of single-crystal mats has been shown to increase and then decreases reversibly upon the addition and subsequent removal of solvents,[20] implying the existence of an amorphous zone capable of being swollen by solvent. A more detailed study[21] has shown that the magnitude of the increase in low-angle spacing depends upon the molecular weight of the polymer and the crystallization conditions, from which it may be inferred that the nature of the amorphous surface layer depends upon these variables. A slightly rough fold surface could account for part of the density deficiency, but it cannot account for all of the other observations.

The above is sufficient to establish the necessity of considering the existence of an amorphous layer of some type on single crystals of polyethylene. Studies on other polymers are sufficient in scope to establish the phenomenon as a quite general one. For the sake of brevity, we have omitted evidence based on relaxation measurements (NMR, dielectric, and mechanical).

Consider now the evidence for fairly 'regular' folding in polymer single crystals. By 'regular' we mean folding with mostly adjacent re-entry and without a substantial fraction of long, loop-like folds. Certainly some 'mistakes' such as occasional non-adjacent re-entry in the growth plane must be allowed, as well as some defects such as row vacancies and cilia Figures 1(b) and 1(c). Also, a slight roughness of the fold surface including a few 'buried folds' may be permitted, but nothing to the extent shown in Figure 1(a) is intended under the definition of 'fairly regular'.

Geil[22] has summarized the crystallographic and morphological evidence of many workers, notably Bassett, Frank, and Keller, and Reneker and Geil, and Niegisch and Swan, for essentially regular folding in polyethylene single crystals of various types. Although many cases are discussed, the evidence is epitomized by his discussion of hollow pyramidal single crystals of polyethylene. The best way to explain the existence of these well-known sectorized single crystals is the staggered regular packing of chain folds by analogy with a terracotta roof. It would seem practically impossible to explain the details of the crystal structure and morphology of such crystals with any of the models shown in Figure 1(a). There is nothing inconsistent with the explanations summarized by Geil in case the rather regular fold surface were subject to an adsorbed layer of polymer molecules: all that would be added would be the amorphous layer suggested by other measurements.

Further evidence for rather regular chain folding with predominantly adjacent re-entry lies in the nature of the fracture in different sectors of the same crystals[23] and the slight difference in crystallographic spacing in adjacent sectors of the same crystal.[24] Clear evidence for mostly adjacent re-entry in polyethylene single crystals has been given by Bank and Krimm[25] based on an analysis of the IR spectra of mixed crystals of deuterated and hydrogen-bearing chains. Finally, we mention the work of Holland and Lindenmeyer,[26] who found dislocation networks between polyethylene single crystals, and the more extensive work of Sadler and Keller,[27] who showed the phenomenon to be specific for molecular weights below 3000. Dislocation networks provide clear evidence for crystallographic contact between layers and hence regularity of the folds — in this case it is clear that no substantial amorphous layer intervenes between the crystals. (Subsequently, we shall suggest possible reasons for the virtual absence of an amorphous layer in material of sufficiently low molecular weight.) In any case, it is clearly not admissible to ignore the sum of the evidence for fairly regular folding in polyethylene crystals formed from dilute solution.

Accordingly, if the experimental evidence for an amorphous layer and for fairly regular folding is simultaneously accepted, one is led to the model shown in Figure 1(c) which depicts a crystal with a fairly regular fold surface and an independent surface layer composed of physically adsorbed polymer. We may now ask if there is any evidence for such a model. Direct experimental evidence for the existence of a thin mechanically detachable layer on single crystals of several polymers has been found by Jones (née Breedon) and Geil.[28,29] From this work there is good reason to believe that single crystals of a number of polymers of normal molecular weight possess a mechanically detachable layer on their fold surfaces. Accordingly, it is believed that the model shown in Figure 1(c) is a reasonable representation of the situation. The adsorbed molecules on the fold surface provide the amorphous component suggested by heat of fusion and density measurements, among other results, and simultaneously provide the material that strips off the fold surface during fracture. The underlying fold surface itself could be fairly regular (including mostly adjacent re-entry) in keeping with the requirements of crystallographic, morphologic, and spectroscopic evidence. The proposed representation is a true 'two-phase' model with a quite abrupt phase boundary between the two phases (as suggested by Fischer et al.[18] and as shown experimentally by Strobl and Müller[19]).

3. THEORETICAL ASPECTS OF THE ADSORBED LAYER THEORY

We turn now to certain practical and theoretical considerations relating to adsorption of polymer molecules on surfaces. Considerable insight may be obtained by sketching out what is known about polymer adsorption on foreign substrates, since little revision is needed to adapt this information to the case of adsorption of molecules on chain folds consisting of the same monomer units.

Theoretical calculations by various workers[30-33] show that an isolated polymer molecule will form something that roughly resembles a two-dimensional or an appreciably flattened three-dimensional random coil on a foreign substrate. Each single surface contact has an energy taken to be ϵ. Traverses or 'loops' intervene between runs of adsorbed segments. Defining θ as the fraction of polymer segments adsorbed for each molecule, statistical mechanical calculations have been made giving θ as a function of ϵ/kT for various chain lengths. In general, the results show an increasing θ with increasing values of ϵ/kT, as shown schematically in Figure 2(a). Although somewhat dependent on the specifics of the model chosen [33] it is characteristic of such calculations to show a low value of θ for $|\epsilon/kT| \ll 1$, and a relatively high value of θ for values of $|\epsilon/kT| \gtrsim 1$. In the case of a monomer unit attaching itself at some appropriate point on a chain fold, where this point represents a surface site, values of ϵ that are roughly comparable to kT do not seem unreasonable. Even a low θ, arising from a low value of ϵ/kT, can involve a

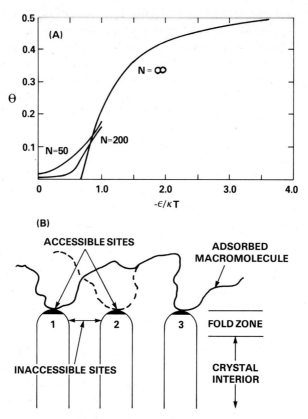

Figure 2. Physical adsorption of macromolecules on surfaces (schematic). (A) Fraction of chain segments θ adsorbed at the surface as a function of $-\epsilon/kT$. The quantity ϵ is the energy of attachment per chain segment and N is the number of segments in the chain (adapted from DiMarzio and McCrackin[31]). (B) Model showing possible origin of reduced mobility of molecules adsorbed on fold surface.

considerable amount of amorphous polymeric material being associated with the fold surface; in this case most of the mass of the amorphous material will be in loops between occupied surface sites. The total amount of the material on the surface will depend on the molecular weight and the crystallization conditions, including the solvent from which the crystal was formed, so some variation in the average thickness of the adsorbed layer is to be expected. In broad aspect, this probably underlies the differences noted earlier in the amorphous fraction detected in polyethylene single crystals crystallized from different solvents[9] and the dependence of the amount of swelling on molecular weight and temperature of crystallization.[20,21]

Experiments on the rate of adsorption and desorption of polymer molecules on foreign substrates[34,35] show two important facts that are relevant to the suggested model shown in Figure 1(c). First, the rate of adsorption is rapid compared to the rate of desorption. In the case of polystyrene on a chromium surface the rate of desorption is extremely slow: in some cases 80% of the original adsorbed polymer still remains after three weeks. The basic reason for the low rate of desorption is that, even when ϵ is rather small, all the points of contact have to be removed in the period of time required for diffusion away from the surface, which is a statistically improbable event.[34] As applied to the present situation, these results imply that the adsorbed layer on the chain fold surface will very likely be highly persistent for material of moderate and high molecular weight. This leads us to the second point, which is that the rates of adsorption and desorption are a function of molecular weight.[35] When expressed as mass per unit area, the lower molecular weight, polymer adsorbs more slowly and desorbs more rapidly than does the higher molecular weight. Accordingly, the adsorbed surface layer on a fold surface may be tenuous in specimens of low molecular weight. (In such cases cilia are expected to be more numerous, and contribute to the 'amorphous' character of the crystal.) In moderate and high molecular weight material, the adsorbed layer will have more segments in contact with the surface and desorb much more slowly. This may explain the fact that dislocation networks, which require intimate contact between two crystals over a considerable area, are found in low molecular weight polyethylene single crystals, but not in ones of moderate to high molecular weight.[27]

It has been noted earlier[2] that the adsorbed layers on the *lateral* surface are apparently mobile, and move across that surface to the site of crystallization. We must now raise the question as to why the material postulated to exist on the fold surface does not share the same fate as that adsorbed on the lateral surface, namely, rapid crystallization. We speculate that a reduced surface mobility of the polymer molecules on the fold surface may be explained as follows. Referring to Figure 2(b), it is seen that it is reasonable to propose that a considerable distance intervenes between 'allowed' sites for attachment of the adsorbed polymer molecule; movement which results in the molecule going from site 1 to site 2 requires total desorption because of the intervening 'inaccessible' sites resulting from spatial considerations. Because the sites are much closer together on the lateral surface, the barriers to be surmounted for mobility could be considerably smaller on this surface. (The mobility on the lateral surface in bulk polymers varies as $\exp[-U^*/R(T - T_\infty)]$, where $U^* \cong 1500$ cal (6275 J) and $T_\infty \cong T_g - 30°$.)[2] We note that our basic concept here would eventually permit migration to a lateral surface and some crystallization of the layer adsorbed on the fold surface. However, it is reasonable to expect this to be very slow (except in low molecular weight material) because of the postulated reduced mobility of the molecules adsorbed on the fold surface, and the large distances over which transport of molecules would have to take place to reach the lateral surface where crystallization would take place. Such a slow migration followed by crystallization may account for the increase of density found in aged preparations of single crystals.[4] It should be mentioned that the presence of long cilia[36] could also account for many of the observations described above, but one would not expect such a layer to be easily detached as in the fracture experiments nor removable by prolonged storage in solvent.

Although it is necessarily highly speculative, some comments on how the adsorbed amorphous surface layer might be reduced or even removed are in order. It is already clear from the work of Jones and Geil that portions of the layer can be removed in fracture experiments, but the area deprived of the amorphous layer is small in these cases, and we therefore concentrate on techniques that might reduce or remove the layer on a larger scale. It has already been remarked that certain solvents yield higher density crystals, implying the presence of less amorphous material. Crystallization from extremely dilute solution for long periods of time might also yield crystals with less adsorbed material. Beginning with such crystals, it might be possible, by long extraction in proper solvents below the dissolution temperature of the crystal, to dissolve the amorphous layer and remove it or eventually crystallize it by prolonged storage. It is conceivable that the presence of appropriate liquids may tend to render the amorphous layer on the fold surface more mobile, and allow it to crystallize. Progress in removing the amorphous layer could be detected by density or heat of fusion measurements or, in extreme cases, by looking for dislocation networks. Removal of the adsorbed layer by chemical attack is problematical, since one must expect the strained bonds in the folds also to be ruptured at a rather early time. However, the molecular weight of the adsorbed layer might be reduced by brief chemical attack in a manner sufficient to allow its removal by other means.

Consider now the properties to be expected of the adsorbed layer on the fold surface. It need not be of constant amount and density in all preparations but may depend on molecular weight, crystallization temperature, solvent used, and time of storage. In the presence of liquids of good solvent power, the layer may be solvated and expanded. In the presence of poorer solvents, it should be less expanded and rather flattened onto the fold surface. In the dried or contracted state, the layer is expected to have a density not far from that of the corresponding subcooled liquid or glassy state. In relaxational studies, the 'loops' between the point of contact should exhibit the mobility and chain motions typical of short-range motions at low temperatures in the glassy state (i.e. the γ_a transition),[37] but it is impossible to state at this juncture whether or not a true glass transition (i.e. a β relaxation) with a typical T_g should appear; we would anticipate that any T_g that did appear would be weak and have abnormal characteristics. Few, if any, single crystal preparations show a β relaxation in dynamic mechanical or dielectric measurements.

A slow increase of density and heat of fusion on prolonged storage in the presence of solvent may be found, resulting from a low but finite surface mobility with subsequent crystallization. In general, the hypothesis of an adsorbed surface layer on a fairly regular fold surface offers a vehicle for alternative explanations of a number of phenomena observed for polymer single crystals.

Although this paper deals primarily with the possibility of an adsorbed layer on polymer single crystals, a number of the considerations may apply to polymers crystallized from the melt. This possibility is discussed in a recent review,[2] where it is brought out that an adsorbed layer may account for a considerable portion of the amorphous content $\alpha = 1 - \chi$ in highly crystalline specimens ($\chi \cong 0.85$ to $\chi = 0.90$). Amorphous material in concentrations higher than $\alpha = 0.1-0.15$ or so may be ascribed at least in part to the presence of interlamellar entanglements that prevent crystallization on a large scale. DiMarzio and Rubin[32] have carried out calculations for the adsorption of molecules between two parallel plates. A long molecule frequently adsorbs on both plates, producing an amorphous link between them. The analogous situation may occur in polymers crystallized from the melt, leading to an increased amorphous content.

REFERENCES

1. J. I. Lauritzen, Jr., and J. D. Hoffman, *J. Appl. Phys.* **44**, 4340 (1973).
2. J. D. Hoffman, G. T. Davis, and J. I. Lauritzen, Jr., in *Treatise on Solid State Chemistry*, ed. N. B. Hannay (Plenum Press, New York, Vol. 3, Chapter 6, 1976).
3. T. Kawai and A. Keller, *Philos. Mag.*, **8**, 1203 (1963).
4. G. M. Martin and E. Passaglia, *J. Res. Nat. Bur. Stand. (U.S.)*, **70A** *(Phys. and Chem.)* No. 3, 221—224 (May-June 1966).
5. D. A. Blackadder and P. A. Lewell, *Polymer,* **9**, 249 (1968).
6. R. K. Sharma and L. Mandelkern, *Macromolecules,* **2**, 266 (1969).
7. E. W. Fischer and R. Lorenz, *Kolloid Z. Z. Polym.*, **189**, 97 (1963).
8. R. Kitamaru and L. Mandelkern, *J. Polym. Sci.*, A-2, **B**, 2079 (1970).
9. A. Nakajima, S. Hayashi, T. Korenaga, and T. Sumida, *Kolloid Z. Z. Polym.*, **222**, 124 (1967).
10. G. T. Davis, J. J. Weeks, G. M. Martin, and R. K. Eby, *J. Appl. Phys.*, **45**, 4175 (1974).
11. H. G. Zachmann, *Z. Naturforschg.*, **19A**, 1937 (1964).
12. A. Keller, E. Martuscelli, D. J. Priest, and Y. Udagawa, *J. Polym. Sci.*, A-2, **9**, 1807 (1971).
13. F. C. Frank, referred to by A. Keller, *Kolloid Z. Z. Polym.*, **231**, 386 (1969).
14. P. J. Flory, *J. Am. Chem. Soc.*, **84**, 2857 (1962).
15. E. W. Fischer and G. Hinrichsen, *Polymer,* **7**, 195 (1966); *Kolloid Z. Z. Polym.*, **213**, 93 (1966).
16. H. Hendus and K. H. Illers, *Kunststoffe,* **57**. 193 (1967).
17. L. Mandelkern, A. L. Allou, Jr., and M. Gopalan, *J. Phys. Chem.*, **72**, 309 (1968).
18. E. W. Fischer, H. Goddar, and G. F. Schmidt, *J. Polym. Sci.*, **B5**, 619 (1967).
19. G. R. Strobl and N. Müller, *J. Polym. Sci. Polym. Phys. Ed.*, **11**, 1219 (1973).
20. Y. Udagawa and A. Keller, *J. Polym. Sci.*, A-2, **9**, 437 (1971).
21. E. Ergöz and L. Mandelkern, *J. Polym. Sci.*, **B10**, 631 (1972).
22. P. H. Geil, *Polymer Single Crystals*, Interscience, New York, pp. 125—139 (1963).
23. P. H. Lindenmeyer, *J. Polym. Sci.*, **C1**, 5 (1963).
24. D. C. Bassett, *Philos. Mag.*, **12**, 907 (1965).
25. M. I. Bank and S. J. Krimm, *Polym. Sci.*, A-2, **7**, 1785 (1969).
26. V. F. Holland and P. H. Lindenmeyer, *J. Appl. Phys.*, **36**, 3049 (1965).
27. D. M. Sadler and A. Keller, *Kolloid Z. Z. Polym.*, **239**, 641 (1970) and **242**, 1081 (1970).
28. J. B. Jones and P. H. Geil, *J. Res. Nat. Bur. Stand. (U.S.)*, **79A** *(Phys. and Chem.)*, No. 5, 609—611 (Sept.—Oct. 1975).
29. J. E. Breedon, Crack Formation in Polymer Single Crystals, Masters Thesis, Case Western Reserve Univ. (1975).

30. A. Silberberg, *J. Chem. Phys.*, **46**, 1105 (1967).
31. E. A. DiMarzio and F. L. McCrakin, *J. Chem. Phys.*, **43**, 539 (1965).
32. E. A. DiMarzio and R. J. Rubin, *J. Chem. Phys.*, **55**, 4318 (1971).
33. R. J. Rubin, *J. Chem. Phys.*, **43**, 2392 (1965).
34. R. R. Stromberg, W. H. Grant, and E. Passaglia, *J. Res. Nat. Bur. Stand. (U.S.),* **68A**, (*Phys. and Chem.*), No. 4, 391–399 (July-Aug. 1975).
35. W. H. Grant, L. E. Smith, and R. R. Stromberg, *Discuss. Faraday Soc.,* **59**, 209 (1975).
36. I. C. Sanchez and E. A. DiMarzio, *J. Chem. Phys.*, **55**, 893 (1971).
37. J. D. Hoffman, G. Williams, and E. Passaglia, *J. Polym. Sci.,* **C14**, 173 (1966).

14

Subjective and Objective Assessment of Surfaces

B. J. Tighe

University of Aston in Birmingham, Department of Chemistry, Gosta Green, Birmingham

1. INTRODUCTION

Probably the most common type of polymer surface encountered in everyday life is that presented by paints and varnishes — known collectively as surface coatings. It is not surprising therefore that it is in this field of study that most of the effort attempting to relate objective characterization of polymer surfaces to their subjective assessment has been made.

The most obvious and arguably the most important subjectively discerned properties of a surface of this type are its colour and gloss. At normal temperatures (i.e. when the surface is appreciably below red heat) our perception of the colour and gloss of an object arises by the action of rays of light that fall on the object from a source such as the sun or an electric light and are then reflected into the eyes. The stimulus received by the retina produces impulses which are transmitted along the optic nerve to the sight centre in the brain, where they are experienced as a perception of the object. Aspects such as the shape and position of the object, together with its movement in relation to its surroundings, will be perceived together with the properties that are important in the present context, namely the colour of the object and the perception of the gloss of its surface. As further consideration will show, gloss is not a simple physical property, and in assessing the glossiness of a surface we make a judgement as to its texture. This subjective assessment is assisted by the sense of touch which provides an impression of surface rugosity, frictional properties, and hardness.

The point about a subjective impression of this type is, of course, that it contains a great deal of information and is virtually instantaneous. On the other hand, storage and transfer (as from one individual to another) of the information is in any absolute and quantitative sense, impossible. In terms of comparison or matching of two surfaces, however, subjective assessment has many advantages, and in the surface coatings industry, for example, a man experienced in these skills is invaluable. It is with the extent to which these skills can be matched or improved on by instrumental techniques and the form in which such information is obtained and stored that this chapter is concerned.

2. COLOUR

The question of colour is of limited relevance to polymer surfaces in themselves, since this property is largely attributable to pigments that are added to polymers for decorative and/or protective purposes in coatings. Some brief comment is appropriate, however. The radiation from a surface that gives rise to the perception of colour is called a 'colour stimulus', being the agent that stimulates the eye to produce the visual perception of colour. Such stimuli can be registered and measured by physical instruments, but the perception of colour can only be experienced by a living creature. In human beings, of course, it is well known that colours give rise to emotional and psychological reactions. There are other less well-known factors that affect the way in which an individual subjectively perceives a given colour stimulus. These include the regional variation which causes colour perception to change with the position on the retina where the colour is formed, after-image and related phenomena which cause the perception of colour to be affected by the sight of another colour viewed immediately prior to the colour in question, and the problems experienced by the eye in adapting to very high and low levels of illumination which both cause changes in colour perception (at extremes in the ability even to distinguish colours).

The preceding points are important not only in their own right, that is, in relation to colour perception and measurement, but because they illustrate the type of limitation encountered in subjective assessment in general. Perception and comparison of colour is a matter of everyday experience and therefore these points are more easily appreciated than similar points relating to the assessment of gloss. Although the eye and brain assess the reflectance properties of the surface as a whole, including colour (bearing in mind the limitations mentioned previously), instruments do not, and questions relating to the angle of illumination and viewing must be discussed.

Because of the limited relevance of colour measurement in its fullest sense to polymers, as distinct from polymer composites, this subject can command little space. Briefly, the geometry of the optical system plays an important role, as will become apparent. Illumination at an angle of $45°$ to the surface, together with measurement of diffuse reflection at $0°$, is common although other systems are in use. The two most important types of instruments are reflection spectrophotometers and tristimulus colorimeters. In the former, monochromatic light is obtained over a wavelength range by means of prisms or diffraction gratings, whilst in the latter only three bands of light (corresponding to specified primary colours) are used and the amount of each needed to match the colour determined. It is, of course, much more difficult to visualize the colour of a surface from a reflectance spectrogram than from tristimulus data. A good general account of colour characterization is given by Nylen and Sunderland[1] whilst various authors have reviewed and discussed more recent developments in colour measurements.[2-6]

3. GLOSS

Although in subjective terms it is impossible to separate colour from other aspects of surface assessment, questions of gloss and texture are of greater obvious relevance in the study of polymer surfaces. As far as the relationship between gloss and colour is concerned it appears that gloss affects the possibility of colour measurement more than vice versa. The relationship is not fully understood, although it is true to say that visual inspection does not indicate any essential dependence of gloss on the colour of the illuminating light. This proposition is, of course, important in relation to the choice of wavelength of illuminating light in the design of instruments for gloss measurement. It has been suggested[7] that the work of Andronow and Leontowicz[8] provides a theoretical basis which supports this observed lack of dependence.

The term 'gloss' (sometimes referred to as sheen or lustre) is used to describe the result of a visual appraisal of a surface. No precise definition is possible to describe what is in fact a complex subjective phenomenon, although various attempts have been made to define it and its various angular qualities.

Figure 1(a) illustrates some of the relevant terms used in this field. Also illustrated in Figure 1(b) is idealized diffuse or matt reflection (which in practice is extremely rare), whilst Figure 1(c) displays the shape of the polar curve for a matt surface having a greater intensity of reflected light in the direction of specular reflection. When an individual makes a subjective visual assessment of the characteristics of a surface he does so using both eyes in fairly arbitrary and (perhaps surprisingly) independent movement. In so doing he forms an impression based both on the sharpness and contrast of images reflected to the eye from the surface, and the diffuse reflectance of the film. Although it is possible to relate the different quantities involved to intensities of reflection at various angles, this does not allow the visual appearance to be precisely defined in terms of physical quantities. Harrison[9] after extensive studies in which the ability of individual observers to rank a range of surfaces according to gloss was compared with instrumental measurements, has concluded the gloss of surfaces is not a simple physical property, but a psychological *Gestalt*, that is an appraisal of the physical situation taken as a whole. He writes:

> Gloss is not a single simple sensation but a complex of at least three simpler sensations. These were found to be: sharpness of mirror image, variations in the brightness of the surface when viewed at different

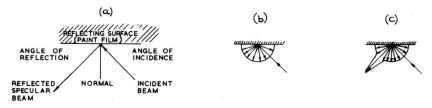

Figure 1. Reflectance phenomena: (a) definition of terms; (b) spatial distribution of light reflected from an ideal matt surface; (c) spatial distribution of light reflected from a semi-matt surface.

angles and the parallactic effect in which we seem to be looking at one surface through another. Gloss is a combination of all three but not in fixed proportions.

It has been found that in making subjective assessments of the relative gloss of surfaces observers tend to fall into one of two categories. In the first (termed the integrating type) observers judge gloss with reference to the total amount of light reflected from the surface, whilst the other (the analytical) judge according to the sharpness of the images seen via the reflecting surface. Thus, an individual can repeat his ranking of surfaces fairly well, but his judgement may differ sharply from that of another observer. The situation is further complicated by the fact that the response of the eye to the intensity of specularly reflected light is non-linear. There is in fact a sound basis (the Weber–Fechner law) for the belief that human observers make visual assessment on a logarithmic rather than a linear scale. The validity of this was demonstrated by Kosbahn[10] who carried out experiments in which a rotating series of strips whose gloss varied in a defined manner was used. A linear differentiation in the reflection properties of the strips resulted in poor visual discrimination. When a logarithmic progression was used the steps between adjacent strips were visually estimated as equidistant.

Despite these complicating factors considerable progress has been made in recent years in characterization of the reflectance properties of surfaces by instrumental techniques, and in the correlation of such measurements with subjective assessment. The fundamental laws governing the reflection of light have been known for a considerable time, and the phenomenological approach to scattering of light from surfaces has attracted the attention of physicists and mathematicians for many years. Theoretical relationships governing the intensity distribution of light scattered from surfaces have been discussed by, for example Barkas[11] and Nimerof.[12] In contrast, the analyses of experimental results in the field of surface characterization by reflectance techniques have been of necessity largely empirical, although related to the fundamental work. A parallel might be drawn with the way in which infrared spectroscopy used as a practical tool for the characterization of organic compounds relies more heavily on empirical comparison than on the fundamental laws governing the vibrational spectra of simple molecules.

Although it is generally recognized that gloss is a complex phenomenon associated with the physical nature of a surface and the way in which it affects the spatial distribution of reflected light, and coupled with this a broad measure of agreement existed amongst earlier workers when describing the factors that affected gloss, there has been continual uncertainty (or difference of opinion) as to the best way in which it can be measured. The types of instrument used can conveniently be described under two headings, namely gloss meters and goniophotometers. A comprehensive review and classification of the former system has been prepared by Konig[13] and a more general account, including reference to the literature on both types of instrument up to 1971, by Zörll.[7]

4. GLOSSMETERS – INSTRUMENTATION

The simplest form of gloss measuring device is the so-called 'specular glossmeter' (also referred to as a 'reflectometer'). It consists essentially of three parts: namely, an incandescent light source, a port at which the specimen is presented, and a photosensitive receiver such as a photomultiplier. These three components are usually in fixed positions to each other, with the light source and the photomultiplier being set in positions corresponding to the incident and reflected beams, respectively, in Figure 1(a) (i.e. the light reflected from the surface is received at the specular angle). Readings taken from the instrument are related to a polished black tile which is assigned an arbitrary 'gloss value' of 100. Various 'fixed-head. instruments are available which differ mainly in the geometry of illuminating and receiving systems. It is also possible to obtain 'variable-head' instruments in which the geometry may be varied over a limited range.

It is now 40 years since the first specular glossmeter was described by Hunter[14] which led to the adoption in 1939 by the American Society for Testing Materials of the A.S.T.M. D523–53T Tentative Method of Test for Specular Gloss. Further work by Hunter[15] showed that a specular angle of 60° appeared to be the most useful for the examination of films of intermediate gloss, although the 45° glossmeter has also been claimed to be satisfactory for a wide range of surfaces.[16] For particularly high gloss values (i.e. over 70), however, 20° specular geometry was preferred, whereas for low gloss values (below 30) 85° geometry was found to give better discrimination. This highlights one of the major limitations of specular glossmeters in comparison to the more sophisticated goniophotometers (see section 5). Indeed, a considerable amount of work[17,18] was carried out on the question of dimensions and tolerances for the source and receiver apertures in order that any error contributed by these features might be limited to a maximum of one gloss unit on any point of the scale, even when measured on different instruments. Despite this, a more recent 'round-robin' evaluation of a set of standard surfaces indicated[19] a general lack of agreement between gloss readings obtained from different laboratories using the specular glossmeter system, attributable mainly to poor standardization and maintenance of the instrument.

In spite of their apparent shortcomings in terms of both accuracy and reproducibility, the advantages of specular glossmeters such as their instrinsic simplicity and ease and speed of analysis are such that this type of instrument is widely used in industry and questions of design and use command regular attention.[20-25]

An alternative but less widely used form of glossmeter is the 'contrast glossmeter' which attempts to accommodate the visual evaluation that a human observer makes of the distinctness of an image observed at the sample surface, rather than the intensity of the reflected light. The principle of this type of instrument involves the use of the sample as a mirror reflector just as in the case of the specular glossmeter, but the degree of the contrast at borderlines between dark and light fields is assessed. Contrast glossmeters have been designed in which the fineness of detail that can be seen by specular reflection at a fixed angle is estimated, or in which a pattern can be reflected specularly at various angles and the angle closest to normal to the sample surface at which the pattern can be seen is taken as a measure of gloss. Although evaluation of the gloss of a surface with this type of instrument is in principle related to a specific model of subjective gloss assessment, the fact that human evaluation is also involved inevitably makes the standardization of such devices difficult. Notable contributions in this field have been made by Hunter,[26] Frier,[27] Randell,[28] and Bohmann.[29] In the present context the most important features of contrast glossmeters when compared to their specular counterparts are their extended range of applicability to surfaces in the medium and low range of reflectance[28] and the fact that they have provided support for the view that gloss should be linked with the sharpness of an image as perceived by an observer.[29]

5. GONIOPHOTOMETERS – INSTRUMENTATION

The most informative and precise technique for measurement of the gloss of surfaces is undoubtedly goniophotometry – the measurement of reflected light intensity as a function of viewing angles of incidence. Although this technique is used solely for gloss characterization by several workers, it is capable of giving a great deal more information in the field of characterization and deterioration of polymer surfaces. Because commercial instrumentation does not always correspond to the requirements of particular researchers several custom-built goniophotometers have been constructed, frequently based on pre-existing optical systems. The essential features of a goniophotometer are shown in Figure 2, although this does not illustrate factors that dictate the precision and accuracy of results obtained, for example, sensitivity and stability, collimation of light source, receptor angle of measuring photocell, and range of illuminating and viewing angles.

The first reference to a goniophotometer appears to have been made over 50

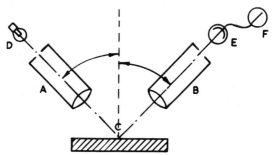

Figure 2. Essential features of goniophotometer. Components: A. Collimating lens system for incident beam from light source D. B. Collimating lens system for reflected beam. E and F represent the detector and read-out meter respectively. The incident and reflectance angles may be varied about point C.

years ago,[30] although it is only during the last decade that interest in the design and use of high-performance goniophotometers has grown. Thus, in 1966 Hoffmann and Kosbahn[31] described a relatively simple design of goniophotometer which was limited only to three angles of incidence (30°, 45°, and 60°) and used a convergent beam with an angle of convergence of about 8°. This has the advantage of obviating the need for a collimator on the viewing side and enabling an inexpensive photocell to be used. In addition the instrument was equipped with an automatic curve-plotting facility.

In the same year a new commercial instrument the GP2 was described by Heinz Loof of the Carl Zeiss Company.[32] Although manual in operation, the Zeiss GP2 has found widespread use in practice because it combines sufficient accuracy of measurement for most purposes with an optical system that is limited to the essential elements. Further instrumental developments have been concerned with such questions as the range of obtainable geometries of illumination and viewing and the relationship of resolution with collimation.[33-37] During this period several interesting pieces of work relating to the modification of existing instruments have been reported. Thus Carr[38] has described an automated version of the Zeiss GP2 goniophotometer, Colling et al. the modification of a Unicam SP500 spectrophotometer[39] and Loof[40] the modification of the Zeiss DMC25 spectrophotometer, all for goniophotometric use. Both Billmeyer and Davidson[41] and Quinney and Tighe[42] have described the modification and use of the Brice – Phoenix light-scattering photometer, which is in many ways ideally suited for this type of work. The last three named instruments[39-41] are, in strict terms, spectrogoniophotometers since they are capable of measuring reflectance as a function of wavelength as well as angle. Although few surfaces are 'goniochromatic' that is to say the colour changes with geometry of illumination, or viewing, or both, the phenomenon (also referred to as geometric metamerism) is well recognized.

It is with goniochromaticity that Billmeyer's work[43-47] has been principally concerned and his instrumentation includes an extensive wavelength range of incident light. Other workers have used goniophotometers principally for the study of phenomena related to surface defects and pigmentation.[39,48-51]

In the writer's laboratory an automated goniophotometric technique has been used, in conjunction with other surface techniques, for the study of the deterioration of polymer surfaces in various natural and artificial environments.[42,52-55] Finally, in this section, it is appropriate to mention that Edwards[56] has described the construction of a simple high-resolution 'laser gonio-reflectometer'.

6. PRESENTATION AND INTERPRETATION OF GONIOPHOTOMETRIC DATA

In contrast to the simple representation of the principle of the goniophotometer shown in Figure 2, Figure 3 illustrates the essential features of the high-performance goniophotometer obtained, in this case, by simple modification of the Brice—Phoenix light-scattering photometer (series 2000). A light beam from lamp (A) passes through a preselected wavelength filter (B) and shutter unit (C) into a collimating tube (E), incorporating a removable polarizer and slit aperture (F) the

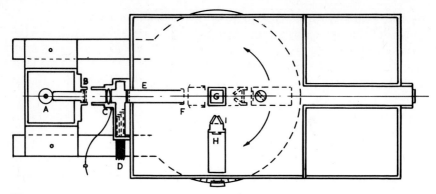

Figure 3. Essential features of a high-performance goniophotometer, based on a Brice–Phoenix light-scattering photometer.

width of which may be varied. The light impinges on the sample mounted at (G) and is thereby reflected, scattered, or transmitted, its intensity being measured as a function of the scattering angle by a photocell (H). The intensity of the light beam is controlled by a series of neutral filters (D). The photocell which may be rotated through an angle of 270° has a removable analyser and variable slit aperture (I). The signal from the photocell is amplified and registered on a galvanometer or chart recorder.

More details of operation, including the design of specimen mounts,[42] the use of a motor to drive the photocell, and the construction of beam-narrowing devices[54] are described in the literature.

Both unpigmented and pigmented polymer films may be used on any flat substrate. Microscope slides are very convenient, especially for the deposition of films of polymer available only in small quantities. Reduction in the size of beam is limited principally by the sensitivity of the instrument, but a 1 mm radius circular beam gives adequate sensitivity and enables small areas of the polymer surface to be examined. For studies such as ageing, which involve successive examinations of the same sample an alignment mark on the mount and sample support facilitate reproducible mounting of the specimen.

Typical goniophotometric curves obtained using the above instrumentation are shown in Figures 4 and 5. These were obtained under identical illumination conditions and are plotted on the same scale. Although they were obtained on black pigmented films the same principles apply to unpigmented samples. The sensitivity of the technique is illustrated by the difference in maximum intensities between the two figures and the fact that well-resolved curves are obtained in both cases. One problem that is immediately apparent is the difficulty of representing intensities as different as 896 units (Figure 4) and 4.5 units (Figure 5) on the same plot. In addition the reflected light normal to the surface (angle of reflectance 0°) is much lower in intensity, but is still an important parameter. One convenient way of overcoming this difficulty is the logarithmic presentation. Thus, Figure 6 shows the curves for 60° angle of incidence from Figures 4 and 5, together with the

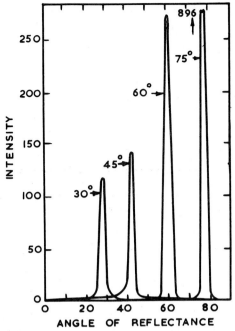

Figure 4. Typical goniophotometric curve for black pigmented polymer film.

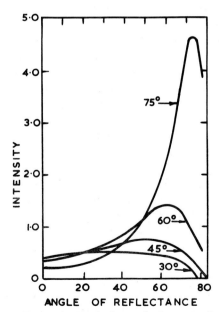

Figure 5. Typical goniophotometric curve for black pigmented polymer film.

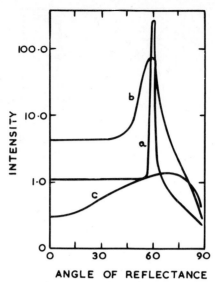

Figure 6. Goniophotometric curves (logarithmic presentation) for black pigmented polymer films (b and c corresponding to Figures 4 and 5 respectively) and a white pigmented film (a).

corresponding curve for a white pigmented polymer film plotted on a logarithmic scale. Both the specular intensity (I_s) and diffuse reflectance (I_D, the intensity of reflected light at $0°$) can be readily determined. It is appropriate to notice that the white sample has a very much higher diffuse reflectance than either of the black pigmented samples. This, of course, affects the image contrast and is one cause of differences between subjective and instrumental ranking of gloss when both colours are included in the range or surfaces compared.

The information that can be deduced from the shapes of goniophotometric curves is very important and must be considered subsequently. However, it is apparent that if several samples are to be compared some method or methods must be devised for simplifying or summarising the information. The simplest value is the peak height or I_s value for a given angle of incidence, but this is reducing the technique to the level of a sensitive specular glossmeter. Although the I_s value is significant it has been shown that neither this nor a peak area measurement give significantly better correlation with subjective assessment than does a simple gloss meter[57].

One useful way of summarizing the information from curves for various angles of incidence is seen in Figure 7 in which ($I_s - I_D$) for each recorded angle of incidence is plotted against $90°$ − angle of incidence. The value corresponding to $90°$ − angle of incidence is, of course, the intensity of the incident beam. The information shown corresponds to two of the specimens referred to in Figure 6,

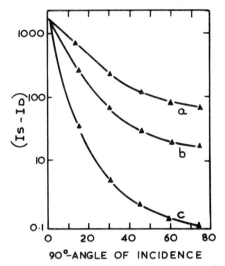

Figure 7. Plot of $(I_s - I_D)$ versus $90°$ − angle of incidence for two of the pigmented films referred to in Figure 6 together with data for a partially degraded specimen of one of them.

together with a partially degraded specimen of one of them. This form of presentation does illustrate the rationale in choosing specular glossmeters having smaller angles of incidence to measure glossy surfaces and greater angles of incidence for matt surfaces. It will be obvious by inspection that this tends to produce less difference between the specular intensities from the different surfaces. Thus, the choice of small angles of incidence for glossy surfaces produces an intensity towards the right-hand axis (Figure 7), while larger angles of incidence will compensate to some extent for the lower reflectance of the matt surface by observing it at a point towards the right-hand axis. The extremes recently recommended by the technical committee working within the framework of the International Organization for Standardization are a $20°$ angle of incidence for high-gloss surfaces and $85°$ for matt surfaces.[58]

Although the method of presentation shown in Figure 7 is useful in comparing goniophotometric curves for different surfaces it does have limitations. Whilst it shows clearly[52,53,55] the progress of deterioration in polymer films for instance, it is neither sufficiently informative as to the nature of the surface changes, nor yet concise enough for use in kinetic work. In order to meet these requirements it is necessary to make use of the peak width at half height $(W_{1/2})$. This enables some account to be taken of the sharpness of the image and is used in calculation of the so-called 'gloss factor' $(I_s - I_D)/W_{1/2}$ which is the most useful single parameter that can be derived from goniophotometric curves. Its magnitude will of course depend, for a given instrument, on the angle of incidence chosen (usually $45°$ or $60°$) and the intensity of the incident light beam. The particular value of the gloss factor lies in its convenience, coupled with the fact that it does reflect the changing nature of the surface, a point that will become apparent in the subsequent discussion. It provides a very useful way of monitoring the change in reflectance properties of a film during ageing or deterioration studies, when the bulk of data accumulated makes it impossible to use the types of presentation shown in Figures 4—7.

7. REFLECTANCE PROPERTIES AND THE PHYSICAL NATURE OF SURFACES

Before considering the information that can be deduced from the shapes of goniophotometric curves it is necessary to mention the relationship between gloss and surface profile, together with the types of technique that are helpful in this respect. The fact that increasing the roughness of a surface will cause a change in the ratio of specular to diffuse reflectance can readily be appreciated from the fundamental point of view, but in terms of surface characterization it is important to consider attempts to correlate results obtained by various experimental techniques.

There are two distinct types of technique that are useful in this respect. The first involves direct physical measurement of the surface profile with instruments such as the Profilometer,[59] Talystep,[51] or Talysurf.[52,55] The principle involved is that of a tracer point which moves over the measured surface transmitting its motion to a coil which moves in the field of a permanent magnet, this producing a small voltage which fluctuates according to the frequency and height of the surface irregularities. The instrument may either print an amplified profile on to chart paper or produce a figure for the average height of roughness irregularities (AHRI) relative to the centre line average (CLA). It is obviously necessary to take a substantial number of readings for each sample to be measured. The second technique is electron microscopy which, since the advent of stereoscan instruments, has proved of great value, especially when used in conjunction with goniophotometry. The greatest single advantage of stereoscan over optical microscopy in this type of work is its great depth of focus. Several workers have demonstrated its value in observing the progressive deterioration of coatings which is usually, but not always, associated with a loss of gloss.[60-64]

The fact that the deterioration and loss in specular reflectance of polymer surfaces do not necessarily proceed in parallel is an important observation and one that is dependent upon the relationship between surface rugosity and reflectance.

A general correlation between surface profile and gloss-meter measurements has long been accepted,[59] but such measurements were obtained on samples that progressively became both rougher and less glossy. This is not always the case since the combined physical and chemical effects of outdoor weathering can produce seasonal changes which result in behaviour that is almost cyclic. Thus, the effects of physical erosion tend to make a surface smoother, even though it is continuing to deteriorate. Examination of a considerable number of such samples[52,55] led to the conclusion that there is a quantitative relationship between the gloss factor $[GF = (I_s - I_D)/W_{1/2}]$ and the AHRI, as illustrated in Figure 8.

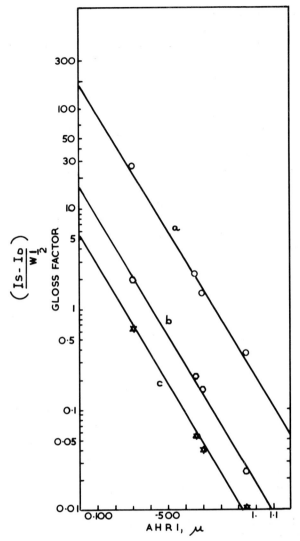

Figure 8. Plot of gloss factor versus average height of roughness irregularities for three polymer films.

Although a linear relationship between AHRI and log *GF* was always found, the relationship was modified by two constants, one of which was characteristic of the surface and the other of the angle of incidence of the light beam.

Additionally and independently of this relationship, it was observed that at each angle of incidence, when the AHRI value exceeded the wavelength of the incident light, the angle of maximum reflectance began to decrease. Although a quantitative relationship between the angular deviation and the AHRI value was observed[52] the number of experimental points was small; this phenomenon must be investigated further before a firm conclusion as to the nature of the relationship can be drawn.

8. INTERPRETATION OF GONIOPHOTOMETRIC CURVES

In the light of the foregoing discussion it will be apparent that the precise nature of surface rugosity will affect the pattern of scattered light and thus the shape of goniophotometric curves. One of the first papers to suggest the use of this in the characterization of paint films was that of Colling et al.[39] Their most valuable contribution lay in the use of interferometry to characterize surface imperfections and the recognition that 'micro' defects which are not greater than the wavelength of the incident light beam remove energy from the beam by diffraction or Mie-type scattering. Unfortunately, they did not reproduce the goniophotometric curves, and the sketches and descriptions provided do not correlate well with the curve shapes obtained with high-performance instruments on a range of surfaces.[42,49,52-54] In particular, none of their sketched curves show the asymmetry that becomes apparent as the size of irregularity increases.

The effect of surface profile on the shape of the resultant goniophotometric curve is not well documented in the literature. The situation is best considered in terms of the I_s value, the I_D and $W_{1/2}$ values, peak asymmetry, and angular divergence. A plane mirror surface will obviously produce, in principle, a perfectly sharp line corresponding in width to the incident light beam. In practice, however, the angle of acceptance of the receiving photomultiplier provides the limitation in most instruments and very sharp, almost triangular, peaks are obtained. The types of curve shown in Figure 4 are typical of those produced from polymer films that have been melt pressed or cast. Any visually discernible imperfections in the surface (such as 'orange peel' effects or 'brush' marks, for example) will produce some distortion of the goniophotometric curves, but the peaks will resemble in shape (i.e. height, width, and symmetry) those obtained from a plane film of the same polymer. The most obvious difference will be a random variation between the angle of incidence and the angle of maximum specular reflection. Since the imperfections are readily discernible and usually related to the technique of film preparation rather than the inherent nature of the surface, such samples are usually discarded.

Taking a plane polymer film that gives sharp narrow peaks (similar to Figure 4) as a starting point, and considering the types of deterioration or surface modification that can occur, the first noticeable change is usually a reduction in the I_s value with no discernible change in I_D and no increase in $W_{1/2}$. This corresponds to an increase in the surface rugosity, both in terms of increased population of imperfections and increased size. At this stage, however, the imperfections are small compared to the wavelength of the incident light ($\simeq 0.5 \ \mu m$). The term 'micro' imperfections is usually used in describing the nature of the surface at this stage and the peaks are often referred to as 'sharp triangular', although this is not strictly accurate as Figure 4 shows.

As the mean size of the imperfections approaches the wavelength of the light beam, peak broadening is observed. This is most readily detected at this stage by the increase in $W_{1/2}$; I_s will continue to fall, but increases in I_D may not be large and the peaks are essentially symmetrical. With a continuing increase in rugosity the increase in $W_{1/2}$ becomes more marked, I_D increases and the peaks become very asymmetric. When this occurs the surface irregularities or defects are often spoken of as being 'macro', being somewhat larger than the wavelength of the incident light beam. The curves shown in Figure 5 illustrate a fairly extreme form of this type of surface. The logarithmic presentation (Figure 6) emphasizes the fact that no curves obtained from polymers are, in fact, absolutely symmetrical. It also usefully illustrates the fact that changes in $W_{1/2}$, I_D, and curve asymetry do occur independently of each other and provide a means of 'fingerprinting' surfaces. Finally, if the instrumentation is sufficiently precise in its location of the sample and the measurement of the angle of incidence, the divergence between this and the angle of maximum specular reflection indicated the extent to which the mean size of surface imperfections exceeds the wavelength of the incident light beam. The size distribution of imperfections obviously plays some part in governing the transitions referred to above, as does the exact profile (e.g. steepness) of the imperfections. No analysis of the effects of variations of this type has been made in the literature.

9. APPLICATIONS OF GONIOPHOTOMETRY

The major fields of use of goniophotometry have been summarized previously, and although the most extensive area of work has been concerned with pigmented coatings, particularly the assessment of gloss and the examination of surface imperfections, the technique is capable of much wider use. Clear specimens can, for example, be examined as readily as pigmented, and since small areas of thin films are studied the technique is ideally suited to small quantities of experimental polymers. In this context goniophotometry has been used[65] to study the relative UV stability of poly-α-esters containing pendant aromatic and fluoroaromatic groups. The technique similarly enables the surface rugosity of polymer surfaces to be assessed as a preliminary step in their examination by other procedures where rugosity can affect the results obtained. This is particularly useful in the determination of surface energetics by contact-angle techniques such as those of Owens and Wendt or Hamilton (cf. Reference 66).

Goniophotometry is an extremely valuable tool in the study of the surface degradation, deterioration, or ageing of polymers in various natural or artificial environments. In this, as in many other potential applications, it is most usefully used in conjunction with other techniques. In this context it has been employed with stereoscan electron microscopy and direct surface profile measurements which have been used to study the physical nature, and multiple internal reflectance IR spectroscopy and surface free energy measurements, aspects of the chemical nature, of polymers during ageing.[52,53,55,67] Other techniques that are potentially valuable in this respect include ellipsometry,[68] ray deflection mapping,[69] and schlieren techniques[70] which can provide more precise information relating to the physical profile of the surface and photoelectron spectroscopy (ESCA)[71] and X-ray probe microanalysis (EDXRA)[72] which give chemical information.

The use of the sense of touch in subjectively assessing the hardness, rugosity, and frictional properties of a surface has been previously mentioned. The objective assessment of surface rugosity is discussed in preceding sections and frictional properties are extensively discussed elsewhere in this volume.[73] The question of objective assessment of surface hardness is a difficult one. Many semiquantitative techniques exist, but a comparison of them would be outside the intended scope of this chapter since the dividing line between the influence of surface and bulk properties is in this case diffuse. There is one technique that is particularly valuable, however, and relevant at this point since it provides an excellent and quantitative complement to goniophotometry and the related techniques mentioned here. Like goniophotometry it requires only small (*ca.* 1 cm^2) areas of relatively thin (*ca.* $10^{-3}-10^{-1}$ cm) film and is therefore ideally suited to the characterization of experimental polymers available only in small quantity, provided that they can be solution cast or melt pressed on to a suitable (glass or metal) substrate. This technique is termed 'micropenetrometry'. The prototype design of Monk and Wright[74] (now commercially available[75]) is capable of providing three types of information. The profile of the indentation curve indicated the viscoelastic nature of the specimen,[74] use of a spherical indenter with various loads enables Young's modulus and the rigidity modulus to be determined,[76,77] and a flat-ended indenter with a calibrated load indicates the resistance of the surface to deformation under the particular experimental conditions chosen (a facility exists on the instrument to vary the temperature between -40 and $90°C$). In this way the effect of eyelid load (*ca.* 2.6×10^4 dynes cm^{-2}) on specimens of materials for use in soft contact lenses and of similar thickness to the final lens (*ca.* 10^{-2} cm) may be determined.[77] The upper and lower limits of acceptability are related to the question of comfort as the eyelid moves over the surface, at one extreme, and the retention of visual acuity after blinking, at the other. The comparison of *in-vitro* examination with *in-vivo* clinical testing provides an interesting, if rather unusual example of the comparison of subjective and objective assessment.

ACKNOWLEDGEMENTS

I am grateful to Robert Molloy for many useful discussions and to Miss M. Devine for typing the manuscript.

REFERENCES

1. P. Nylen and E. Sunderland, *Modern Surface Coatings*, Interscience (1965).
2. R. S. Sinclair and W. D. Wright, *Appl. Optics*, **8**, 751 (1969).
3. J. Beresford, *J. Oil Chem. Assoc.*, **53**, 9, 800 (1970).
4. R. W. G. Hunt, *Rev. Progress in Coloration*, **2**, May, 11–9 (1971).
5. E. I. Stearns, *Textile Chemist and Colorist*, **6**, 2, 38 (1974).
6. F. W. Billmeyer, Jr., E. D. Campbell, and R. T. Marcus, *Appl. Optics*, **13**, 6, 1510 (1974).
7. U. Zörll, *Progr. Org. Coatings*, **18**, 113 (1972).
8. A. Andronow and M. Leontowicz, *Z. Physik*, **38**, 485 (1926).
9. V. G. W. Harrison and S. R. C. Potter, *Research*, **7**, 128 (1954).
10. T. H. Kosbahn, *Farbe. u. Lack.*, **70**, 693 (1964).
11. W. W. Barkas, *Proc. Phys. Soc.*, **51**, 274 (1939).

12. J. Nimeroff, *J. Opt. Soc. Am.*, **42**, 579 (1952).
13. W. Konig, *Plaste u. Kautschuk*, **16**, 5, 366 (1969).
14. R. S Hunter, *J. Research NBS*, **18**, 19 (1937).
15. R. S. Hunter and D. B. Judd, *ASTM Bulletin*, No 97, 11 (1939).
16. V. G. W. Harrison, *J. Sci. Instr.*, **26** 84 (1949).
17. H. K. Hammond, III, and I. Nimeroff, *J. Research NBS*, 585 (**1950**).
18. H. K. Hammond, III, *Off. Dig. Fed. Soc. Paint Technol.*, **36**, 471, 343 (1964).
19. H. R. Johnson, *J. Paint Technol.*, **40**, 527, 572 (1968).
20. V. G. W. Harrison, *J. Oil Col. Chem. Assoc.*, **36**, 8, 569 (1953).
21. D. L. Tillard and T. R. Ballett, *J. Oil Col. Chem. Assoc.*, **36**, 545 (1953).
22. 'Symposium on gloss measurement', *Off. Dig. Fed. Soc. Paint. Technol.*, **36**, 471, 343 (1964).
23. O. N. Eqirov and M. A. Slutskaya, *Continental Paint and Resin News*, **8**, 12, 9 (1970).
24. Anon, *Brit. Print*, **85**, 5, 91 (1972).
25. U. Zörll, *Continental Paint and Resin News*, **11**, 11, 3 (1973).
26. R. S. Hunter, *Off. Dig. Fed. Soc. Paint. Technol.*, **36**, 348 (1964).
27. H. J. Preier, *Farbe. u. Lack.*, **73**, 316 (1967).
28. O. Merz and U. Randel, *Farbe. u. Lack.*, **70**, 600 (1964).
29. T. H. Bohmann, *Farbe. u. Lack.*, **76**, 15 (1970).
30. L. A. Jones, *J. Opt. Soc. Amer.*, **5**, 213 (1921).
31. K. Hoffmann and T. H. Kosbahn, *Farbe. u. Lack.*, **72**, 119 (1966).
32. H. Loof, *J. Paint Technol.*, **38**, 501, 632 (1966).
33. M. A. Karagueuzoglou and R. Poisson, *Double Liaison*, **17**, 177, 51/269 (1970); **17**, 178, 53/357 (1970).
34. U. Veiel, *Farbe. u. Lack.*, **73**, 743 (1967).
35. J. Roire and A. Karagueuzoglou, *XIIth FATIPEC Congress*, Garmisch-Partenkirchen, 385 (1974).
36. F. Finus, *XIIth FATIPEC Congress*, Garmisch-Partenkirchen, 393 (1974).
37. E. Ledstedter, *Defazet*, **29**, 6, 258 (1975).
38. W. Carr, *J. Oil Col. Chem. Assoc.*, **57**, 12, 403 (1974).
39. J. H. Colling, W. E. Craker, and J. Dunderdale, *J. Oil Col. Chem. Assoc.*, **51**, 524 (1963).
40. H. Loof, *1st AIC Congress 'Colour'*, 69, Stockholm.
41. F. W. Billmeyer, Jr. and J. G. Davidson, *J. Paint Technol.*, **41**, 539, 647 (1969).
42. P. S. Quinney and B. J. Tighe, *Br. Polym. J.*, **3**, 274 (1971).
43. F. W. Billmeyer, Jr., R. L. Abrams, and J. G. Davidson, *J. Paint Technol.*, **40**, 519, 143 (1968).
44. F. W. Billmeyer, Jr. and R. L. Abrams, *J. Paint Technol.*, **45**, 579, 23 (1973).
45. F. W. Billmeyer, Jr. *J Paint Technol.*, **45**, 579, 31 (1973).
46. F. W. Billmeyer, Jr. and D. G. Phillips, *J. Paint Technol.*, **46**, 592, 36 (1974); *Am. Chem. Soc. Div. Org. Coatings and Plastics Chem.*, **33**, 1, 25 (1973).
47. D. G. Phillips and F. W. Billmeyer, Jr., *J. Coatings Technol.*, **48**, 616, 30 (1976).
48. J. H. Colling, W. E. Craker, M. C. Smith, and J. Dunderdale, *J. Oil Col. Chem. Assoc.*, **54**, 1057 (1971).
49. G. Guillaume, *Paint Technol.*, **33**, 1, 16 (1969).
50. R. Amberg, *J. Oil Col. Chem. Assoc.*, **54**, 211 (1971).
51. L. Gate, W. Windle, and M. Hire, *TAPPI*, **56**, 3, 61 (1973).
52. M. Tahan, *J. Paint Technol.*, **46**, 590, 35 (1974).
53. M. Tahan and B. J. Tighe, *J. Paint Technol.*, **46**, 590, 48 (1974).
54. M. Tahan, *J. Paint Technol.*, **46**. 597, 52 (1974).
55. M. Tahan, R. Molloy, and B. J. Tighe, *J. Paint Technol.*, **47**, 602, 52 (1975).
56. J. B. Edwards, *Paint R.A. Tech. Report*, TR/6/74 (1974).

57. J. H. Colling, W. E. Craker, and J. Dunderdale, *IXth FATIPEC Congress*, Sect. 3, 84 (1968).
58. U. Zörll, *J. Oil Col. Chem. Assoc.*, **59**, 439 (1976).
59. M. Hess, *J. Oil Col. Chem. Assoc.*, **39**, 185 (1956).
60. S. H. Bell, *J. Oil Col. Chem. Assoc.*, **43**, 466 (1960).
61. S. Wilska, *J. Oil Col. Chem. Assoc.*, **50**, 911 (1967).
62. R. D. Marley and H. Smith, *J. Oil Col. Chem. Assoc.*, **53**, 4, 292 (1970).
63. L. H. Prinan and F. L. Baker, *Am. Chem. Soc. Div. of Coatings and Plastics Chem., Abs. of Papers*, 161st meeting, Los Angeles, Ab. 73 (1971).
64. R. L. Dasai and W. A. Cote, Jr., *J. Coatings Technol.*, **48**, 614, 33 (1976).
65. N. Kosolsumallamas, Ph.D. Thesis University of Aston (1976).
66. A. Barnes, Ph.D. Thesis, University of Aston (1976).
67. R. Molloy, Ph.D. Thesis, University of Aston (1977).
68. W. E. J. Neal and R. W. Fane, *J. Phys (E)*, **6**, 5, 409 (1973).
69. D. M. Howell, *J. Oil Col. Chem. Assoc.*, **58**, 41 (1975).
70. H. J. Frier, *Farbe u. Lack.*, **79**, 307 (1973).
71. D. T. Clark, this volume.
72. S. S. Labana and M. Wheeler, *App. Polym Symposia*, No. 23, 61 (1974).
73. B. Briscoe, this volume and D. Dowson, this volume.
74. C. J. H. Monk and T. A. Wright, *J. Oil Col. Chem. Assoc.*, **48**, 520 (1975).
75. Research Equipment Limited, London.
76. R. J. L. Morris, *XIth FATIPEC Congress*, **1** (1972).
77. B. J. Tighe, *Br. Polym. J.*, **8**, 71 (1976).

15

The Examination of Polymer Surfaces by Infrared Spectroscopy

H. A. Willis and Veronica J. I. Zichy

I.C.I. Plastics Division, Welwyn Garden City

1. INTRODUCTION

Infrared spectroscopy is one of the most informative techniques in the study of the physical and chemical nature of polymers. With the realization that in many polymer applications, and particularly in the case of organic films, great interest attaches to the characterization of the polymer surface, considerable thought has been given to devising methods which will enable the infrared spectrum of the surface to be determined without interference from the bulk of the sample.

Two different methods of measuring the infrared spectra of polymer surfaces have been devised. Although both might be applied to polymer surfaces in general, they are most applicable to film surfaces. One method, described by Johnson[1] involves removing the surface layer by abrasion; the infrared spectrum of the material so removed being measured in a normal transmission experiment.

Alternatively, some form of reflection spectroscopy may be used, since a reflection spectrum of a sample is more representative of the nature of the reflection surface than of the bulk. If determined in the simple and obvious way, by measuring the radiation specularly reflected from the air/sample interface, the spectrum is more dependent upon refractive index change than upon absorption. This spectrum has the appearance of the first derivative of an absorption spectrum, and infrared spectroscopists, therefore, find it very difficult to understand. It, is, furthermore, very weak because the refractive index changes in organic materials are comparatively small in the infrared region.

However, reflection spectroscopy gained rapidly in popularity when it was realized that if the reflection was measured from the interface between the sample surface and a high refractive index optical element (Figure 1) rather than between the sample surface and air, the reflection spectrum not only becomes much more intense but it also becomes very similar to an absorption spectrum.

This form of spectroscopy was suggested almost simultaneously by Fahrenfort[2] and Harrick.[3] In the experiment usually described as ATR (attenuated total

288

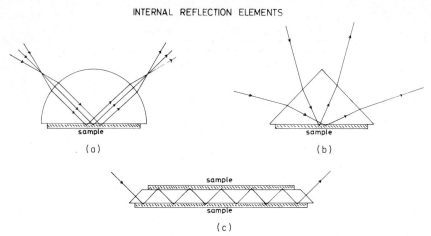

Figure 1. Internal reflection elements for infrared reflection spectroscopy. (a) Hemispherical ATR unit. Beam is well collimated on sample surface. (b) Isosceles prism ATR unit. Wide angular spread of beam on sample surface. (c) MIR reflection element.

internal reflection) spectroscopy, the spectrum is measured when the radiation beam has suffered a single reflection from the surface of the sample.

In the examination of surfaces, it is important to restrict the effective depth of penetration of the radiation into the sample. This is governed[4] by equation (1):

$$d_p = \frac{\lambda_0}{2\pi n_1 (\sin^2\theta - n_{21}^2)^{1/2}} \tag{1}$$

where d_p is the distance required for the electric field amplitude to fall to e^{-1} of its value at the surface of the sample, θ the angle of incidence between the radiation beam and the normal to the surface,

$$n_{21} = \frac{n_2}{n_1} = \frac{\text{refractive index of sample}}{\text{refractive index of the optical element}}$$

and λ_0 is the wavelength of the radiation. Examination of this equation shows:

(a) The depth of penetration increases linearly with the wavelength of the radiation. Thus, in relation to an absorption spectrum, bands become increasingly strong with increasing wavelength (decreasing frequency).

(b) The depth of penetration decreases as the angle of incidence of the radiation increases, as the refractive index of the optical element increases, and as the refractive index of the sample falls. Thus, all these changes reduce the intensity of the spectrum.

The only uncontrolled parameter in (b) is the refractive index of the sample. The refractive index of the sample varies in the vicinity of absorption bands, falling below the mean level on the high-frequency side and rising above the mean level on the low-frequency side of the bands. In extreme cases this is seen as a 'negative

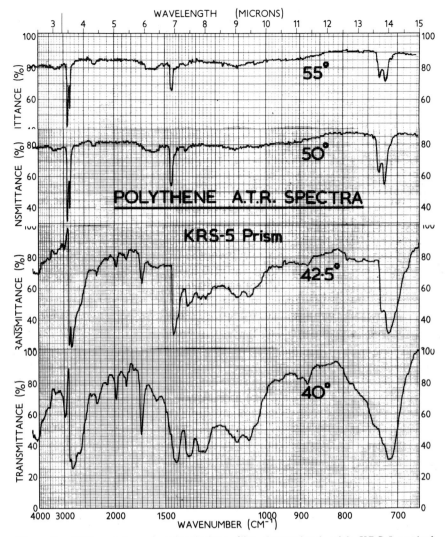

Figure 2. ATR spectra of polyethylene film determined with KRS-5 optical element with different angles of incidence. Note improved quality of spectrum, but decreased intensity, with increasing angle of incidence.

band' in the spectrum, for example at 2900 cm^{-1} in the polythene spectra in Figure 2 recorded at low angles of incidence. The corresponding effect of broadening on the low-frequency side of the bands is also evident in this figure, the distortion of the 720 cm^{-1} band being particularly noticeable.

Returning to equation (1), it may be seen by inspection that increasing θ or increasing n_1 will reduce changes in

$$\sin^2\theta - n_{21}^2$$

when n_2 changes. The improvement in the quality of the spectrum is evident in Figure 2 at the higher angles of incidence ($50°$ and $55°$). A similar effect results from an increase in the refractive index of the ATR element.

Thus, producing an undistorted spectrum, and restricting the spectrum to surface features are achieved simultaneously, but at the expense of intensity. This can be accommodated by ordinate scale expansion of the spectrum, but expansion may result in loss of minor spectral features by increasing the noise level. For this reason, the MIR (multiple internal reflection) method is often used (Figure 1(c)) in which attenuation is increased by causing the beam to be reflected successively from the sample surface. There is inevitably an increasing loss of radiation in the MIR experiment due to defocusing of the radiation beam as the number of reflections is increased, so a compromise must be made. Commercially, apparatus to give nominally 5, 9, or 25 reflections is available and the spectrum strength can be reduced, with a particular element, by progressively reducing the area of the element face which is contacted by the sample.

The effective penetration of the radiation into the sample as dealt with in detail by Harrick,[4] but generally a penetration of 10 000 Å or more is usual. This is very large compared with what may be achieved by some other spectroscopic techniques; nevertheless observations of considerable practical importance may be obtained by the infrared reflection method.

Johnson originally suggested[1] that his abrasion method might refer to a layer of 25 Å depth. This calculation was based on the weight of abraded material, and assumed uniform abrasion of the surface. Unfortunately, this is not the case; parts of the surface are deeply scored and other parts untouched. Subsequently, Johnson suggested[5] that a few hundred Ångstroms' penetration was more likely. This figure may be somewhat optimistic and perhaps 1000 Å would be more reasonable. Even so, the penetration in the ATR or MIR experiment is at least 10 times as great as that in the Johnson surface abrasion method. It is reasonable to ask why reflection measurements are favoured whilst surface abrasion is rarely done. The answer lies in the relative cost and convenience of the two methods. Apparatus for surface abrasion is not commercially available, but would cost realistically perhaps £1000 to build. Apparatus to measure ATR or MIR spectra costs £200 to £300, and is available from a number of accessory manufacturers as a unit which may be conveniently and quickly fitted to, and removed from, any commercially available infrared spectrometer. Measurement of infrared spectra with an ATR or MIR unit certainly requires a degree of skill greater than that necessary to measure transmission spectra, but when the operator has acquired this skill, reflection spectra may be obtained in as short a time as transmission spectra of comparable quality. Conversely, surface abrasion is a tedious experiment. Rather than the 10–15 minutes required to measure the reflection spectrum, one or two days would be considered reasonable to complete a successful experiment. Furthermore, the physical form of the surface is likely to be destroyed during abrasion so the method is usually applicable only where chemical features are of interest. Even so, the possibility of interaction between chemical groups in the surface layer, and the abrading powder (usually potassium bromide) must be considered in interpreting the infrared spectrum of the abraded material.

We propose, in this survey, to give examples to illustrate the wide range of problems which can be tackled by infrared spectroscopic examination of surfaces. The measurements reported have been obtained by simple spectroscopic methods, and do not reflect the improvements in sensitivity which may be expected to result from the use of a computer in conjunction with either a ratio recording infrared spectrometer or an interferometer.

2. EXPERIMENTAL

Spectra reproduced in all the figures except those measured with polarized radiation were measured on the Grubb—Parsons GS2A spectrometer, the Perkin—Elmer 457, or the Perkin—Elmer 580 spectrometers. All these are grating spectrometers.

For polarized infrared spectra, to study molecular orientation, polarized light was generated by means of a Cambridge Consultants Ltd wire grid polarizer* which we find to be a great improvement on reflection or multiplate transmission polarizers. One polarizer is mounted in the common beam of a double-beam instrument, that is in such a position that both beams pass through it (e.g. immediately before or after the exit slit). This gives much flatter backgrounds than other arrangements. Spectra with polarized light were measured on the Perkin—Elmer 157P or Hilger H800. In these prism instruments the inherent instrument polarization is small and almost constant with changing frequency, as opposed to grating instruments, in which the polarization is large and changes rapidly with change of frequency.

To avoid confusion, we refer only to the plane of the electric vector of the polarized radiation beam, and describe this as vertical or horizontal by reference to the axes of the spectrometer. There is no practical ambiguity here, since spectrometers are invariably operated with the active surface of the dispersing element in a vertical plane, and it is not practical to use diagonal illumination with internal reflecting elements.

ATR and MIR units from a number of manufacturers were used. We found that best-quality ATR spectra were obtained with a micro-ATR unit with a hemicylinder as the high refractive index element, the sample being held in contact with the hypotenuse face (Figure 1(a)). The explanation may be that suggested by Harrick[4] that, when correctly focused, the hemicylinder produces a well-collimated beam covering a small sample area. The internal reflection elements used were of thallium bromo-iodide (KRS5), (refractive index ≈2.4), or germanium (refractive index ≈4.0). Good contact between sample surface and prism surface is essential. A portion of the film is cut out to fit precisely the surface of the reflection element to

---

*Grid polarizers are marketed by Cambridge Consultants Ltd., Cambridge, and Perkin—Elmer, Inc. Norwalk, U.S.A.

be used. Film samples tend to adhere well to the surface, but we ensure good contact by placing a backing pad of butyl rubber behind the sample for ATR experiments. For MIR experiments we use three or four layers of filter paper, again cut to the size of the element surface.

3. THE SURFACE ABRASION METHOD

This method, in its original form, is described in detail and with some very clear figures, in Johnson's paper.[1] Our experience of his method suggested, however, that some minor modifications led to a considerable improvement in its effectiveness. We found that the steel wool abrasion disc, which Johnson describes, cut deep trenches in the surface, as observed under the optical microscope. Although this is probably irrelevant when some additive to the polymer such as an antioxidant or slip agent is to be removed from the surface, it is not an efficient method of sampling when it is required to abrade from the surface a layer of the polymer which has been modified by oxidation, acid treatment, or corona discharge. We retained the assembly much as described by Johnson, but replaced the steel wool pad by a potassium bromide disc; this was prepared following the usual practice in a disc press, but with potassium bromide purified by reprecipitation and drying. This potassium bromide disc was mounted in a steel frame. This enables the disc to be rotated by means of a drive magnet, mounted under the film surface, following Johnson's original method. We encountered considerable difficulty through contamination of the surface with dust from the atmosphere, and found it necessary to enclose the apparatus in a 'Perspex' box. After the abrasion operation had been completed we collected the potassium bromide powder from the surface with a camel-hair brush and pressed this, together with the potassium bromide abrading disc, into a sample for measurement by transmission spectroscopy. In our experiments we abraded an area of film 12 inches (300 mm) square, and finished with a potassium bromide disc about 10 sq. mm area. Thus, a compression of the surface is achieved by a factor of 10^4.

4. OBSERVATIONS ON CORONA-DISCHARGE TREATMENT OF POLY-OLEFIN FILMS IN SURFACE ABRASION

Johnson describes[5] abrasion of polyethylene film before and after treatment to enhance adhesion by what is evidently corona discharge.[6] Whilst we agree qualitatively with his data, we disagree with his interpretation. The change in the 7.3 μm ($\sim$1370 cm^{-1}) region is in our opinion not due to an intensification of the methyl group band, but arises from the appearance of a new band at 7.225 μm (1384 cm^{-1}) on the side of the methyl band. We believe this new feature is due to the nitrate ion. This is presumably generated from the nitrogen of the air under the discharge. The species involved is potassium nitrate formed by double decomposition with the potassium bromide. In addition to the nitrate ion, we find evidence also of nitrate ester.

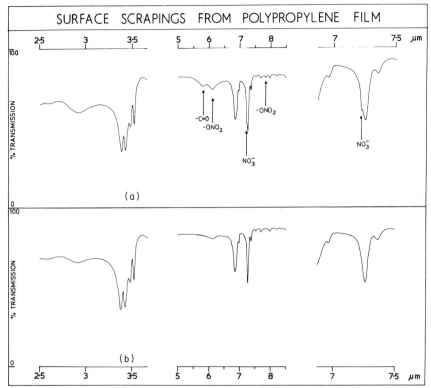

Figure 3. Spectra of material abraded from surfaces of corona-discharge treated polypropylene film. (a) Treated surface. (b) Opposite (non-treated) surface. Trace on the right is expanded, to show NO_3^- band.

A similar reaction occurs in the corona-discharge treatment of polypropylene film, and we show our findings in Figure 3. The two spectra were obtained by abrasion sampling of the corona-discharge treated surface, and the non-treated reverse side, of a polypropylene film. Absorption bands due to the various nitrogen-containing groups are evident and our assignment is marked on the figure. Our film did not contain amide additives, hence the carbonyl species formed during the treatment are readily seen. The hydroxyl absorption near 3 μm (3300 cm^{-1}) is we believe genuinely due to hydroxyl groups removed from the surface, as we took great precautions to purify and dry the potassium bromide. We believe that Johnson's observation that the corona-treated surface of a film containing an amide slip agent is relatively deficient in amide is a misinterpretation of his data, and we shall discuss this later.

Thus, our examination reveals both oxygen- and nitrogen-containing groups on the surface of polyethylene and polypropylene after corona-discharge treatment, but clearly much more work would be needed to establish the relevance of this surface chemistry to the adhesion process.

5. APPLICATIONS OF INTERNAL REFLECTION SPECTROSCOPY TO SURFACE CHEMICAL PROBLEMS

(i) Introduction

Internal reflection spectroscopy is, experimentally, much simpler than the surface abrasion technique and is applicable to the investigation of both chemical and physical features of a polymer surface. Thus Shimada and Hoshino[7] studied the effect of fluorine gas on the surface of a number of synthetic polymer films, whilst Blais *et al.*[8] have followed the photodegradation of the surface of polyethylene terephthalate film. Luongo and Schonhorn studied the crystallinity of the surface of polyethylene film nucleated against high- and low-energy surfaces,[9] and Blais *et al.*[10] report that surface oxidation of polypropylene film results in chain scission of the polymer, which is followed by an increase in crystallinity of the surface layer.

In our experiments, we have applied surface spectroscopy to observe chemical changes due to oxidation and acid etching, and reactions such as cross-linking occurring in thin coatings on a polymer surface. We have also observed physical changes resulting from attack on the surface by solvents, and the orientation which occurs when a film is stretched. We have noted also that the nature of the surface may influence the physical state of additives which 'bleed out' from the composition on to the polymer surface. In section 2(ii) we shall give examples in each of these areas.

(ii) Surface oxidation of polyethylene

For some packaging applications, where gas and water permeability must be restricted, aluminium foil is used. A layer of polyethylene ~30 μm thick is applied by melt extrusion to one surface of the foil. This coating enables heat-sealed packages to be made. The stability of the container depends upon producing a strong bond between the extruded polyethylene coating and the aluminium foil, and occasional failures at this interface led to an investigation of the effect of the more obvious variables in the process, namely the temperature of the extruded polyethylene melt curtain and the linear speed of the foil during the extrusion operation.

Initially, we tried to measure the spectrum of the polyethylene by flattening an area of the laminate and attempting to measure a 'double-transmission' spectrum of the polyethylene layer by reflecting the radiation beam from the aluminium surface. This gave no useful information. We then dissolved away the aluminium from samples of the laminate by immersion in sodium hydroxide solution and measured the transmission spectrum of the polyethylene film which remained. In all cases we measured only a spectrum of unmodified polyethylene, no differences being observed which related to the variables of extrusion temperature or line speed.

We then measured MIR spectra from each of the two surfaces of the polyethylene and found that both the surface in contact with the aluminium and the outer surface were oxidized, and both to about the same extent. This meant that we did not need to remove the aluminium, and could record the surface spectrum from the outer polyethylene surface because the polyethylene layer was so thick that the radiation beam, in the MIR measurement, did not penetrate to the polyethylene–aluminium interface.

With this simplified form of experiment, we were able to record surface spectra easily from a large number of samples, and some representative spectra are given in Figure 4. Either increasing the temperature of the melt curtain, or reducing the line speed, led to an increase in the intensity of the pair of bands at 1720 and 1735 cm^{-1}, which we ascribe to carbonyl groups in the polyethylene surface.

The lower frequency band we consider to arise from ketonic and acidic carbonyl groups. The higher frequency band we believe arises from the ester group, which results from a cross-linking reaction between carboxylic acid and hydroxyl groups.[11]

In order to relate the degree of surface oxidation to the peel strength of the polyethylene to aluminium laminate, we needed to measure these MIR spectra quantitatively.

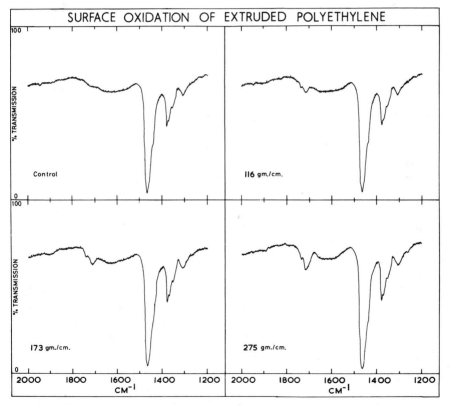

Figure 4. MIR spectra with KRS-5 element of polyethylene extruded on to aluminium foil. Samples of higher carbonyl content show increased peel strength of polyethylene/aluminium seal (see Figure 5).

The quantitative application of MIR spectra raises some difficult theoretical points, and the subject is discussed in detail by Harrick.[4] As Harrick points out, for weak bands, and with the depth of penetration restricted so that the spectrum approximates to an absorption spectrum, band intensities can be regarded as following the Beer–Lambert law, so long as there is good contact between the sample and the reflection element. However, completely reproducible contact between sample and ATR or MIR element cannot be assumed when changing from one sample to another. Thus, it is not possible to relate absorbances directly in different spectra. We have adopted the method of internal normalization. This involves dividing absorbances of bands of interest in a particular spectrum by the absorbance of a reference band. This will usually be a feature of the spectrum arising from a group which should be at the same concentration in all samples. Absorbances, normalized in this way, against a common reference, may now be compared from one spectrum to another. Even so, these reflectance measurements cannot be related directly to those measured on a normal absorption spectrum, because band intensities in the reflection spectrum increase with decreasing frequency (equation (1)).

In the quantitative measurements described in this paper, all absorbances are measured against linear backgrounds as in conventional transmission spectroscopy, and in all cases absorbance ratios are quoted against a reference band.

In the present case a further difficulty was presented because the bands due to ketonic, acidic, and ester carbonyl groups overlap, and they almost certainly all have different extinction coefficients. Since there is no simple way of resolving this problem, we added together the absorbances at 1720 and 1735 cm^{-1}. As a reference we used the absorbance of the methyl group at 1375 cm^{-1}. This absorbance ratio is plotted against peel strength in Figure 5. The excellent relationship shows that surface oxidation is the predominant factor in adhesion between these two surfaces.

In the manufacture of the aluminium foil/polyethylene laminate, both sides of the polyethylene layer are oxidized to the same extent. Increased peel strength between polyethylene and aluminium is limited by the increasing difficulty of sealing the polyethylene layers together as their surfaces become increasingly oxidized, so a compromise must be reached. This infrared test is of particular value in establishing the optimum conditions for making this laminate, enabling plant operating conditions for a satisfactory product to be established quickly, and for routine control purposes in maintaining product quality.

Corona-discharge treatment is only one of the methods used for increasing the strength of the bond between a polyethylene surface and other resins; a number of etching reagents have been suggested, including chromic acid.[12] We have examined a number of samples prepared by Brewis and co-workers by immersing polyethylene film in chromic acid solution $[K_2Cr_2O_7 : H_2O : H_2SO_4 :: 4.4 : 7.1 : 88.5]$ at $70°C$ for quite short periods (up to 60 sec). During this time the adhesive bond to other materials increases rapidly in strength (Blais et al.[13]).

Surface spectra of polyethylene film etched in this way are shown in Figure 6. After 30 sec etching at $70°C$ we observe new structure at 1215 and 1260 cm^{-1}.

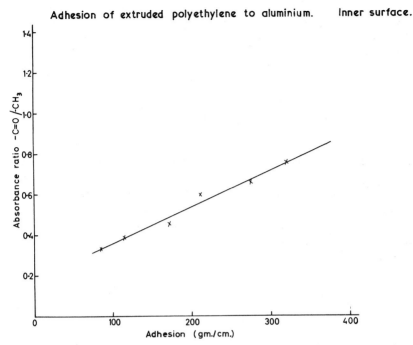

Figure 5. Peel strength of polyethylene/aluminium laminate as a function of the carbonyl content of the polyethylene surface.

This band pair could well arise from the presence of an alkyl sulphate group on the surface. After 60 sec or more exposure to the reagent, our results are similar to those reported by Blais *et al.*[13], the significant new bands occurring at ~1200 cm^{-1} ($vS=O$) and ~1050 cm^{-1} ($vS—O$). These correspond to the appearance of an alkly sulphonate group. Bands near ~3300 and 1700 cm^{-1}

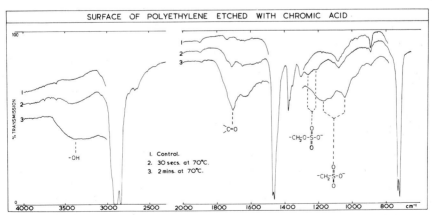

Figure 6. MIR spectra with KRS-5 element of polyethylene etched with chromic acid solution, composition $K_2Cr_2O_7 : H_2O : H_2SO_4 : : 4.4 : 7.1 : 88.5$ by weight.

298

indicate the appearance of hydroxyl and carbonyl groups on the polyethylene surface. Blais *et al.*[13] suggest that these polar groups are responsible for the increase in adhesive bonding, but there appears to be no direct evidence to support this view.

(iii) Reactions occurring on film surfaces

Many adhesives and coatings are applied to a polymer surface in the form of liquid constituents which then react to form a solid product. Infrared spectroscopy is a simple method of following the chemical reactions which occur in these materials. For example,[14] the curing of an epoxy resin can be followed in infrared transmission spectroscopy by observing the reduction in intensity of the band at 915 cm^{-1} due to the epoxy ring, which decomposes in the cross-linking process. This reaction can also be followed by reflection spectroscopy in a thin layer of the

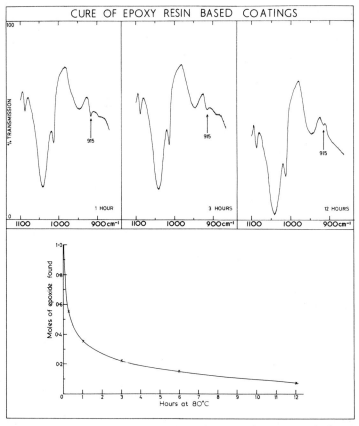

Figure 7. MIR spectra with KRS-5 element of epoxy resin layer (~3 μm thick) on a polymer film, showing disappearance of epoxy group band on curing at 80°C.

resin (~3 μm thick) which has been applied to the surface of a polymer film (Figure 7). In this case the resin was cured at 80°C, and it is evident from the spectra that the cure is virtually complete after 12 hr.

Polyester film, vacuum coated with aluminium, and with a layer of coloured lacquer covering the aluminium surface, is used for decorative purposes. The lacquer is frequently based on a polyurethane resin, and it is important that the resin cures adequately, otherwise the wear resistance is poor. The cure of a polyurethane resin, *in situ* on the aluminium surface, may be conveniently followed by MIR spectroscopy, as shown in Figure 8. In this case, the band at 2280 cm^{-1}, due to the isocyanate group, disappears as the resin cures. The reaction is followed by plotting the absorbance ratio N=C=O to NH as a function of time (Figure 8).

Epoxy and polyurethane resins form the base of a large number of lacquers, surface finishes, and adhesives, and it is of considerable interest to be able to follow the chemistry of the cross-linking process when such a material is *in situ* on a polyolefin or polyester film.

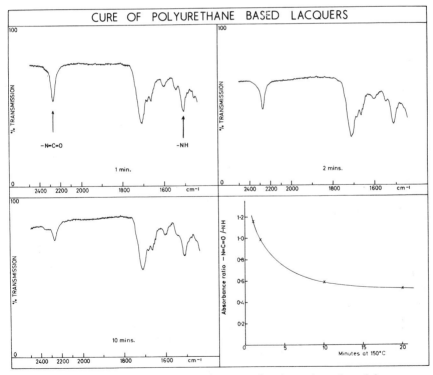

Figure 8. MIR spectra with KRS-5 element of polyurethane-based lacquer on the aluminized surface of polyester film. Owing to high intensity of spectra, only one side of the element was covered by the sample. Curing reaction followed at 150°C.

(iv) Additives on film surfaces

Most additives to films (slip agents, antioxidants, antistatic agents) slowly bleed out of the bulk of the film and on to the surface. The majority of these compounds are polar, and may be detected on the surface either by the abrasion method, as described by Johnson,[1,5] or by ATR or MIR spectroscopy.

Aliphatic esters such as glyceryl stearates may be observed by means of their $\nu C=O$ and $\nu C-O$ bands near 1730 and 1180 cm^{-1}, respectively, and in alkyl ethers such as polyethylene oxide the $\nu C-O-C$ band near 1100 cm^{-1} is distinctive. Tertiary amine additives, on the other hand, are almost impossible to detect by these methods as they have no strong characteristic absorption bands.

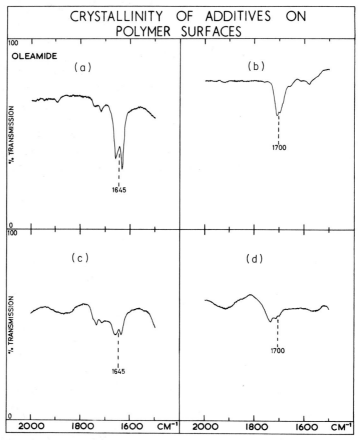

Figure 9. MIR spectra with KRS-5 element of oleamide on the surface of polypropylene film. (a) Transmission spectrum of crystalline oleamide. (b) Transmission spectrum of oleamide in dilute solution in hydrocarbon oil. (c) Spectrum of untreated polypropylene film surface. (d) Spectrum of corona-discharge treated polypropylene film surface.

The detection of long-chain primary amides, such as oleamide and stearamide, added as slip agents is of some interest. Johnson[5] showed that in the case of polyethylene film containing a long-chain amide slip agent, the carbonyl band due to the amide group at ~ 1650 cm^{-1} was relatively weaker on the corona-discharge treated side of the film. Our own measurements (Figure 9) confirmed this observation, which is surprising, as there is no evident reason why corona-discharge treatment should suppress the migration of additives to the surface.

Further investigation shows, however, that while the doublet band at about 1645 cm^{-1} is weak or absent, new bands appear close to 1700 cm^{-1}, as shoulders on the carbonyl band at 1720 cm^{-1}. The latter arises from the reaction with oxygen during the corona-discharge treatment. Reference to the comparison transmission spectra in Figure 9 show that the doublet near 1645 cm^{-1} is present in the spectrum of the amide when it exists as a crystalline solid, but when in solution in a hydrocarbon solvent the band moves to about 1700 cm^{-1} and decreases in strength. The intermolecular hydrogen bonding which exists in the crystal is destroyed when the material is dissolved in a relatively non-polar solvent, as a result a substantial increase occurs in the carbonyl frequency. A similar effect is observed on the carbonyl frequency of carboxylic acids between the solid and dilute solution states.

We suggest, therefore, that the transport of the long-chain amide to the corona-discharge treated surface is not suppressed, but that the additive which has reached this surface is, in effect, dissolved in the surface layer. It seems possible that the increased polarity of this surface enhances the solubility of the additive.

6. PHYSICAL EFFECTS AT POLYMER SURFACES

(i) Introduction

Although we have so far emphasized the chemical effects which may be observed at polymer surfaces by infrared spectroscopy, the ability to use this technique to study the physical nature of surfaces is of considerable interest. We now give examples in which surface treatment affects the conformation, orientation, or crystallinity of film surfaces.

(ii) Solvent etching

Solvent etching is occasionally used to promote adhesion to polymer films (e.g. McGowan[11]). We thought it would be of interest to see if any difference could be found after a film surface had been etched.

In Figure 10 we show the surface spectrum of a polyethylene terephthalate film (biaxially drawn and heat set). The two spectra are those of the opposite sides of film, but one surface has been etched with trichloracetic acid. The differences between the spectra are most marked. The polyethylene terephthalate molecule exists in the extended or *trans*-form, and the bent, or *gauche* form. The *trans*-isomer can be present in either crystalline or amorphous regions. The *gauche* is present only

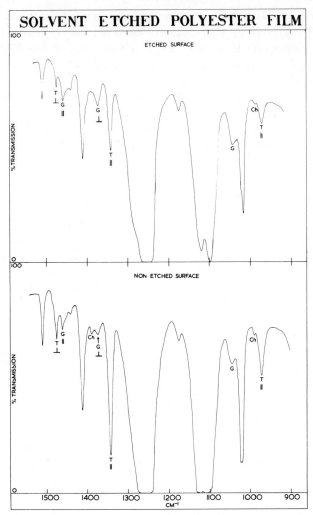

Figure 10. MIR spectra with germanium element of polyethylene terephthalate film spectra of etched and untreated surface. Owing to strength of spectrum, only one side of the element was covered by the sample. T = band due to *trans*-isomer, G = band due to gauche isomer, Ch = band due to chain fold in crystalline lamellae.[22]

in amorphous regions. Bands appear in the infrared spectrum from each of these forms (usually known as *trans*- and *gauche* bands).[15] Comparing the spectra in Figure 10, it is evident that all the bands due to the *trans*-isomer have decreased significantly in intensity in the etched surface. This applies to both parallel and perpendicular bands of the *trans*-isomer, so the effect is not merely due to reorientation of the surface layer. Bands marked Ch have also decreased in intensity. According to Koenig and

Hannon[16] these latter bands arise from chain folds in the crystalline regions of the material. It may be said with confidence, therefore, that the solvent has destroyed crystalline regions of the surface; this observation is confirmed by the relative strengthening in the spectrum of the etched surface, of bands due to the gauche isomer, marked G.

These observations, although well founded, are surprising, especially as Blais *et al.*[10] found that oxidative attack on polypropylene increased surface crystallinity. It is possible that trichloracetic acid acts by hydrolysis of the film surface and the chain folds in crystalline regions may be particularly prone to attack, thus destroying the crystalline lamellae.

(iii) Measurement of surface orientation in polypropylene

The measurement of orientation in polymer films by transmission spectroscopy with polarized radiation is well known.[17] Bands in the absorption spectrum are classified as parallel if they appear more strongly when the plane of the electric vector is parallel to the molecular axis. In the case of one-way drawn film this is predominantly in the direction of draw (often known as the machine direction). Conversely, these bands which are stronger when the plane of the electric vector is at right angles to the molecular axis are classified as perpendicular bands. The polypropylene molecule is ordinarily present in the polypropylene film in a threefold helix form, i.e. a turn of the helix is completed in three monomer repeat units.[18] In a simplified view, therefore, the one-way drawn polypropylene film may be thought of as an assembly of long thin cylinders, rather like a packet of spaghetti, with their long axes all parallel and in the draw (machine) direction.

Molecular motions along the molecular axis result in parallel bands, molecular motions at right angles to the molecular axis result in perpendicular bands. Because the molecule has cylindrical symmetry, different directions perpendicular to the axis cannot be distinguished, i.e. all perpendicular motions are equivalent (in spectroscopic terms they are degenerate). These considerations are of importance in interpreting the spectra of the surfaces of oriented polypropylene film.

The oriented film was examined by being placed in contact with the faces of an MIR element (Figure 1(c)). Portions of the film were cut so that the molecular axis was vertical (MD vertical) and examined with the radiation beam polarized vertically, and then with the beam polarized horizontally. These pieces of film were then removed from the element, and replaced with pieces cut so that the draw direction was horizontal (MD horizontal). This assembly was examined with vertically polarized light, and then with horizontally polarized light. The MD and polarization directions are indicated by the arrows in Figure 11. Additionally, spectra are given in this figure measured with the MD vertical and horizontal, but with no polarizer in the spectrometer. Whereas the pair of spectra with no polarizer are almost identical, and show both parallel and perpendicular bands, the four views with polarized light show pronounced differences.

When the MD is vertical, if the sample is well oriented, vertical polarization can only activate parallel bands and, indeed, in this view only parallel bands are seen.

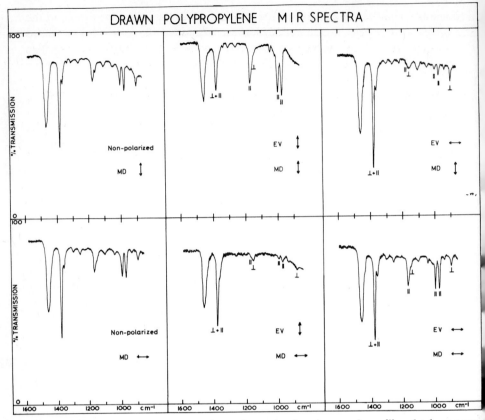

Figure 11. MIR spectra with germanium element of polypropylene film 6 : 1 one-way drawn. Direction of draw (MD) and plane of electric vector (EV) indicated by arrows.

The band at 1375 cm^{-1} (symmetrical methyl deformation motion) has both parallel and perpendicular components, and hence appears in all views. However, the perpendicular component is stronger, so in the fully parallel view the 1375 cm^{-1} band appears at its weakest, whilst the purely perpendicular bands at 899 and 1150 cm^{-1} are almost completely absent. When the polarizer in the instrument is rotated through 90° so that the plane of polarization is horizontal, only perpendicular bands should be activated, and indeed in this view the parallel bands are extremely weak. The residue probably arises through imperfection of orientation.

Perpendicular bands are absent from the parallel view, whereas parallel bands are seen weakly in the perpendicular view. Presumably this is merely because the parallel bands are inherently much stronger than the perpendicular bands in the region of the spectrum examined.

When the sample is rotated through 90° (MD horizontal) and illuminated with a beam in which the electric vector direction is vertical, again only perpendicular bands should be activated since no component of a parallel molecular motion exists in the vertical plane. The spectrum is clearly very similar to the other 'crossed' view in that perpendicular bands are strong and parallel bands very weak, but the entire spectrum is weaker. Presumably the reason is that the depth of penetration of the radiation into the sample is less, by about a factor of two under the present conditions, when the electric vector is vertical (see Harrick,[4] p. 44).

The other view, in which both EV and MD are horizontal, is different from those discussed above in that activation of both parallel and perpendicular bands would be expected. Indeed, this EV horizontal, MD horizontal view is remarkably similar to that obtained without a polarizer.

This work on one-way drawn polypropylene agrees well with that reported by Flournoy.[19] It is clear that, in general, the behaviour of parallel and perpendicular bands in ATR or MIR spectroscopy is in line with that which would be observed in the more usual transmission experiment.[20]

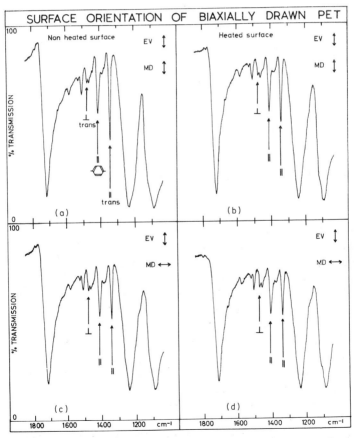

Figure 12. ATR spectra with KRS-5 element, of polyethylene terephthalate film biaxially drawn after being heated from one side only; (a), (b), (c), and (d) views are described in the text.

(iv) Surface orientation of polyester film

Having established the validity of the treatment, we can now proceed to a practical problem. When polyester film is biaxially drawn, it is necessary to heat it first. This may be done by heating the film from both sides (e.g. in an air oven) or from one side by radiant heating. If heated from one side, it seems possible that the two sides of the drawn film may have different orientation.

In biaxially drawn polyethylene terephthalate film, there is usually a greater degree of axial alignment of the molecules in one direction; and this direction may be found by ascertaining the direction of highest refractive index in the plane of the surface. This is marked MD in Figure 12, and is in the same direction on both sides of the film. This axial alignment is evident from the figures, as in both cases the parallel bands are stronger, and the perpendicular bands weaker when EV is parallel to MD. On the other hand, comparing corresponding pairs on the heated and non-heated sides, parallel bands are stronger in (a) than in (b), and are also stronger in (c) than in (d).

This effect arises from the planar orientation of the film. In addition to the axial orientation, the biaxial draw draws the plane of the molecules into the plane of the film. When the planar orientation increases, parallel bands increase in strength. It is apparent that the planar orientation of the non-heated surface (spectra (a) and (c)) is higher than that of the heated surface (spectra (b) and (d)).

It is clear, therefore, that MIR or ATR spectroscopy can reveal the orientation of biaxially drawn film more completely than is possible by normal transmission measurements, in which any variation of orientation through the film is averaged and not detected.

7. CONCLUSIONS

We have endeavoured, in this survey, to indicate the many features of surfaces, both chemical and physical, which can be revealed by infrared spectroscopic studies. The examples have been chosen so as to indicate the important and practical problems which can be investigated by these methods.

ACKNOWLEDGEMENTS

We would like to thank our co-workers in the Spectroscopy Group at ICI Plastics Division for recording the spectra which we have used in this survey, and to E. L. Zichy for providing information on his modifications of the Johnson surface abrasion technique. We thank J. M. Chalmers for preparing the figures.

REFERENCES

1. W. T. M. Johnson, *Official Digest of the Oil and Colour Chemists' Association,* **32**, 1067 (1960).
2. J. Fahrenfort, *Spectrochimica Acta,* **17**, 698 (1961).
3. N. J. Harrick, *Phys. Rev. Letters,* **4**, 224 (1960).

4. N. J. Harrick, *Internal Reflection Spectroscopy*. Interscience, New York (1967), pp. 30, 44.
5. W. T. M. Johnson, *Official Digest of the Oil and Colour Chemists' Association*, 33, 1489 (1961).
6. P. Blais, D. J. Carlsson, and D. M. Wiles, *J. Applied Polymer Science*, 15, 129, (1971).
7. J. Shimada and M. Hishino, *J. Applied Polymer Science*, 19, 1439 (1975).
8. P. Blais, M. Day and D. M. Wiles, *J. Applied Polymer Science*, 17, 1895 (1973).
9. J. P. Luongo and H. Schonhorn, *J. Polymer Science*, A-2, 6 1649 (1968).
10. P. Blais, D. J. Carlsson, and D. M. Wiles, *J. Polymer Science*, A-1, 10, 1077 (1972).
11. J. McGowan, private communication.
12. D. M. Brewis, *J. Material Science*, 3, 262 (1968).
13. P. Blais, D. J. Carlsson, G. W. Csullog, and D. M. Wiles, *Journal of Colloid and Interface Science*, 47, 636 (1974).
14. R. E. Burge and B.P. Geyer *The Analytical Chemistry of Polymers*, ed. G. M. Kline, Interscience, New York, Pt. 1, p. 123 (1959).
15. P. G. Schmidt, *J. Polymer Science*, A, 1, 1271 (1963).
16. J. L. Koenig, and M. J. Hannon, *J. Macromol. Sci. (Phys)* B1, 119 (1967).
17. S. Krimm, *Advances in Polymer Science*, 2, 51 (1960).
18. G. Natta, P. Corradini, and M. Cesari, *Atti. Accad. Naz. Lincei, Rend. Classe Sci. Fis., Mat. Nat.*, 21, 365 (1956).
19. P. A. Flournoy, *Spectrochimica Acta*, 22, 15 (1966).
20. J. H. Schachtschneider and R. G. Snyder, *J. Polymer Science*, C7, 99 (1964).

16

The Investigation of Polymer Surfaces by Means of ESCA

D. T. Clark

University of Durham

1. INTRODUCTION

Since solids communicate with the rest of the universe by way of their surfaces, it is a truism that structure and bonding (in the chemical sense) of the surface of solids is of fundamental importance in any detailed discussion at the molecular level of many important phenomena. Thus, the characterization of polymer surfaces is clearly central to a detailed understanding of both surface polymerizations (synthesis) and the physical, chemical, electrical, and mechanical properties of polymer samples. Despite the obvious importance of the nature of the surface and immediate subsurface of polymer samples there are few techniques currently available for routinely delineating the aspects of structure and bonding pertaining to these regions. In Chapters 14 and 15 consideration has been given to the techniques which have traditionally been employed in answering some of the important questions relating to the characterization of polymer surfaces.[1] Thus, with due allowance for surface topographical effects the measurement of surface free energies can provide indirect evidence on the structure of the immediate surface of a polymer sample, whilst multiple attenuated total internal reflectance spectroscopy[2] can provide valuable information on the outermost few hundreds of Ångströms. The available information levels from these and other traditional techniques is somewhat limited, however, and it is not generally possible to elaborate details of structure and bonding in the immediate surface and subsurface of a sample, much less tackle the more difficult problem of the investigation of inhomogeneous samples, such as those that arise from the initial interaction of a polymer surface with a given agent followed by diffusion into the sample. This circumscribes many important areas such as the initial stages of weathering and chemical modification and degradation in general.

Fortunately, the past few years have witnessed the emergence of electron spectroscopy for chemical applications (ESCA) as a spectroscopic tool *par excellence* for studying in considerable detail aspects of structure and bonding in the surface regions of polymers.[3] Indeed, the available information levels from a single experiment are such as to allow the elaboration of details of the surface regions of inhomogeneous samples and since, as will become apparent, the typical sampling depth is < 100 Å the technique nicely complements those discussed in the preceding chapters.

The application of ESCA to the study of polymers has been the subject of numerous reviews over the past five years, and the number of contributed papers on this and related topics in the symposium attests to the increasing awareness of the great potential of the technique.[3] Indeed, with the current background of data accumulated from the study of well-authenticated systems it is possible to study quantitatively systems of considerable complexity. Indeed, it will become apparent that there are areas of study in which the required information can only at present be derived from ESCA studies, whilst in others the technique nicely complements the more established spectroscopic tools. In general, however, ESCA provides data at a much coarser level than most other spectroscopic tools and information pertaining, for example, to conformational effects may only be inferred rather indirectly. In many areas of application ESCA does not compare favourably in terms of resolution, sensitivity, etc. with more established spectroscopic tools. The fact remains, however, that this is more than compensated by the great range of information available from a single ESCA experiment, such that in the future one can envisage that ESCA will be the technique of choice for any initial investigation of a polymer sample. The ability to provide information straightforwardly on uncharacterized samples is unique to ESCA and gives the technique great potential (already exploited in some areas) for tackling not only academic problems but those of an applied 'trouble-shooting' nature.

In this chapter we briefly consider the available information levels from the ESCA experiment, together with examples taken from recent work which should serve to illustrate to the uninitiated the important role that the technique has in general in the development of the surface science of polymers.

2. THE ESCA EXPERIMENT

ESCA involves the measurement of binding energies of electrons ejected by interactions of a molecule with a monoenergetic beam of soft X-rays.[4] For a variety of reasons the most commonly employed X-ray sources are $Alk\alpha_{1,2}$ and $MgK\alpha_{1,2}$ with corresponding photon energies of 1486.6 and 1253.7 eV, respectively. In principle, all electrons from the core to the valence levels can be studied, and in this respect the technique differs from uv photoelectron spectroscopy (UPS) in which only the lower energy valence levels can be studied.[5] The basic processes involved in ESCA are shown in Figure 1.

With the conventionally employed X-ray photon sources, cross-sections for core levels for most elements of the periodic table are within two orders of magnitude of that for the C_{1s} levels, and the technique thus has a convenient sensitivity range for all elements.[6] The cross-sections for photoionization of core levels is generally considerably higher than for valence levels, and this taken in conjunction with the fact that core orbitals are essentially localized on atoms and therefore have binding energies characteristic of a given element, means that in ESCA the predominant

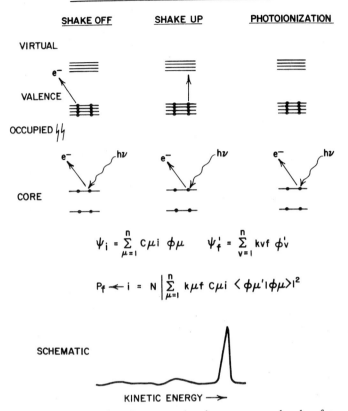

Figure 1. Schematic of core and valence energy levels of a molecule, illustrating the realationship between direct photo-ionization, shake-up, and shake-off phenomena. (*The shake-up states may be viewed as monopole excited states of the core-ionized system and their probability of excitation depends on the overlap integral between the occupied orbital of the initial system and the virtual orbital of the core-ionized system.*)

emphasis is on the study of core levels. Although core electrons do not take part in bonding they monitor closely valence electron distributions, and it is this particular feature which endows the technique with such wide-ranging capabilities.

The removal of a core electron (which is almost completely shielding as far as the valence electrons are concerned) is accompanied by substantial reorganization of the valence electrons in response to the effective increase in nuclear charge. This perturbation gives rise to a finite probability for photoionization to be accompanied by simultaneous excitation of a valence electron from an occupied to an unoccupied level (shake-up) or ionization of a valence electron (shake-off). These processes giving rise to satellites to the low kinetic energy side of the main

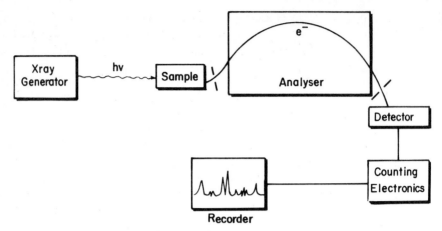

Figure 2. Schematic illustrating basic experimental set-up for ESCA.

photoionization peak, follow monopole selection rules and their measurable parameters (intensities and separation with respect to the direct photoionization peaks) enormously broadens the scope of the technique, as will become apparent in the ensuing discussion.

The basic experimental set-up is shown in Figure 2, and is largely self explanatory.

The most flexible of the commercially available designs employ double-focusing electrostatic analysers with retarding lens systems.† This allows ready access to the sample for in situ preparation or modification and also allows angular studies to be made straightforwardly.

To maintain the integrity of the sample during the time taken for measurement and to obviate scattering of electrons entering the analyser, samples are maintained in the spectrometer source at pressures of 10^{-7} Torr or better. The relatively low sticking coefficient for most small molecules comprising the extraneous atmosphere for polymers contrasts with the situation for evaporated metal films, where pressures in the range 10^{-10} Torr would typically be required to accommodate the very high sticking probabilities. Conventional cold-trapped diffusion pumps backed by two-stage rotary pumps are therefore normally employed, and the vacuum requirements are such that it is feasible to use pre-pumped insertion locks. Samples may thus be introduced from atmosphere into the spectrometer and be ready for investigation at the requisite operating pressure in a matter of minutes. Samples may conveniently be studied as films or powders mounted on a sample probe which may be taken into the spectrometer from atmosphere via insertion locks and valves. Provision is usually made to enable samples to be heated or cooled *in situ,* and ancillary equipment may be mounted directly on to the source of the spectrometer for *in-situ* preparation or pre-treatment (e.g. argon-ion bombardment, plasma

---

† For example the ES200 and 300 series of electron spectrometers manufactured by Kratos (A.E.I. Scientific Apparatus Ltd.) Manchester, England.

synthesis, electron bombardment, UV irradiation, chemical treatment, etc.). Addition of a quadrupole mass spectrometer facilitates many sample treatment studies and allows close control to be kept of the extraneous atmosphere in the sample region.†

The typical X-ray fluxes employed in commercially available spectrometers are such that there are relatively few systems for which appreciable radiation damage occurs during the time taken to record a spectrum. Polythiocarbonyl fluoride depolymerizes rather rapidly, whilst polyvinylidene fluoride slowly eliminates HF and cross-links. By contrast the dose rates involved in conventional Auger spectroscopy (employing an electron beam for excitation) are several orders of magnitude larger so that radiation damage poses severe problems.[7] The surface chemistry of polymers may therefore only be conveniently studied by ESCA, and Auger spectroscopy is not a viable alternative. This contrasts with the situation for inorganic systems. The sample requirements are modest and the surface sensitivity of the technique is such that the technique samples approximately the outermost 100 Å or so of sample and, depending on spectrometer design, ~0.2 sq. cm in area.

One of the most distinctive features of ESCA as a spectroscopic tool which sets it apart from any other is the large range of information levels that are available in a single experiment and these are set out in Table 1 (for a fuller discussion see Clark *et al.*[3]).

Table 1 Hierarchy of information levels available in ESCA

---

(1) Absolute binding energies relative peak intensities, shifts in binding energies. Element mapping for solids, analytical depth profiling, identification of structural features, etc. Short-range effects directly, longer-range indirectly.

(2) Shake up—shake off satellites. Monopole excited states; energy separation with respect to direct photoionization peaks and relative intensities of components of 'singlet and triplet' origin. Short and longer range effects directly (analogue of U.V.).

(3) Multiplet effects. For paramagnetic systems, spin state, distribution of unpaired electrons (analogue of E.S.R.).

(4) Valence energy levels, longer-range effects directly.

(5) Angular dependent studies. For solids with fixed arrangement of analyser and X-ray source, varying take off angle between sample and analyser provides means of differentiating surface from subsurface and bulk effects. Variable angle between analyser and X-ray source, angular dependence of cross sections, asymmetry parameter β, symmetries of levels.

(6) Sample charging phenomena. Additional means of monitoring structure and bonding in the surface regions. Electrical properties of polymer films from biasing experiments.

---

† This also provides the potential for investigating polymer surfaces by a secondary ion mass spectrometer (SIMS) — a field which has been little developed to date.

Table 2 ESCA applied to polymers

A. *Aspects of structure and bonding (static studies)*
    (i) Gross chemical compositions
        (a) elemental compositions
        (b) % incorporation of comonomers in copolymers
        (c) polymeric films produced at surfaces
    (ii) Gross structural information; e.g. for copolymers, block, alternating, or
        random nature. Domain structure in block copolymers
    (iii) Finer details of structure
        (a) structural isomerisms
        (b) experimental charge distributions in polymers
    (iv) Valence bands of polymers
    (v) Identification of polymers, structural elucidation
    (vi) Monopole excited states

B. *Aspects of structure and bonding (dynamic studies)*
    (i) Surface treatments, e.g. CASING, plasma modification
    (ii) Thermal and photochemical degradation
    (iii) Polymeric films produced at surfaces by chemical reaction; e.g. fluorination
        (including the use of ESCA for depth profiling and quantitative measurement
        of film thickness)
    (iv) Chemical degradation of polymers; e.g. oxidation, fluorination, etc.

C. *Electrical properties*
    (i) Mean free paths of electrons as a function of kinetic energy
    (ii) Photoconductivity of polymers
    (iii) Statics and dynamics of sample charging
    (iv) Triboelectric phenomena

The areas of application of ESCA in relation to polymers which have already been delineated are shown in Table 2. Further details are available in Clark *et al.*[3]

3. MEAN FREE PATHS OF ELECTRONS AS A FUNCTION OF KINETIC ENERGY IN POLYMERIC MATERIALS

In section 1 it was inferred that ESCA as a technique had excellent surface sensitivity and also provided the capability of differentiating surface from subsurface phenomena. Both these features are a consequence of the extremely short mean free paths (escape depths) of electrons in solids. Thus, in general, the ESCA spectrum of a given core level consists of well-resolved peaks corresponding to electrons escaping without undergoing energy losses, superimposed on a background tailing to lower kinetic energy arising from inelastically scattered electrons, as is evident from the wide-scan spectra shown in Figure 3. (These also show how the direct photoionization and Auger peaks may readily be distinguished on changing the photon energy, since the kinetic energy for the latter are independent of how the initial core hole is created.)

For the commonly used X-ray sources the mean free path of the photons is typically $\sim 10^4$ Å which is many orders of magnitude larger than the typical mean free paths of photoemitted electrons, so that in the applications which are outlined

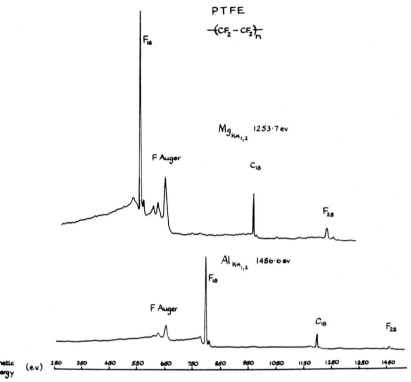

Figure 3. Wide-scan ESCA spectra of PTFE measured with $Mg_{K\alpha_{1,2}}$ and $Al_{K\alpha_{1,2}}$ photon sources.

below we may assume that the X-ray beam is essentially unattenuated over the range of surface thickness from which the photoelectrons emerge.

The two situations of common occurrence are for a bulk homogeneous material for which the intensity of the elastic peak is given by equation (1) and for an overlayer of thickness d on a bulk homogeneous substrate for which the corresponding equations for photoemission from overlayer and substrate core levels are equations (2) and (3).†

$$I_i^\infty = F\alpha_i N_i K\lambda_i \tag{1}$$

where F is the X-ray flux, α_i the cross-section for photoionization, N_i the number of atoms A per unit volume, K a spectrometer-dependent term which includes geometric and instrumental response terms, and λ_i the electron mean free path or escape depth.

---

† It should be noted that these expressions refer essentially to a given angle between X-ray source and analyser (this is fixed in most commercially available spectrometers, and a fixed take-off angle θ for the electrons with respect to the sample).

For an overlayer of thickness d,

$$I_0 = I_0^\infty[1 - \exp(-d/\Lambda_0)] \qquad (2)$$

For substrate

$$I_s = I_s^\infty \exp(-d/\Lambda_s) \qquad (3)$$

We can regard photoemission from core and valence energy levels as a means of producing electrons of accurately known kinetic energy, and the study of the relative intensities of the elastic peaks for photoemission from polymer overlayer and substrate levels provides one of the most versatile techniques for the direct measurement of electron mean free paths in the energy range typically 0–1500 eV.

Extensive studies have been made of electron mean free paths in a variety of polymeric materials by the substrate overlayer technique.[8] Thus, by producing films of known thickness by *in-situ* polymerization of paraxylylene precursors produced in a pyrolysis chamber[9] or by polymerization of reactive species produced in inductively coupled RF plasmas excited in appropriate monomers[10] it is possible to follow the attenuation of substrate and overlayer core levels as a

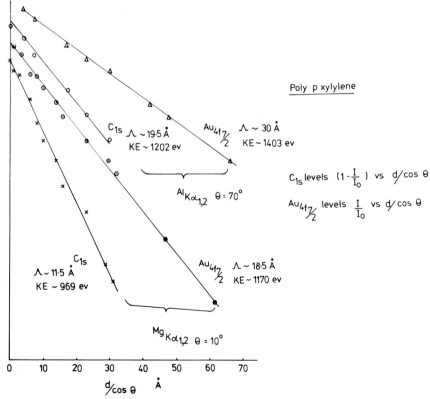

Figure 4. Representative substrate overlayer data for polyparaxylylene films of known thickness deposited on gold.

function of film thickness for either linear Paralene† -type polymers or for cross-linked polymers produced in plasmas. Independent evidence may be adduced for the uniformity of such films deposited, for example, on gold, and typical data are shown in Figure 4 where the study of the C_{1s} levels of a polyparaxylene (Paralene N) overlayer and $Au_{4f_{7/2}}$ levels of the gold substrate, with both $MgK\alpha_{1,2}$ and $AlK\alpha_{1,2}$ photon sources, provide the information that mean free paths are ~12, 20, and 26 Å for photon energies of ~970, ~1200, and ~1400 eV, respectively.[9]

It is worth while emphasizing at this stage that sampling depth and escape depth should not be confused. For escape depths of 12, 20, and 26 Å roughly 95% of the signal intensity of the elastic peaks derive from the topmost 36, 60, and 78 Å of sample in each case.

The results for cross-linked, plasma-produced polymer films are very similar,[10] and indeed it may be shown that electron mean free paths measured directly for polymer films correspond closely to those for typical metals and semiconductors, this being gratifyingly in agreement with the available crude theoretical models.[11]

Since many of the important chemical, physical, electrical, and mechanical properties of thin films are determined by the structure of the outermost few tens of Ångströms, any technique capable of differentiating surface from subsurface and bulk is of some considerable importance. One means for example of depth profiling a sample is to study core levels corresponding to different mean free paths for the photoemitted electrons. This is illustrated schematically in Figure 5, which also shows the general form of electron mean free paths as a function of kinetic energy.

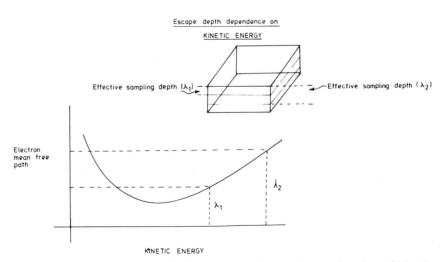

Figure 5. Generalized form of electron mean free paths as a function of kinetic energy. Also illustrated is the difference in *sampling depths* (λ_1 and λ_2) for electrons of differing kinetic energy and hence mean free path.

---

†Union Carbide patented designation for polyparaxylenes.

318

Effective
sampling
depth d

Effective sampling
depth d/cos θ

Figure 6. Illustration of the use of angular-dependent studies to enhance surface sensitivity for a thin surface film of thickness d on a homogeneous bulk material.

Surface, subsurface, and bulk may also be differentiated in principle by angular-dependent studies. The basic philosophy here is that we can specifically enhance surface features by studying electrons emerging at grazing exit from the surface, this being illustrated schematically in Figure 6.

By studying different levels of the same and different elements in a sample as a function of take-off angle it becomes possible in principle to determine whether the outermost ~100 Å of a sample is representative of the bulk, and whether in this region the sample is homogeneous, laterally inhomogeneous, vertically inhomogeneous, or if the sample is completely inhomogeneous in this region; this is illustrated in Figure 7

It should be emphasized here that although the depth resolution of ESCA is unsurpassed the lateral resolution is quite poor. By appropriate angular-dependent studies, however, it is still possible to obtain information on laterally inhomogeneous samples.[1][2]

4. SAMPLE CHARGING AND THE ELECTRICAL CHARACTERISTICS OF POLYMER FILMS

The study of essentially insulating samples as polymer films, or indeed samples of metals insulated from the spectrometer probe by an insulating film, results in an apparent shift in kinetic energy scale arising from sample charging, i.e. under the conditions of X-ray irradiation insufficient charge carriers are available to maintain the overall electroneutrality of the sample. The dynamic equilibrium which is established under X-ray irradiation invariably produces an overall positive charge on the sample and hence a retardation of the electrons leaving the sample surface, the net effect being a shift to lower kinetic energy. The study of sample charging

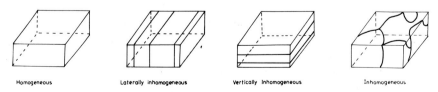

Homogeneous   Laterally inhomogeneous   Vertically Inhomogeneous   Inhomogeneous

Figure 7. Classification of polymer surfaces into homogeneous, laterally inhomogeneous, vertically inhomogeneous, and inhomogeneous systems.

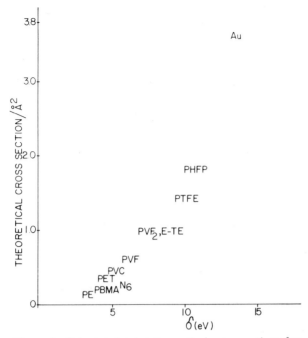

Figure 8. Calculated total theoretical cross-sections for normalized unit areas for various polymers and gold versus experimentally determined equilibrium charging shift (δ) for samples irradiated with a $Mg_{K\alpha_{1,2}}$ X-ray photon source (12 kV, 15 mA) and electrically isolated from the spectrometer.

provides an interesting means of studying photoconductivity in polymeric films. Since the total electron flux leaving the sample surface is a function of the total cross-section for photoionization, we might anticipate that under a given set of instrumental conditions (e.g. samples, either metal or polymer mounted in such a way as to be insulated from the spectrometer), the normalized cross-sections should show a relationship to sample charge measured by ESCA. Such a relationship is shown in Figure 8.[13]

The surface sensitivity of sample-charging phenomena is readily demonstrated by deliberately allowing a hydrocarbon film to build up upon samples, and the data displayed in Figure 9 also illustrate that the cross-section for photoionization from this overlayer is close to that of polyethylene.

This illustrates in quite a striking manner the relationship between surface structure and sample charging, and indeed the investigation of such phenomena provides direct information on surface structure, a particularly striking example being provided by the investigation of surface modification phenomena is discussed in section (7v)

The characteristic nature of the C_{1s} binding energy arising from deliberate surface contamination by hydrocarbon of polymer samples provides a ready means

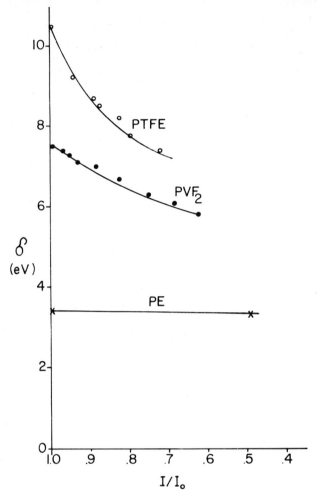

Figure 9. Sample charging for PTFE, polyvinylidene fluoride (PVF$_2$), and polyethylene (PE) as a function of hydrocarbon contamination.

of calibration of the energy scale which is highly reproducible.† Figure 10, for example, shows the shift in binding energy between the hydrocarbon as reference and appropriate core levels for gold and PTFE samples as a function of the absolute shift in core binding energy which amply demonstrates this point. Although sample charging in most designs of commercial instrumentation employing unmonochromatized X-ray sources is typically in the range 0–20 eV, with monochromatized X-ray sources the removal of bremstrahlung as a source of secondaries

---

†Although this must be established independently, the binding energy for hydrocarbon contamination is usually 285.0 ± 0.01 eV.

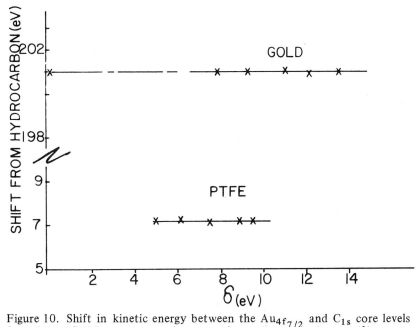

Figure 10. Shift in kinetic energy between the $Au_{4f_{7/2}}$ and C_{1s} core levels for gold and PTFE samples as a function of overall sample charging (δ)

can lead to severe sample-charging problems and, depending on the design, this can typically be as large as several hundred electronvolts shift in the kinetic energy scale. In such cases it is possible to use flood guns and low-powered, low-pressure UV lamps as sources of electrons to obviate sample charging, further details are available elsewhere.[13]

For thin polymer films < 1000 Å thick, sufficient charge carriers are available such that sample charging is insignificant and the Fermi levels of the spectrometer and sample are effectively the same. This may readily be shown by applying a bias voltage to the sample and observing the shift in the kinetic energy scale for core and valence levels. Appropriate bias voltages may also be employed to study the secondary electron distribution to provide a direct measurement referenced to the Fermi level. Interesting situations arise if thicker polymer films are studied. In the range 10–100 μm, for example, if the polymer film is in good contact with the spectrometer probe, sample charging occurs, but none the less core and valence levels and the secondary electron distribution follow an applied bias, but not exactly. If thick enough films are produced, however, an applied bias voltage does not effect the kinetic energy of the photoemitted electrons, this being shown schematically in Figure 11. ESCA, therefore, provides a convenient means of studying electrical characteristics of thin films under the conditions of X-ray irradiation, and considerable work in this area has already been completed. It is inappropriate in an article such as this to dwell on this topic and further details are available elsewhere.[13] It is clear, however, that ESCA can shed new light on this important area.

322

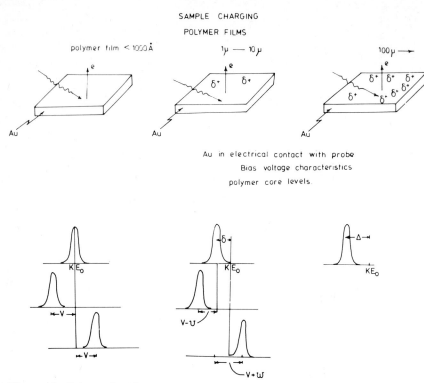

Figure 11. Schematic illustrating typical biasing characteristics for polymer films of different thickness.

5. INFORMATION DERIVED FROM ABSOLUTE AND RELATIVE BINDING ENERGIES AND RELATIVE PEAK INTENSITIES

We consider briefly in this and succeeding sections the sorts of information which may be obtained on polymeric systems using the hierarchy of information levels outlined above.

The first levels of information available derive from the measurement of absolute and relative binding energies and relative peak intensities. The distinctive nature of core levels means that identification of elements is straightforward (cf. Figure 3). With appropriate calibration, the relative intensities and shifts in binding energy for components of a given core level may be used to identify structural features and repeat units. Figure 12, for example, shows high-resolution ESCA spectra for two polymer samples, and from the absolute and relative binding energies and relative peak intensities these may be identified as PVC and polyisopropyl acrylate.

Previous studies of substituent effects on core levels in simple monomeric systems has shown that these are highly characteristic for a given substituent and follow simple additivity models.[3] The results may be quantified by detailed non-empirical calculations, and this forms a sound basis for understanding the

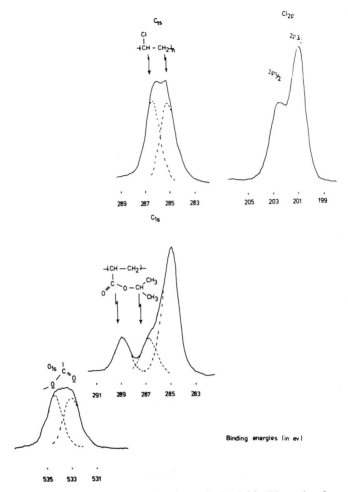

Figure 12. Core-level spectra for polyvinylchloride and poly-isopropyl acrylate films.

electronic factors determining both absolute and relative binding energies. This has enabled computationally inexpensive models based on an all valence electron CNDO/2 SCF MO formalism to be developed which can be extended quantitatively to describe polymers. A large amount of data have previously been reviewed which relates to fluorocarbon-based polymers.[3] However, a systematic study of a large number of homopolymers of simple vinyl monomers provides a compilation of substituent effects on C_{1s}, N_{1s}, O_{1s}, F_{1s}, Si_{2p}, P_{2p}, S_{2p}, and Cl_{2p} levels.[3] Figure 13, for example, shows some of the data pertaining to substituent effects on C_{1s} levels in polymers.[14]

The characteristic nature of many substituent effects can now be used as a fingerprint, much in the same manner as one might use infrared or NMR data.

It is sometimes the case that isomeric species have core-level spectra which are

324

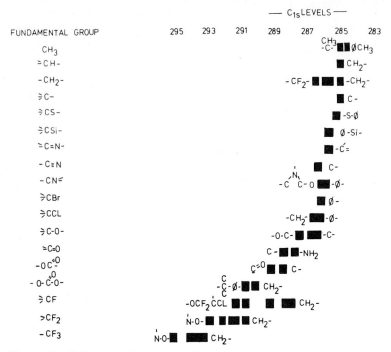

Figure 13. Shift range for commonly encountered substituents for C$_{1s}$ levels of polymeric systems. (In each case the typical range of binding energies encountered is indicated by the shading.)

virtually identical, since the factors which determine core-level shifts are very short range in nature. Although the spectra may therefore be used to identify the gross structure, they may not allow a distinction to be drawn between isomeric species. The fingerprint nature of the valence energy levels, however, may be utilized to good advantage in such cases, and this is illustrated in Figure 14.[15] Comparison with appropriate model systems allows an unambiguous assignment of particular structural isomers of the polybutyl acrylates.

Although there are well-developed techniques for studying chemical compositions and features of structure and bonding pertaining to the bulk of polymer samples, until the advent of ESCA information with regard to surface compositions could only be inferred rather indirectly by, for example, surface free energy measurements. Since any solid communicates with the rest of the world primarily by way of its surfaces, such information is important in many areas. ESCA may therefore be conveniently employed to answer the question, 'Is the surface composition typical of the bulk'? This is central to the investigation of polymer samples synthesized *in situ* by photopolymerization and plasma techniques, for example, to the study of surface oxidation and weathering phenomena, and to surface modification and degradation in general. With the background of data on absolute and relative binding energies and relative peak areas it is possible to use ESCA as a competitive technique for the routine non-destructive analysis of

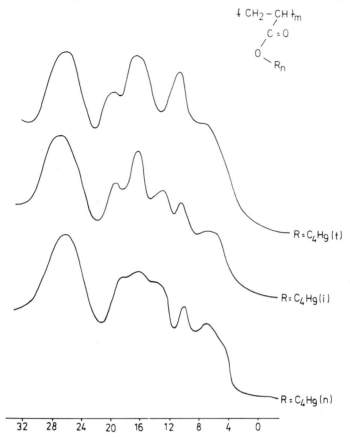

Figure 14. Valence-level spectra for a series of isomeric poly-butyl acrylates, illustrating their distinctive fingerprint nature.

samples for which the surface composition corresponds to that of the bulk. In addition, for relatively low molecular weight materials where the structural features corresponding to end groups have a distinctive nature, it becomes possible to obtain DPs directly. As illustrative examples we consider examples taken from the fluoropolymer field. As a simple example, Figure 15 shows the core-level spectra for a series of ethylene–tetrafluoroethylene copolymers. The higher binding energy component of the C_{1s} levels arise from CF_2 groups whilst the lower binding energy component is appropriate to CH_2 components. The shifts in binding energy between the two component peaks also establishes that the copolymers are largely alternating in character. Much of our early work in ESCA applied to polymers involved the study of homopolymers of simple fluorocarbon monomers, and this has allowed the accumulation of an extensive background of information on absolute and relative binding energies as a function of structural feature and on the relative intensities of core levels.[3] By comparison with these simple homopolymers

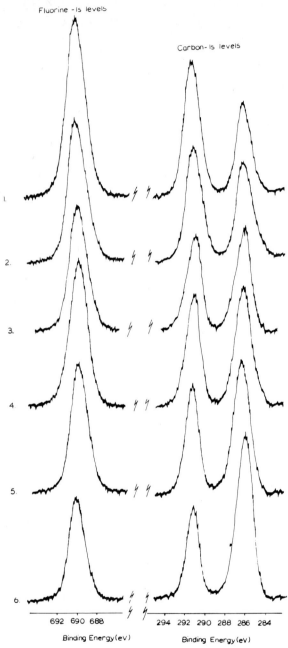

Figure 15. Core-level spectra for a series of ethylene–tetrafluoroethylene copolymers.

Table 3 Analysis of ethylene/tetrafluoroethylene comonomer incorporations

Sample	Composition monomer mixture (mol.% C_2F_4)	Copolymer composition (mol. % C_2F_4)				
		Predicted from monomer reactivity ratios	Calc. from C analysis	Calc. from F analysis	Calc. from area ratio: C_{1s} peak: F_{1s} peak	Calc. from C_{1s} (CH_2 peak) C_{1s} (CF_2 peak)
1	94	63	61	61	63	62
2	80	53	52	54	52	52
3	65.5	50	49	48	47	46
4	64	50	47	45	44	45
5	35	45	41	40	42	40
6	15	36	–	–	32	31

the compositions of the copolymers may readily be established by two independent means.

Firstly, from a comparison of the integrated area ratios for the F_{1s} and C_{1s} levels, and secondly from the individual components of the C_{1s} levels.[16] This readily establishes that the materials are copolymers of ethylene and tetrafluoroethylene which are largely alternating in character and that the outermost surface sampled by ESCA is identical in composition to the bulk. This is shown in Table 3 where a comparison is drawn with compositions determined by standard microanalysis (carbon by combustion, fluorine by potassium fusion). ESCA is highly competitive as a routine means of establishing compositions for fluoropolymers in particular, in terms of accuracy, non-destructive nature, and speed. The homogeneous nature of the samples is also demonstrated by the fact that the intensity ratios F_{1s}/C_{1s}, F_{1s}/F_{2s} are independent in each case of the electron take-off angle (θ) with respect

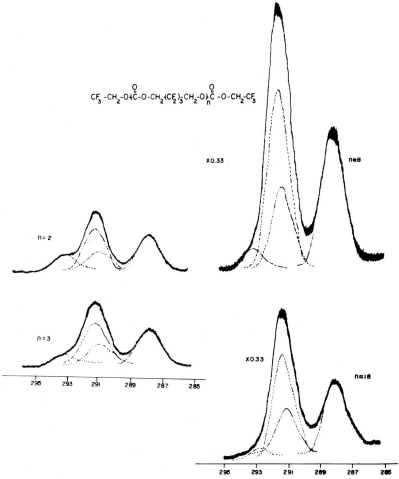

Figure 16. C_{1s} levels for a series of fluorocarbonate polymer samples.

to the sample surface, a point which is dealt with in the section dealing with the plasma modification of these systems (section 7v).

As we have previously noted it is often the case that particular structural features may be characteristic of the end groups of a given polymer system. The direct detection of such end groups by means of their characteristic binding energies provides a convenient means of establishing DPs in relatively low molecular weight material. A particularly favourable situation arises for systems for which the terminal groups involve $\underline{CF}_3$ residues. If due care is taken to ensure that ESCA

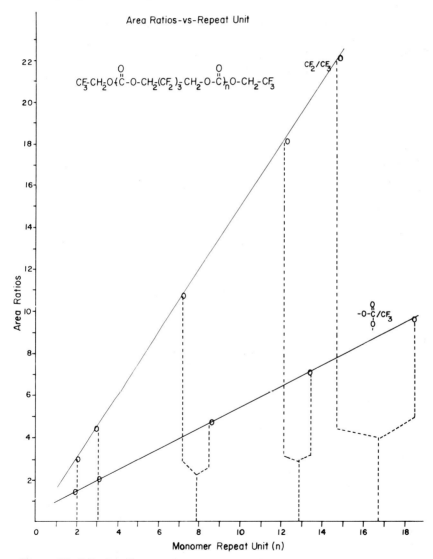

Figure 17. DPs for fluorocarbonate polymers obtained from CF_3/CF_2 and $CF_3/O-C\overset{\nearrow O}{\underset{O}{\diagdown}}$ core level intensities.

statistically samples the repeat unit (by, for example, considering the relative intensities of different levels of the same element with differing escape depth dependencies), then the comparison of area ratios for chemically shifted components of a given core level may be used to straightforwardly estimate DPs.[17] For example, in a series of fluorocarbonate polymers of the general formulae shown in Figure 16 it may readily be shown that the carbon 1s levels appropriate to carbonate $-O-C{\overset{\nearrow O}{\underset{\searrow O}{}}}$ and $\underline{C}F_2$ environments occur at approximately the same binding energy. The C_{1s} levels for the series of low molecular weight materials shown in Figure 16 fall into three distinct regions, and with appropriate calibration of linewidths and lineshapes for individual components from the study of model compounds the lineshape analysis produces the components indicated by the dotted curves. From the relative areas of the CF_3 carbons to the $\underline{C}F_2$ and carbonate carbon peaks DPs may be elaborated as indicated in Figure 17. The two methods of elaborating DPs give slightly different results which may indicate specific orientation effects, however, the two are within $\sim 10\%$ and show an excellent correlation with DPs determined by vapour pressure osmometry, and this is shown in Figure 17. By contrast DPs determined by ^{19}F NMR do not agree with those determined by these two techniques, although the reason for this discrepancy is not clear. Similar applications may be made in the study of polycarbonates in general which tend to be of relatively low molecular weight compared with other common polymeric systems, and it is therefore possible to establish overall compositions and DPs in the same experiment.[17]

As a more complicated example of how the homogeneity and composition of thin polymer films may be established by ESCA, Figure 18 shows the core-level

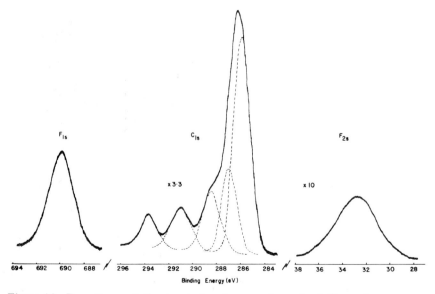

Figure 18. F_{1s}, C_{1s}, and F_{2s} levels for plasma-polymerized films of vinylidene fluoride.

spectra for a thin (~ 200 Å) film produced by plasma polymerization of vinylidene fluoride. The complex lineshape for the C_{1s} levels in which components appropriate to $\underline{C}F_3$, $\underline{C}F_2$, $\underline{C}F$, and $\underline{C}$ structural features are apparent offer direct evidence for the extensive molecular rearrangements occurring in the plasma polymerization. The stoichiometry may be established both from the components of the C_{1s} levels and from the intensity ratios of the C_{1s}/F_{1s} levels, once it has been established that these are independent of take-off angle θ. The stoichiometry in fact is $(C_2 F_1)_n$, strongly suggesting that the reactive species involved in polymerization derive from fluoroacetylene which may readily be envisaged to arise from effective elimination of HF from vinylidene fluoride. This is an interesting example, since it would be extremely difficult to investigate structure and bonding in this cross-linked polymer systems such as this by any technique other than ESCA. Further examples amplifying this theme are presented in Chapter 9.

Whilst the discussion has emphasized the application to the study of fluoro-polymer systems, which are often difficult to study by conventional techniques, with the careful compilation of data on appropriate model systems, it becomes a straightforward matter to extend the analysis to polymer systems in general. As a typical example we may consider the investigation of a series of polyalkyl acrylates.[15] The basic points of interest being, are the structures of these polymers the same in the surface regions as in the bulk, and are there any specific orientation effects as the length of the alkyl side chain increases?

The core-level spectra for a given series of samples are shown in Figure 19. The O_{1s} levels show a doublet structure (somewhat obscured in the case of polyacrylic acid because of hydrogen bonding effects); the binding energies for the two components being characteristic for an ester group. The C_{1s} levels in each case show a high binding energy component attributable to $-C\overset{\displaystyle \nearrow O}{\underset{\displaystyle \searrow O-}{}}$ type environments shifted ~ 4 eV from the main peak, which in each case arises from CH_3, $\underline{C}H_2$, and $\underline{C}H$-type environments from the backbone carbons and carbons of the alkyl group not attached to oxygen. In going from polyacrylic acid to the polymethylacrylate, a shoulder to the high binding energy side of the low binding energy component develops, shifted by ~ 1.6 eV and attributable to the carbon attached to the ester oxygen. This shoulder gradually decreases in relative intensity with respect to the main peak as the chain length of the alkyl group increases. The assignment of core levels is readily confirmed by reference to simple model compounds for which detailed theoretical analyses have been performed.

A study of the valence levels for simple model compounds reveals that in esters the O_{2s} levels are well separated from the remainder of the valence band and are essentially core-like in nature. The approximate kinetic energies pertaining to photoemitted electrons from O_{1s} and O_{2s} levels using a $MgK\alpha_{1,2}$ photon source are ~ 720 and ~ 1227 eV, respectively, and from a consideration of the generalized curve of escape depth versus kinetic energy these should correspond to significant differences in electron mean free paths. By studying a series of simple oxygen-containing organic molecules as condensed films such that ESCA statistically samples the molecules, it can be shown that with the particular experimental arrangement peculiar to our spectrometer the apparent O_{1s} to O_{2s} area ratio is

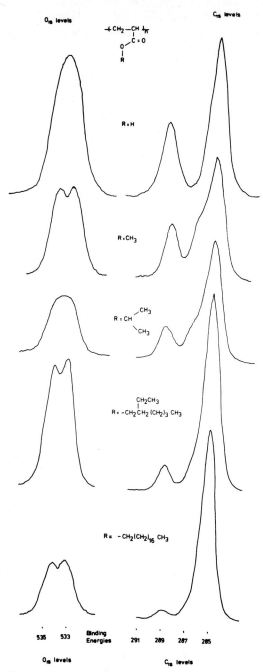

Figure 19. Core-level spectra for a series of polyalkyl acrylates.

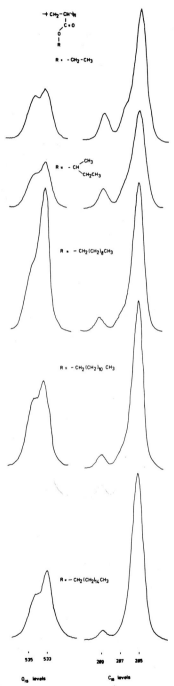

Figure 20. Core-level spectra for a series of sur-
face-oxidized polyalkyl acrylates.

11 ± 1. By measuring the area ratios of the O_{1s} to C_{1s} peaks versus the stoichiometric ratios for these molecules it is also possible to derive the instrumentally dependent apparent sensitivity ratios for oxygen with respect to carbon for unit stoichiometry. Analysis of the O_{1s} to O_{2s} area ratios for the samples corresponding to the core-level spectra illustrated in Figure 19 reveal that these samples are homogeneous on the ESCA depth sampling scale as far as oxygen is concerned. With a knowledge of the relevant sensitivity factors it is a straightforward matter to establish the stoichiometry from a consideration of the total O_{1s} to C_{1s} intensity ratios, and from the ratio of the high binding energy $\left(-C\begin{smallmatrix}\nearrow O \\ \searrow O-\end{smallmatrix}\right)$ component of the C_{1s} levels with respect to the remainder. The two independent means of establishing compositions are in excellent agreement and show that on the ESCA depth sampling scale the samples are homogeneous and correspond exactly to that appropriate to the bulk.

A somewhat more complicated situation obtains for the series of poly-alkylacrylates whose core-level spectra are shown in Figure 20. A distinctive feature clearly evident in all of the spectra is the obvious inequality in intensity of the two component peaks of the O_{1s} levels. A similar analysis to that presented previously in this section can straightforwardly be carried out. Figure 21, for example, shows a plot of the ratio of intensities for the individual components of the O_{1s} levels and also the total O_{1s}/O_{2s} ratios. For comparison purposes the dotted lines indicate the correlations expected for samples which on the ESCA depth profiling scale corresponds to a statistical sampling of the appropriate repeat unit in the polymer. It is clear that there are considerable deviations from such correlations in a direction which overall suggests that the samples are oxidized. If we consider the polydecyl acrylate for example, the O_{1s}/O_{2s} ratio is significantly higher than for the reference compounds, suggesting that since the mean free path for the O_{1s} levels is considerably shorter than for the O_{2s} levels that the oxidation is largely confined to the surface. The absolute binding energies in each case for the O_{1s} component levels which have apparently increased in intensity corresponds to $\underline{C}$=O structural features, as is apparent from a comparison with data for the model systems. It is interesting to note that high-resolution infrared studies revealed no major distinction of the type clearly evident from the ESCA spectra, and the carbonyl region for all of the samples showed only a single peak in the range 1734 ± 6 cm^{-1}, consistent with $-C\begin{smallmatrix}\nearrow O \\ \searrow O-R\end{smallmatrix}$ structural features. This is readily understandable since the infrared data pertains essentially to the bulk. Detailed examination of the C_{1s} spectra for the samples shows that the overall line profiles can only be quantitatively fitted with the addition of a small peak appropriate in binding energy to isolated carbonyl features as might arise from oxidation. We will consider further aspects of sample inhomogenities which may be investigated on the tens of Ångströms scale by means of ESCA in section (7). At this stage, however, we consider some aspects of shake-up phenomena which provide an important level of information, particularly in those systems for which the span in molecular core binding energies is very small.

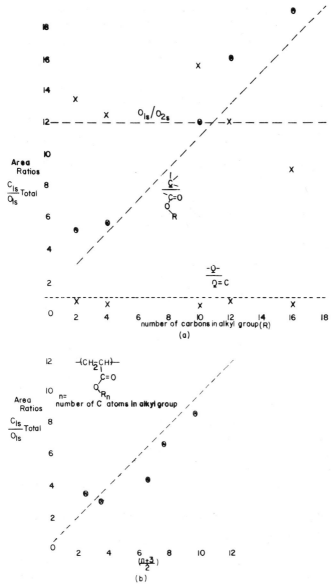

Figure 21. (a) Plot of the intensity ratios for the individual components of the O_{1s} levels, and also the O_{1s}/O_{2s} ratios for the series of polyalkyl acrylates shown in Figure 20. (b) Plot of the C_{1s} and O_{1s} area ratios as a function of the number of carbons in the alkyl groups. (Dotted lines are the correlations corresponding to statistical sampling of the repeat units.)

6. INFORMATION FROM THE INVESTIGATION OF SHAKE-UP PHENOMENA

The removal of a core electron (which is almost completely screening as far as the valence electrons are concerned) is accompanied by reorganization (relaxation) of the valence electrons in response to the effective increase in nuclear charge. This perturbation gives rise to a finite probability for photoionization to be accompanied by simultaneous excitation of a valence electron from an occupied to an unoccupied level (shake-up), or ionization of a valence electron (shake-off). These processes giving rise to satellites to the low kinetic energy side of the main photoionization peak, follow monopole selection rules and considerably extend the scope of ESCA as a technique, as will become apparent. The relationship between direct photoionization, shake up, and shake off are shown schematically in Figure 1. The relationship between relaxation of the valence electrons which typically stabilize the core-hole state by ~ 12 eV for C_{1s} levels,[18] and the intensity and energy separation with respect to the direct photoionization peak of the shake-up and shake-off satellites is well understood:[19] however, the scope of this review does not allow the development of this interesting aspect of chemical physics at this point.

The main conclusions from this analysis, however, are that the shake-up transitions most readily observed are those whose energies are quite close to that for the relaxation energy, and that the transitions of highest intensity are to shake-up states of singlet parentage in the doublet manifold.[20] There are, in fact, similarities in the investigation of the monopole excited states of core-ionized species and the corresponding dipole excited states of the neutral system. Since the shake-up states follow monopole selection rules the intensity of the transitions is determined by the overlap between the highest occupied valence orbitals of the initial system and the lowest unoccupied orbitals of the core-ionized species; this is presented analytically in Figure 1. It should be evident from this that shake-up spectra should be highly characteristic of a polymer system. It should perhaps be pointed out that for solids the investigation of shake-up and shake-off states is complicated by the presence of the general inelastic tail (arising from photo-emission from a given core level followed by energy loss by a variety of scattering processes) which provides a broad energy distribution, usually peaking for polymeric systems ~ 20 eV below the direct photoionization peaks. This generally obscures any underlying high-energy shake-up or shake-off processes, such that it is only for systems exhibiting relatively high-intensity, low-energy shake-up peaks that information derived from this source can conveniently be exploited. Fortunately, such a situation generally obtains for polymer systems which contain either unsaturated backbones or pendant groups, since low-energy $\pi \rightarrow \pi^*$ shake-up transitions are available.[3]

As a typical example Figure 22 shows the C_{1s} spectra for typical saturated polymers polyethylene (high density) and polydimethylsiloxane, and polystyrene and polydiphenylsiloxane which represent prototype systems with saturated backbones and unsaturated pendant groups. For the latter, well-developed shake-up structures are apparent which clearly distinguishes them from the saturated systems, although the differences in lineshape and linewidths for the main

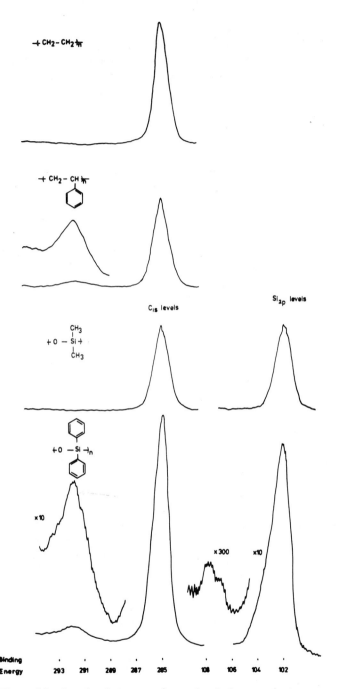

Figure 22. Core-level spectra for polyethylene, polystyrene, polydimethyl siloxane, and polydiphenyl siloxane showing distinctive low-energy $\pi \rightarrow \pi^*$ shake-up satellites for the aromatic systems.

338

photoionization peaks are closely similar. (For the siloxanes, of course, the relative intensities of the C_{1s} with respect to the O_{1s} and Si_{2p} levels may be used to effect a ready distinction between the two siloxanes.) It will become apparent that the transitions giving rise to the satellite structures in PS and PDPS are due to $\pi \rightarrow \pi^*$ transitions, as might indeed be inferred from the much smaller shake-up peak associated with the Si_{2p} levels and the lack of any low-energy shake-up structure accompanying the O_{1s} levels for PDPS. It may readily be shown from both experimental and theoretical studies that the low-energy satellite structures arise

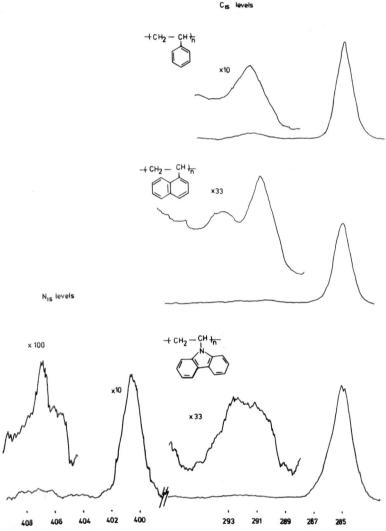

Figure 23. Core-level spectra for polystyrene, polyvinylnaphthalene, and polyvinylcarbazole showing distinctive low-energy $\pi \rightarrow \pi^*$ shake-up satellites.

from transitions involving the two highest occupied and lowest unoccupied orbitals of the pendant phenyl groups.[21]

Low-energy shake-up satellites may be used to remove ambiguities in the interpretation of the primary sources of ESCA, namely absolute and relative binding energies and relative peak intensities, and are highly characteristic of the unsaturated system from which they derive.[21] Figure 23, for example, shows the core-level spectra for polystyrene, polyvinylnaphthalene, and polyvinylcarbazole.

Clearly, the distinctive nature of shake-up satellites can add a new dimension to ESCA data for systems in which the shifts in core levels are negligible. As one example Figure 24 shows the C_{1s} and O_{1s} levels for poly-n-hexylmethacrylate and polyphenylmethacrylate.[22]

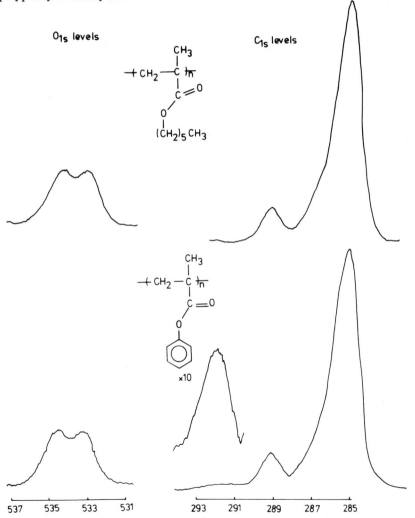

Figure 24. Core-level spectra for poly-n-hexylmethacrylates showing the shake-up structure for the latter.

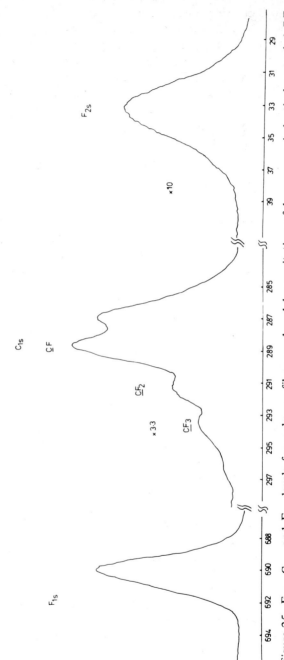

Figure 25. F_{1s}, C_{1s}, and F_{2s} levels for polymer films produced by excitation of low-power, inductively coupled RF plasmas in perfluorobenzene.

A further example is shown in Figure 25 which illustrates the C_{1s} core-level spectra obtained for polymer films produced from RF plasma polymerization of perfluorobenzene. The extended nature of the core levels illustrates the considerable complexity of the rearrangements occurring in the plasma. The appearance of a component to the high binding energy side of that assigned to $-\underline{C}F_3$ structural features is clearly not attributable to a primary photoionization peak, and detailed analysis shows that this component at ~ 296 eV arises from $\pi \rightarrow \pi^*$ excitations accompanying core ionization of CF structural features, thus showing that the polymer possesses a certain degree of unsaturation involving the π systems.[23]

7. SAMPLES FOR WHICH THE SURFACE IS DIFFERENT FROM THE BULK

(i) Introduction

Although, as we have previously noted, the surface of a polymer sample is often identical in composition to that of the bulk (indeed ESCA provides the only spectroscopic tool for establishing this), the technique really comes into its own in elaborating details of structure and bonding in inhomogeneous samples. Since solids communicate with the rest of the world by way of their surfaces the investigation of inhomogeneous samples forms a necessary prerequisite to any definite study of such diverse areas as weathering, oxidative degradation, surface modification, surface contamination, and migration phenomena. The technique has improved the level of understanding at a fundamental level in many of these areas and we briefly consider a few representative examples

(ii) Specific differences in surface structure in polymers

Since as we have indicated electron mean free paths are typically in the range 7–30 Å (dependent on KE) it is possible (for systems involving lengthy repeat units) for ESCA not to statistically sample repeat units and hence to reveal specific differences in surface structure. Two examples which illustrate this most effectively have been discussed in detail elsewhere and concern specific orientation effects in the surface regions of styrene alkane copolymers particularly those involving lengthy alkane components,[24] and the specific segregation of the lower surface energy component in solvent cast films of polystyrene/polydimethylsiloxane AB block copolymers.[25] The investigation of specific surface orientation effects in styrene alkane copolymers originates in the detailed examination of shake-up phenomena which is specific for the styrene component. The study of AB block copolymers is particularly interesting since, depending on the overall composition and block lengths, it is possible to obtain a considerable range of bulk morphologies. From the relative intensities of core-level spectra and from shake-up satellites it is a straightforward matter to show that the general situation involves a substantially different surface structure from that of the bulk, being such that the surface of lowest surface energy is produced.

(iii) Surface contamination

The fact that polymer surfaces are often contaminated in the surface regions has many important ramifications, since unless this is recognized investigations of many different phenomena may proceed on the false premise that the surface composition is well understood. Thus, the surfaces of mouldings often involve specific segregation of release agents, whilst polymers produced by emulsion polymerization often show evidence of segregation of entrained emulsifying agents at the surface.[3,26] Solvent cast films, on the other hand, may involve the segregation of additives, stabilizers, etc. to the surface. Segregation of low molecular weight plasticizers at surfaces is also common, whilst the formation of, for example, epoxy resins may be accompanied by migration of accelerators and/or curing agents to the surface.[26] Polymers which possess hydrophilic centres ($-NH_2$, $-COHN_2$, $\geq C=O$, etc.) may well become specifically contaminated at the surface by hydrogen-bonded water, and this may readily be detected by ESCA since the binding energy for the O_{1s} levels for the contaminant is distinctive.[3,15] Polymer films produced by pressing may be contaminated by mass transfer from the pressing matrix and this may readily be detected by ESCA. Most polymer surfaces have a relatively low sticking probability for extraneous contaminants present in the average laboratory environment, none the less it is usually a straightforward matter to study low levels of surface contamination which previously would have gone unnoted. Contamination arising from the handling of samples may also be studied in some detail.

One area where knowledge of the precise nature of the surface is of critical importance is in delineating models for the interpretation of data relating to triboelectric phenomena, and this serves as a specific example of how the technique may be employed to study surface contamination. Contacting polymer films from opposite ends of the so-called triboelectric series results in charge transfer such that there is a considerable build-up of static charge on each component of the contacting pair. Such charge transfer could conceivably occur via electron transfer from the material of low to that of higher work function, thus equalizing the Fermi levels, or alternatively by mass transfer (transfer of ions) between the two components of the contacting pair. Even more likely is that both processes are of importance; however, the two possible mechanisms are not entirely separable since the propensity for adsorption of ions at the surface of a given polymer will undoubtedly be a subtle function of its electronic structure, as will the work function. It is known that triboelectro-phenomena, in general, are explicable in terms of a charge density of the order of 1 in 10^4 of surface sites, which is probably an order of magnitude lower than can currently be detected by ESCA. Nonetheless, in contacting polymer samples, it is of interest to see whether or not there is mass transfer between the components. If mass transfer is observed it certainly does not resolve the problem of how charging occurs, but it certainly allows one to say that mass transfer cannot be ruled out as a possible mechanism.

We have therefore studied the surfaces of a variety of polymer films both before and after contacting events.[2] The great advantage of such an investigation by means of ESCA is the ability to look at *both* halves of a contacting pair. As an example,

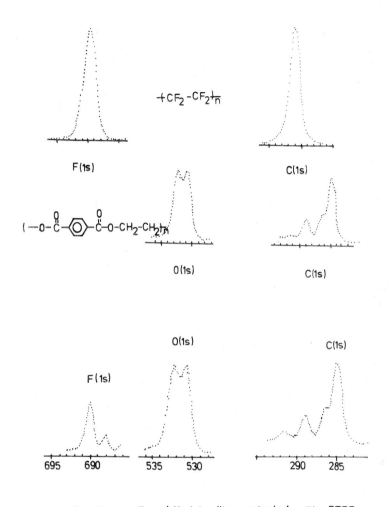

Polyethylene Terephthalate film contacted with PTFE

Figure 26. Core level spectra for PTFE and PET and for the PET component after lightly contacting the two polymer films.

Figure 26 shows the core-level spectra for PTFE, and PET films, which are highly characteristic for each polymer. In addition, there is also shown the core-level spectra for the polyethylene terephthalate component after lightly contacting the PTFE film. Since the polymers are from the opposite ends of the triboelectric series the films show a strong tendency to adhere to one another, even on a light contact.[27]

The observation of the high binding energy component in the C_{1s} spectrum and the observation of the F_{1s} levels shows that some PTFE transfers to the PET surface. It may be estimated that this represents fractional monolayer coverage. Of particular interest is the fact that in the F_{1s} peak, in addition to the major high binding energy component associated with covalent CF_2 linkage, there is a lower binding energy peak attributable to fluoride ion, thus providing evidence for bond cleavage accompanying the mass transfer. Examination of the other half of the component, namely PTFE, shows the presence of PET as evidenced by the characteristic O_{1s} and C_{1s} levels. These simple experiments illustrate the utility of ESCA in this area.

(iv) Chemical modification of surfaces

The chemical modification of surfaces represents one of the most fruitful areas for ESCA investigations. Since reaction is initiated at the surface and is followed by diffusion into the bulk, the initial stages of any reaction produces complex inhomogeneous systems, and in the elaboration of the complexities which typically arise ESCA really comes into its own as a spectroscopic tool.

We have previously alluded to the detection of surface oxidation in polyalkyl acrylates,[15] and in Chapter 9 a discussion has been presented on the RF plasma oxidation of polymers. ESCA has already provided interesting new insights into photo-oxidation and weathering of polymeric systems, and applications in this area illustrate the versatility of the technique in investigations at different levels. The weathering of paints, for example – particularly in relation to the modification of adhesive properties – represents a technologically important topic of great complexity.[26]

The central issues, however, are straightforward, namely what changes in surface chemistry control any adhesive failure and where does such failure occur. The major differences in ESCA spectra for a fresh and weathered surface, and for peeled surfaces which have suffered adhesive failure, can often provide the answers to these questions at a qualitative level and enable corrective steps to be taken to rectify the problem. Since the technique does allow the possibility of gaining an edge over competitive products it is likely that many important technological applications will remain part of technical reports which do not see wide distribution.[26]

As a simple example of how ESCA may be used to investigate oxidative degradation we may briefly consider some aspect of the corona-discharge treatment of polymers. The corona-discharge treatment of polymeric films is one of the most widely used techniques for surface modification which essentially entails selective degradation of the surface. Such treatments have a desired objective in mind, however, the corona charging of polymer-based photoconductors as employed in copying machines represents an area where the overall effects of repeated charging cycles leads to undesirable degradative processes. The study of corona treatments of polymers is therefore of importance in several diverse areas of industry.

A common method of improving the adhesive and printing characteristics of a polymer film is via a corona discharge excited in air. Figure 27(a) shows the O_{1s}' N_{1s}, and C_{1s} regions of the ESCA spectrum of a polyethylene sample after treatment with such a discharge. Several features are immediately obvious. The C_{1s} spectrum of the treated sample exhibits a high degree of structure, largely due to

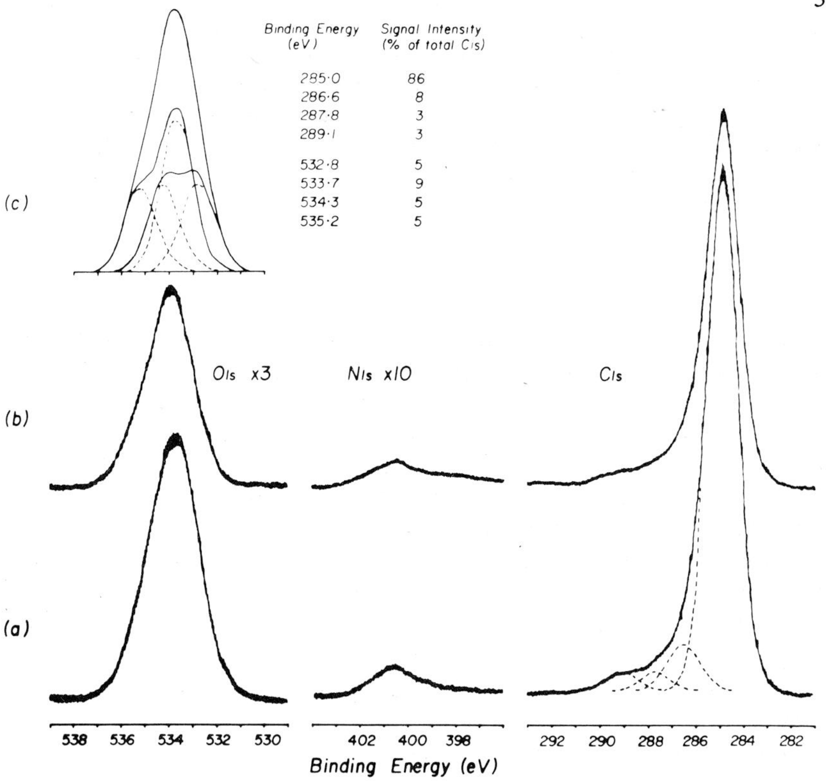

Figure 27. Core-level spectra for corona-discharge treated samples of polyethylene.

—C(=O)O—, >C=O, and ≥C—O structural features at binding energies of 289.1, 287.8, and 286.6 eV, respectively. This observation is supported by the signal in the O_{1s} region which is broad, suggesting several oxygen environments. The N_{1s} signal is of much lower intensity than that for O_{1s}, reflecting the relative concentrations and reactivities of nitrogen and oxygen species in the corona. Figure 27(b) is the spectrum of the same sample taken at an angle corresponding to grazing exit of the photoemitted electrons. The overall O_{1s}/C_{1s} ratio is closely similar for the two sets of data, which is indicative of the fact that the modification extends beyond the ESCA sampling depth. With a knowledge of model systems it is possible to analyse the O_{1s} and C_{1s} lineshapes as shown in Figure 27. The O_{1s} levels, for example, consist of a doublet of equal intensity centred at ∿ 532.8 and 534.3 eV corresponding to —C(=O)O— structural features. The difference arises from two components of intensity ∿ 5% and 9%, respectively, centred at ∿ 535.2 and 533.7 eV. The lower binding energy component may be ascribed to oxygen in alcohol, ether, or ketone structural features, whilst the higher binding energy component may be tentatively ascribed to peroxy features.

Several chemical modifications of surfaces have been proposed to improve adhesive bonding characteristics, including controlled oxidation by pressing films against aluminium foil in air[28] and chromic acid etching,[29] and these have been the subject of ESCA investigation which have showed considerable light on the changes in surface chemistry.

A particularly complex example which has been investigated in great detail is the controlled fluorination of polymers. The activation energy for the rate-determining process in the chain reaction involved is considerably smaller than for diffusion of fluorine into the polymer, and the reaction is therefore diffusion controlled with the fluorine content of the modified polymer decreasing with depth into the sample. Over the past 25 years there has, in fact, been considerable interest in the controlled direct fluorination of polymers, dating from an I.C.I. patent granted to Rudge which claimed that PTFE could be produced by direct fluorination of polyethylene. The work over the intervening period has largely been concerned with the extensive fluorination of polymers, rather than detailed investigations of the initial stages of the reaction. The sensitivity of the technique is shown by the form of the core-level spectra of high-density polyethylene which have exposed to 10% fluorine with nitrogen as diluent for ½ and 30 sec, respectively (Figure 28).

The initial sample exhibited an O_{1s} level signal corresponding to carbonyl structural features at submonolayer coverage. The C_{1s} levels for this initial sample correspond essentially to a single peak centred at a binding energy of ~ 285 eV. After ½ sec exposure, however, a F_{1s} peak appropriate in binding energy to C$\underline{F}$ structural features become evident, and the band profile for the $\underline{C}_{1s}$ levels also reveals the presence of $\underline{C}$F structural features with components appropriate to carbons directly bonded to fluorine and β to fluorine substituents. From a complete

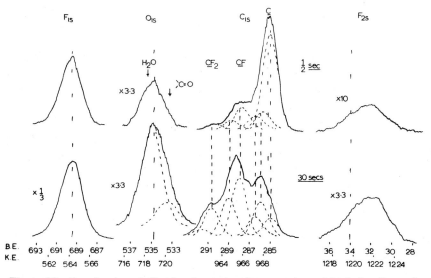

Figure 28. Core-level and F_{2s}-level spectra for samples of high-density polyethylene exposed to fluorine (10% in nitrogen) for periods of ½ sec. and 30 sec., showing development of carbon fluorine structural features.

analysis of the data it is possible to show that after ½ sec exposure, fluorination has proceeded to the first monolayer stage and that the composition of the latter corresponds quite closely to that appropriate to polyvinyl fluoride (namely after ½ sec exposure the composite structure corresponds essentially to a monolayer of polyvinyl fluoride on polyethylene). After 30 sec exposure the lineshape for the C_{1s} levels becomes extremely complex, extending to higher binding energies appropriate to $\underline{C}F_2$ structural features. By reference to model systems it is possible to analyse the lineshape and this, taken in conjunction with the relevant area ratios for the F_{1s} and F_{2s} levels, leads to a straightforward analysis which shows that fluorination extends to a depth of ~ 30 Å, with the overall stoichiometry in this region being $\sim C_1 F_{1.2}$. In this region, however, the surface will be more highly fluorinated than subsurface and bulk. It is possible to unravel the complexities of such a situation and it may be shown that the composition of the first monolayer at this stage approaches that appropriate to polytrifluoroethylene.[30] It should be evident from this that the time dependence of the chemical modification of polymers, particularly in the initial stages, may readily be followed by ESCA.

(v) Modification of polymer surfaces by direct and radiative energy transfer

The interactions of polymers with plasmas, and with discharges in general, forms the basis for several processes of industrial importance. Argon-ion bombardment, for example, forms the basis for the CASING[31] process for improving adhesive bonding. ESCA provides a most convenient means of following the changes in surface chemistry resulting from such treatments. Figure 29, for example, shows the F_{1s} and C_{1s} spectra of a largely alternating ethylene tetrafluoroethylene copolymer (52% TFE) after irradiation with a low-energy (2 kV) beam of argon ions having a beam current of 5 μA for successive periods of 5 sec. The results are quite striking and it is clear that argon-ion bombardment causes extensive modification within the ESCA sampling depth. The main features are as follows. The F_{1s} signal decreases as does the high binding energy component of the C_{1s} levels owing to $\underline{C}F_2$ structural features, with a concomitant increase in intensity of a component of intermediate binding energy in the C_{1s} spectrum owing largely to $\underline{C}F$-type environments. There are in fact close parallels to be drawn with the corresponding RF argon plasma treatment of polymers. It is indeed possible to illustrate the features in common with the direct energy transfer component of the plasma modification, this being illustrated by the kinetic data illustrated in Figure 30.

Since for the argon-ion treatment only direct energy transfer is involved, the rate processes monitored by the attenuation of core-level signals show a linear plot in striking contrast to the curvature evident for the plasma treatment (cf. Chapter 9).

The use of ESCA as a tool for elaborating details of the kinetics and mechanism of the interaction of polymers with plasmas excited in inert gases has been discussed in some detail in Chapter 9. It is worth while concluding this section by illustrating the wealth of data which may be obtained from complete analysis of the ESCA data. It has previously been emphasized that the kinetics and mechanisms of both the direct and radiative energy transfer processes may be investigated in some detail by monitoring the relative intensities and binding energies of the components

348

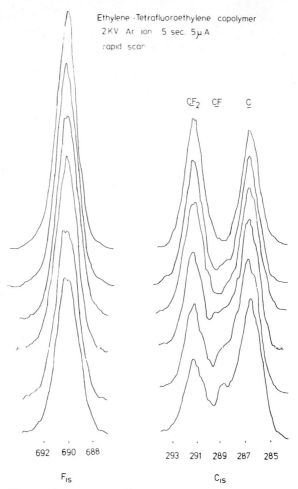

Figure 29. C_{1s}, and F_{1s} core-level spectra for an ethylene—tetrafluoroethylene copolymer as a function of exposure to an argon-ion beam.

of the core-level spectra. A similar analysis may in fact be based on the surface sensitivity of sample-charging phenomena. Thus, the changes in structure and bonding are such that substantial differences are manifest in the equilibrium charging shift which the sample exhibits under a given set of conditions. This is illustrated by the data in Figure 31 which mainly complements that obtained from an analysis of the primary ESCA data.[13]

8. CONCLUSIONS

It should be evident from the brief outline presented here that ESCA has a key role to play in the elaboration of details of structure and bonding of polymer surfaces.

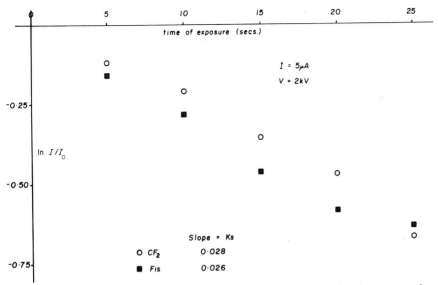

Figure 30. Analysis of rate data for argon-ion treated samples in terms of a pseudo-first-order process.

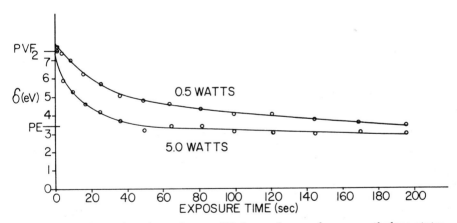

Figure 31. Change in dynamic equilibrium charge for an ethylene–tetrafluoroethylene copolymer (48% : 52%) as a function of modification in an argon plasma. Also indicated are the equilibrium charges for polyvinylidene fluoride (PVF_2) and polyethylene (PE).

The hierarchy of information levels endows the technique with unrivalled capability for studying inhomogeneous samples and, as such, ESCA will clearly have an important role to play in many fundamental investigations of both the synthesis and of the chemical, physical, mechanical, and electrical properties of surfaces.

REFERENCES

1. B. J. Tighe, Chapter 14, this book.
2. H. A. Willis and V. G. J. Zichy, Chapter 15, this book.
3. (a) D. T. Clark 'Chemical applications of ESCA', in *Electron Emission Spectroscopy*, ed. W. Dekeyser, D. Reidel Publishing Co., Dordrecht, Holland (1973). (NATO Summer School Lectures, Ghent, September 1972.) (b) D. T. Clark, *Advances in Polymer Friction and Wear*, Los Angeles, March 1974, ed. L. H. Lee, Vol. 5A, Plenum Press, New York, 1975. (c) D. T. Clark and W. J. Feast, *J. Macromol. Sci. Reviews in Macromol. Chem.,* **C12**, 191 (1975). (d) D. T. Clark 'Structure and bonding in polymers as revealed by ESCA', in *Electronic Structure of Polymers and Molecular Crystals*', ed. J. Ladik and J. M. Andre, Plenum Press, New York, 1975. (NATO Summer School Lectures, Namur, September 1974.) (e) D. T. Clark in *Structural Studies of Macromolecules by Spectroscopic Methods*, Chapter 9, ed. K. Ivin, Wiley, London (1976). (f) R. S. Swingle II and W. M. Riggs, *CRC Crit. Rev. Anal. Chem.*, 5 267 (1975). (g) D. T. Clark, in *Advances in Characterization of Polymer and Metal Surfaces*, New York, April 1976, ed. L. H. Lee, Academic Press, New York, 1977. (h) D. T. Clark, 'ESCA applied to polymers', in *Advances in Polymer Science* (Ed. H. J. Cantow *et al.*) Vol. 24, Springer Verlag, Berlin–Heidelberg. (1977).
4. K. Siegbahn *et al.*, *Nova Acta R. Soc. Sci. Uppsala*, Ser. IV, 20 (1967).
5. D. W. Turner, A. D. Baker, C. Baker, and C. R. Brundle, *Molecular Photoelectron Spectroscopy*, Wiley, New York (1970).
6. J. H. Schofield, *J. Electron Spectry.*, **8** 129 (1976).
7. C. D. Wagner, *Disc. Faraday Soc.*, **60** (1975).
8. (a) J. C. Tracy, in *Electron Emission Spectroscopy*, ed. W. Dekeyser, D. Reidel Publishing Co. Dordrecht, Holland (1973). (b) C. J. Powell, *Surface Science*, **44**, 29 (1974). (c) I. Lindau and W. E. Spicer, *J. Electron Spectry.*, 3 409 (1974).
9. D. T. Clark and H. R. Thomas, *J. Polymer Science, Polymer. Chem. Edn*, **15**, 2843 (1977).
10. D. T. Clark and D. Shuttleworth, *J. Polymer Science*, in press (1978).
11. D. R. Penn, *J. Electron Spectry.*, **9** 29 (1976).
12. C. S. Fadley, R. J. Baird, W. Siekhans, T. Novakov, and S. A. L. Bergstöm, *J. Electron Spectry.*, 4, 93 (1974).
13. D. T. Clark, A. Dilks, H. R. Thomas, and D. Shuttleworth, *J. Polymer Science*, in press (1978).
14. D. T. Clark and H. R. Thomas, *J. Polymer Science*, in press (1978).
15. D. T. Clark and H. R. Thomas, *J. Polymer Science, Polymer Chem. Edn.*, **14**, 1671 (1976).
16. D. T. Clark, W. J. Feast, I. Ritchie, W. K. R. Musgrave, M. Modena, and M. Ragazzini, *J. Polymer Science, Polymer Chem. Edn.*, **12** 1049 (1974).
17. (a) D. T. Clark and H. R. Thomas, *J. Polymer Science*, in press (1978). (b) H. R. Thomas, Ph.D. Thesis, University of Durham (1977).
18. D. T. Clark, I. W. Scanlan, and J. Müller, *Theoretica Chim. Acta* **35** 341 (1974).
19. R. Manne and T. Aberg, *Chem. Phys. Letters*, 7, 282 (1970).
20. D. T. Clark, in *Progress in Theoretical Organic Chemistry*, ed. I. G. Csizmadia, vol. 2, Elsevier, Amsterdam (1977).
21. (a) D. T. Clark, A. Dilks, J. Peeling, and H. R. Thomas, *Disc. Faraday Soc.*, **60** 183 (1975). (b) D. T. Clark, D. B. Adams, A. Dilks, J. Peeling, and H. R.

Thomas, *J. Electron Spectry.*, **8**, 51 (1976). (c) D. T. Clark and A. Dilks, *J. Polymer Sci., Polymer Chem. Edn,* **15**, 15 (1977).

22. D. T. Clark and H. R. Thomas, *J. Polymer Science, Polymer Chem. Edn.,* **14**, 1701 (1976).
23. D. T. Clark, A. Dilks, and D. Shuttleworth, Chapter 9, this book.
24. D. T. Clark and A. Dilks, *J. Polymer Science, Polymer Chem. Edn.,* **14**, 533 (1976).
25. D. T. Clark, J. Peeling, and J. M. O'Malley, *J. Polymer Science, Chem. Edn.,* **14**, 543 (1976).
26. D. T. Clark, A. Dilks, D. Shuttleworth, and H. R. Thomas, unpublished observations.
27. D. T. Clark, A. Paton, and W. Salanek, *J. App. Phys.,* **47**, 144 (1976).
28. (a) D. T. Clark, W. J. Feast, W. K. R. Musgrave, and I. Ritchie in, *Advances in Polymer Friction and Wear,* Los Angeles, March 1974, ed. L. H. Lee, Vol. 5A, Plenum Press, New York (1975). (b) D. T. Clark, A. Dilks, and D. Shuttleworth, *J. Materials Science,* **12**, 2547 (1977). (c) D. Briggs, D. M. Brewis, and M. B. Konieczko, *J. Materials Science,* **12**, 429 (1977).
29. D. Briggs, D. M. Brewis, and M. B. Konieczko, *J. Materials Science,* **11**, 1270 (1976).
30. D. T. Clark, W. J. Feast, W. K. R. Musgrave, and I. Ritchie, *J. Polymer Science, Polymer Chem. Edn.,* **13**, 857 (1975).
31. R. H. Hansen and H. Schonhorn, *Polymer Letters,* **4**, 203 (1966).

17

Defects of the Surface Zone — Their Origin and Inhibition with Particular Reference to PVC

P. L. Clegg and A. D. Curson

I.C.I. Plastics Division, Welwyn Garden City

1. INTRODUCTION

In this paper we present, from an admittedly industrial rather than academic viewpoint, a brief review of the salient features governing the surface and subsurface deterioration of plastics exposed to normal external weathering environments, descriptions of the various types of defect that can occur which are illustrated from our own work, and finally some thoughts on inhibition. Special mention is made of PVC, not so much because of any unique problems it poses as because this was the request of the symposium organizers.

In-service deterioration is common to many industrial and natural products and this has led to the widespread use of renewable protective films. Plastics are not so protected and their initial cost and expected lifetime has to be judged in relation to lack of maintenance. In fact, of course, the plastics industry and their supporting finer chemical suppliers have spent some considerable effort to improve lifetimes by synthesizing, and proving, effective antioxidants and ultraviolet absorbers. Their success is indicated by the enhanced lifetimes for various protected polymers quoted in Table 1; their failure relative to the simplest and most effective protection of all is reminiscent of Henry Ford's dictum, 'You can have any colour you like so long as it's black'.

2. GENERAL MECHANISM OF FAILURE

It is simple to set down conditions that must apply before failure can occur; either (a) the stress (or strain) in the plastic must increase beyond the supportable limit, or (b) there must be a detrimental change in the material's ability to support such stresses (or strains) as do exist. That it is not so simple as it seems is shown by the enormous amount of experimental and theoretical work which it has been necessary to undertake in order to fill in much of the detail; there are at least two notable books[1,2] which provide a wealth of references. A general scheme of failure is shown in Table 2.

Table 1 Expected minimum lifetime (years) under natural weathering conditions in the UK

Polymer	Standard grades	Natural coloured grades designed for outdoor use	Black grades
LDPE (mouldings)	3−5	5−7	20+
LDPE (film)	1−1.5	3	10+
PP	1	2−3	10−20
PVC (plasticized)	1−2	5−10	20+
PVC (rigid)	2−5	5−10	10−20
PMMA	10−15*	10−15*	15−20
ABS	2−3	−	5

*All standard PMMA grades are designed for outdoor use.

Considering first the excess stress which may cause failure, the least interesting case is when the macroscopic stress limit is exceeded, and ductile or brittle failure follows, depending upon the strain rate and the plastic concerned. This type of failure can be avoided quite simply by conventional design to essentially macroscopic stress limits. The phenomena involved in exceeding a microscopic tensile stress (or strain) limit are more complex.

Macroscopic tensile stresses (or strains) well below short-time tensile limits may well, in time, cause minute orthogonal failures, called crazes, to appear and grow until a true crack is initiated; this may be followed fairly rapidly by complete failure. The nature and environment of the stress field[3] as well as the inherent properties of the plastic (see Vincent[1]) will determine the type of failure and the way in which the ductility of the material in the neighbourhood of the growing craze tip is involved.

The main problem has been to explain satisfactorily why material should begin to fail under an apparently simple tensile stress. Many electron-microscopic studies, for example[4,5] have shown clearly that the initial craze failure is formed by submicro cavitation at about 100 Å centres, followed by the drawing of 'proto filaments'[2] from the material between cavities (see Figure 1). Given some initial heterogeneity or pre-existent voiding of molecular structure or the presence of 'flaws' in the form of dust, catalyst residues, etc. it is possible to see how the necessary triaxial stress system could be established that would result in

Table 2 A general scheme of failure

Level of failure	Stress increases beyond limit tolerable		Ability to withstand existing stress decreases	
	Time scale		Time scale	
Macroscopic	Short	External loading stress increases and leads to 'uninteresting' macroscopic failure	Long	Failure at stresses much lower than at short times due to initiation of 'brittle' failure modes
Submicroscopic	Short/med.	Rupture or physical transport of overstressed molecules leading to submicrocavitation and, later, crazing	Longer	(a) Stress field alters at tip of craze and a crack propagates (b) Cyclic loading increases rate of propagation of failure
	Med./long	Local stress increases due to evolution of gaseous by-products of degradation	Med./long	Molecular scission due to degradation processes
			Any	Presence of an 'active' environment

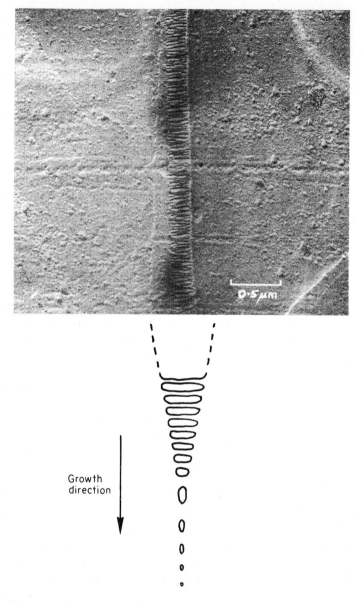

Figure 1. Transmitted electron micrograph of a craze in polystyrene illustrating the protofilament structure, × 32 200. The line drawing shows how this craze structure is developed from microcavitation produced in advance of the craze tip.

submicrocavitation. Evidence for such submicrocavitation, consequent upon the dissociation of overstressed interatomic bonds by thermal fluctuation, is to be found for example in the work of Zhurkov and his collaborators published over the last 15 years, and whose most recent paper[6,7] shows that the surface concentration of such defects is at least three to five times greater than their volume concentration.

One is reluctantly forced to agree with the conclusion of Halpin and Hassall[1] that

> the products that we use generally contain pre-existing flaws or defects which are either inherent in the materials or introduced during a fabrication process. Relatively large flaws may be detected by quality control and inspection procedures which are largely deficient in non-metallic materials and corrected by repair or maintenance procedures. Unfortunately, undetectable or undetected small surface cracks or embedded flaws do grow to critical sizes.

Secondly, one must consider changes in the ability of the initial molecular structure to support the stress field that exists. Discounting normal chemical or biochemical attack, all polymers are subject to degradation because of oxidation, thermal motion, or photon action. The most significant change that occurs is chain scission, and although this may be followed by a cross-linking reaction, the inevitable result of high levels of degradation is a mechanically useless plastic. It certainly seems quite clear that chain scission caused by ultraviolet (UV) or by any other means will locally weaken the molecular structure and increase the probability of a submicrovoid growing under the prevailing stress field. Indeed, the most damaging weathering defects are found in the presence of both stress and polymeric degradation.

Chain scission can be neatly surgical as in the case of PMMA, or rather messy as in the case of PVC[8] which degrades primarily through a dehydrochlorination reaction. The HCl evolved will presumably increase stress levels locally, although, of course, most will permeate through to the surface. It is not unreasonable to postulate that such stresses, in addition to the prevailing stress field, will increase the chance of craze initiation and cracking in the surface structure, especially if this is molecularly weakened by scission.

Of most concern in the context of this paper is the effect of UV photons combined with modest temperatures (a black plastic in bright sun can reach about $100°C$) and, quite frequently in our climate at least, copious water! Because plastics have a high general level of absorption in the UV, photodegradation is most pronounced at the surface. Thermal degradation and oxidation are more likely to be uniformly spread through the thickness of the article.

The effect of 'active' environments — which differ from polymer to polymer — has been well known for 30 or more years, especially for PMMA and polythene, as environmental strain cracking.[9] More recently this phenomenon has been reported in unplasticized PVC,[10] and Faulkner and Atkinson[3] have shown the sensitivity of unplasticized, plasticized, and modified PVC to hexane, hexane-toluene and toluene environments.

In another recent paper, Kitagawa[11] shows that even in an anisotropic glassy polymer, crazes are always perpendicular to the maximum strain direction calculated by taking into account the internal stress, and furthermore that in an 'active' environment crazing may occur even under a pure torsional load. The evidence to date does not indicate, unambiguously, whether the critical criterion for the onset of crazing is one of strain or of stress. Whilst of fundamental significance this distinction is not of great practical importance.

3. DEFECTS OF THE SURFACE ZONE

Phenomenologically, the deterioration of plastic materials by natural weathering is evidenced by their original colour being modified, their clarity impaired, and perhaps by macroscopic failure from cracks initiated by changes in surface structure.

With bulk materials it is often only when a close inspection is carried out that the defects come to light, although in some instances the effect of weathering is made apparent by the fall-off of essential, critical properties in the application to which the material is being put, for example the development of haze in clear materials. In the case of films, however, because the 'surface' accounts for a

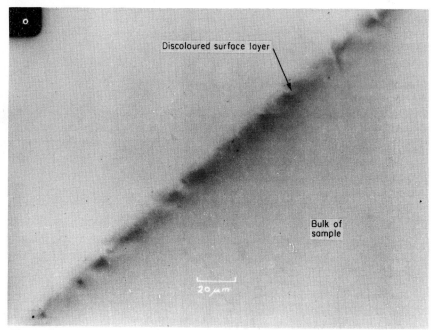

Figure 2. Transmitted common light micrograph of a 5 μm section through the surface of a UPVC sheet discoloured by artificial ageing, ×700.

considerable amount of the bulk of the film, the results of the development of the defects on the continued performance of the material is made more apparent.

Perhaps the most common feature of weathering defects is a noticeable change in the original colour of the material. This may come about as the result of one or more quite different reactions of the plastic to weathering. It is possible to separate these phenomena into those arising from true discoloration of the polymer and those arising from the development of scattering centres leading to masking of the original colour or clarity. The majority of cases fall into the latter category, and these may also be subdivided into phenomena affecting the actual surface and those which are essentially subsurface.

(i) True discoloration

Where true discoloration has occurred there are not necessarily any observable changes in structure of the material. The three main causes of this form of defect are discoloration of the polymer (Figure 2) In PVC this is associated with the formation of conjugated double bonds and the release of HCl, degradation of pigments used in the formulation; and chemical reaction of one or more of the formulation additives with components of the atmosphere (for example the staining of lead-stabilized unplasticized PVC by a sulphurous environment Figure 3).

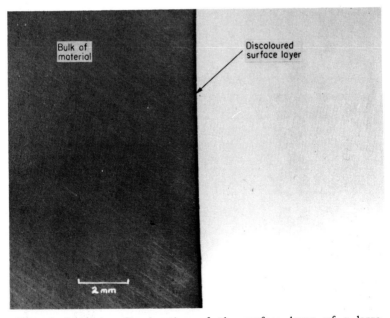

Figure 3. In-service discoloration of the surface layer of a large-diameter UPVC pipe due to sulphurous environment (sewerage). Refelcted light, ×9.

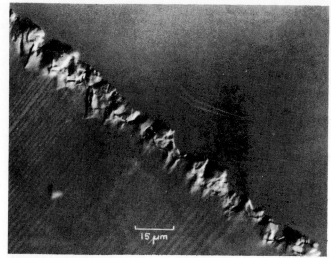

Figure 4. Section through severe surface deterioration of UPVC sheet produced by natural weathering. Transmitted differential interference contrast, ×1000.

Figure 5. Naturally weathered UPVC sheet exhibiting developed haze. The centre circle has both surfaces 'oiled-out' and the outer circles have either surface oiled out. This demonstrates that the haze is the result of a surface defect and is restricted to one surface (the exposed) only. Common light, ×1.5.

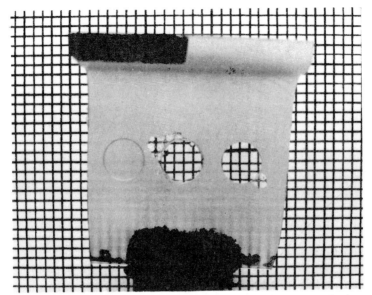

Figure 6. As for Figure 5, but the surface of this sample has deteriorated to the extent that there is total loss of clarity. Common light, x1.5.

Figure 7. Naturally weathered polypropylene showing craze and crack development associated with superficial flow defects. The large disruptions are handling damage arising from the brittleness of the surface. Reflected light, x7.

(ii) Surface defects

Almost totally representative of the first subdivision is the development of what is variously described as surface break-up, chalking, powdering, surface whitening, etc. It occurs with all polymers; glassy (PMMA), crystalline (polypropylene), and granular (PVC) and, depending upon the time of exposure, ranges from the development of surface haze (Figure 5) to a total loss of clarity (Figure 6) in clear materials or loss of original colour in opaque grades. In some applications, particularly those using injection-moulded components, the initial stages of the development of the defect can be seen to be associated with the memory of superficial flow patterns (Figure 7) resulting from the filling of the mould. In others the defect may occur as patches or as a uniform deterioration over the whole exposed surface of the material.

In a very few cases of the development of such a surface defect on a commercial product, we have found the cause to be the migration or leaching out and subsequent degradation of additives at the surface. It is more usual to find that the defect is caused by the physical disruption of the surface layer of the plastic, generally to a depth of no more than about 5 μm (Figure 4). In all known cases this break-up is the conclusion of a progressive deterioration of the surface which, although not necessarily starting with, passes through an earlier stage of craze formation.

Analysis by infrared spectroscopy of the surface layer of exposed surfaces of PVC usually produces no evidence consistent with a change in formulation or of

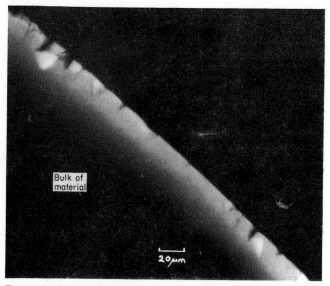

Figure 8. Photomicrograph of a section through the weathered surface of a sample of clear UPVC sheet illustrating the fluorescence layer due to degradation of the polymer. Transmitted fluorescence microscopy, ×500.

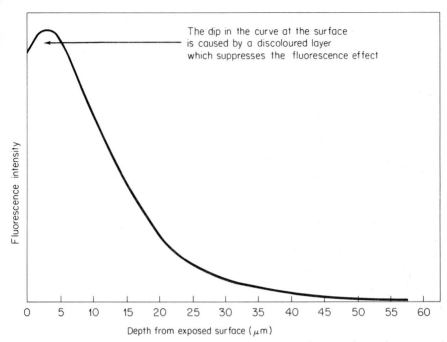

Figure 9. Plot of fluorescence intensity as a function of depth from the exposed surface in weathered UPVC.

additive migration. However, UPVC develops fluorescence quite readily (Figure 8) on exposure to UV radiation, the intensity showing an exponential relationship with the depth of penetration (Figure 9). As already mentioned the accepted sequence of the degradation of PVC is that dehydrochlorination is followed and

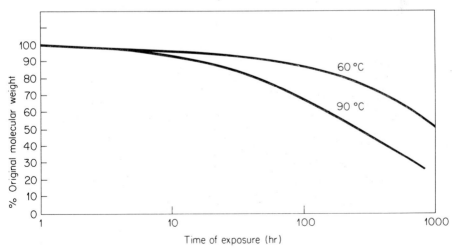

Figure 10. Change in molecular weight of PMMA as a result of exposure to intense UV radiation (450 W Hg discharge lamp without glass envelope).

accompanied by conjugated double bonding.[8] This process would not only give rise to the development of fluorescent sites but is also likely to cause physical disruption, although it should be noted that this does not necessarily imply any decrease in molecular weight of the polymer.

On the other hand, changes in molecular weight of thin (approx. 50–100 μm) foils of polymethylmethacrylate exposed to UV radiation (Figure 10) lead to quite different conclusions in respect of this polymer. Although, admittedly, not at normal ambient temperatures, the results do, however, show that a reduction in molecular weight is brought about by exposure to the radiation. An examination of typical transmission curves of stabilized PMMA at 3650 Å (Figure 11) shows that the rate of absorption is greatest at the surface. It would therefore not be unreasonable to expect that the greatest effect on molecular weight would also be at the surface.

(iii) Subsurface defects

Phenomena which are subsurface are almost entirely associated with voiding in one form or another. The exception being a few isolated instances of internal

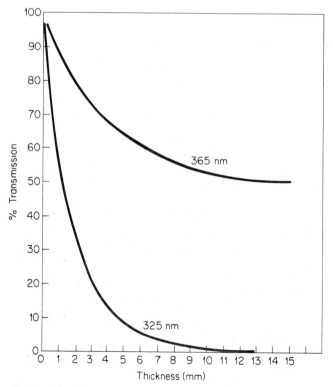

Figure 11. Transmission curves for PMMA containing 0.005% UV absorber.

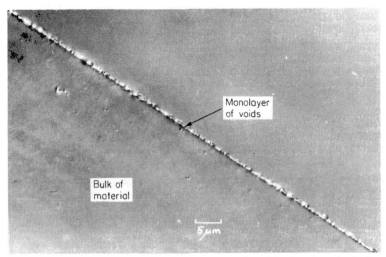

Figure 12. Section through a monolayer of voids developed at the exposed surface of naturally weathered clear UPVC sheet. Transmitted differential interference contrast microscopy, ×2000.

crazing and phase separation. This subsurface voiding should not be confused with the microcavitation associated with crazing, which is typically of the order of 0.005–0.025 μm diameter, but are many times larger, i.e. 0.5–1.0 μm.

Typical of clear unplasticized PVC is the development of a surface monolayer of voids (Figure 12), or of a layer of voids diminishing in concentration to a depth of

Figure 13. Section through the exposed surface of naturally weathered clear UPVC sheet, showing subsurface band of voids. Transmitted differential interference contrast microscopy, ×2000.

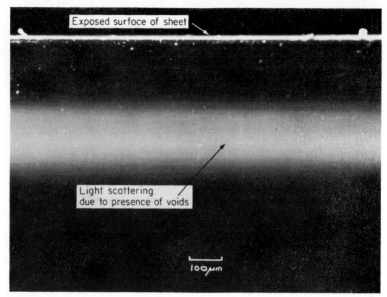

Figure 14. Light scattering caused by a subsurface layer of microvoids in a section through the exposed surface of clear PMMA sheet exposed to high humidity and intense UV radiation. Reflected dark ground microscopy, x126.

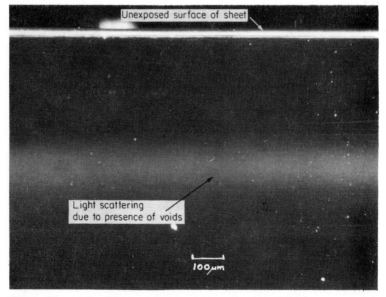

Figure 15. As for Figure 14, but the section this time is through the unexposed surface. The thickness of the sheet was 3 mm.

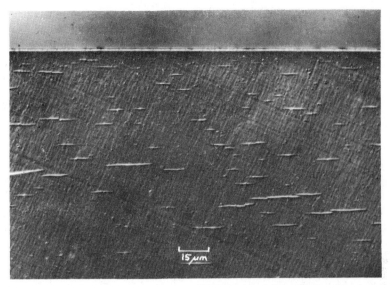

Figure 16. Disc-shaped voids in a PMMA moulding illustrating how voiding can be associated with additive particles. Transmitted differential interference contrast, ×730.

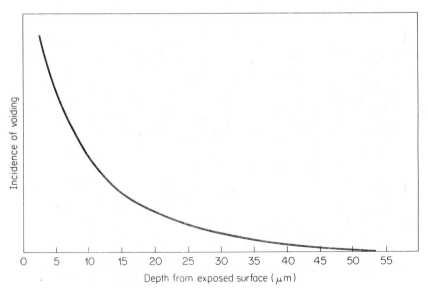

Figure 17. Incidence of voiding as a function of depth from the exposed surface in weathered UPVC.

about 50 μm below the exposed surface (Figure 13). Their spherical appearance suggests that they probably result from the release of gaseous compounds of degradation, presumably HCl. Like fluorescence they show an exponential dependence of concentration on depth from the surface (see Figure 17) indicating a relationship with the absorption of radiation.

Discrete internal layers of voids are more usual in glassy polymers, for example PMMA, and their occurrence has always been associated with conditions of high humidity and intense UV radiation (street lighting bowls) (Figures 14 and 15).

Within pigmented compounds, or those incorporating particulate additives, it can be difficult to say precisely whether or not the voids are associated with the additive particles. More often than not they are, and their formation leads to a diminution of the colouring effect of the pigments by light scattering at the voided interfaces. The disc-shaped voids illustrated in Figure 16 gave rise to a peculiar pearlescent effect: the shape of the voids in this instance resulting from the stress patterns set up around the pigment particles during the shaping of the plastic.

4. IMPORTANCE OF POLYMER MORPHOLOGY

Maintaining our interest in the real world rather than the idealized one of section 2, the importance of the detailed morphology of a plastic subjected to weathering can scarcely be over-exaggerated. The emphasis is apparent when it is fully realized that the integrity of an article will depend upon the degree to which all chains bear an equal part of any external stress and, in the absence of external stress, on the range of local stress variations experienced from one molecular chain to another and along each chain; these factors are dominated by morphology.

Even in apparently ideal situations, for example in the case of an amorphous polymer like PMMA polymerized very slowly indeed, from exceptionally pure monomer, light-scattering studies[12] show significant optical anisotropy, and presumably associated molecular stress variations, on the scale of about 500 Å. Pure amorphous systems polymerized and fabricated under commercial conditions are very substantially more heterogeneous.

Of course, to add to the complexity, many polymers crystallize and the detailed way in which the amorphous chains are left to bear stress will depend on nucleation and crystallization kinetics, on the type of spherulitic structure developed, and on the extent of the interspherulitic regions.

Pure systems, however, barely exist outside the realms of fantasy, and all polymers are contaminated with, for example, 'dirt' and catalyst residues and most contain not merely one but several additives and pigments or dyes of varying degrees of solubility or solidity. All these non-polymeric additives will affect the stress distribution in local molecular chains, and this will be particularly marked where there is non-uniform adhesion between polymer matrix and particulate additives or in regions of local phase separation.

Another factor which may be very significant, is certainly so in the case of PVC[13] or even more exaggeratedly so in the case of PTFE,[14] is the degree to which the polymer particles (granules or powders) are more or less integrated during the melt-compounding operation that usually precedes fabrication. The powder particle of PVC is very complex, being an agglomeration of primary particles, each one of which is composed of subprimary particles about 200 Å in diameter. In a poorly compounded article one can clearly see that original powder particle boundaries still exist, even if deformed; it takes very — one might almost say unusually — good compounding to remove all traces of the subprimaries. Indeed, Frey[13] has stated in relation to PVC that 'a completely isotropic material cannot be realized in practice'.

It is becoming more and more apparent that the complexity and degree of anisotropy of processed UPVC and other plastics is still not fully understood, and the deeper one delves into the structure of plastics the more complex the situation becomes. It is certain that the presence of the unduly stressed and seemingly weak molecular links abounding in poorly compounded polymer are obvious from almost any physical test.

Clearly, too, the state of the solid polymer will reflect the degree of molecular orientation that existed in the melt prior to solidification. This in turn certainly affects spherulitic structure and, to varying degrees, the level of crystallinity achieved in crystallizing polymer. The end result in either amorphous or crystalline polymers may well be a marked anisotropy along and across the direction of orientation.

The effect of strain imposed at much lower temperatures can be even more dramatic and lead ultimately, as in the case of fibre made from film, to virtually zero strength perpendicular to the direction of draw.

Finally, and again most importantly, one must recognize the great significance of the degree of uniformity of dispersion of additives; particularly, in the case of

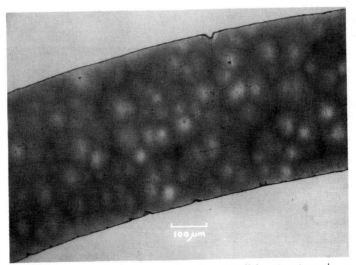

Figure 18. UV photomicrograph of spherulitic structure in a polypropylene moulding containing a UV absorber illustrating rejection of the additive to the spherulite boundaries. UV absorption microscopy (285 nm) x140.

Table 3 Importance of polymer morphology

Factors leading to:

Increased local stress concentrations	Increased likelihood of failure
Polymeric inhomogeneity	Inadequate compounding
Presence of particles	Poor dispersion of particles owing
— pigments	to poor compounding or subsequent
— fillers	segregation during cooling
— residues	
— 'dirt'	
Particle/matrix adhesion	Variable adhesion
State of 'tie' molecules	Stress orthogonal to molecular
	orientation
— amorphous regions	Increasing low m.wt content of
	polymer
— at crystalline/amorphous interface	Any unusually taut molecule

weatherability, of those designed to protect the polymer from the effects of oxidation or UV radiation. Carbon black when used as a protection against UV must be well dispersed[15] on both macroscopic and submicroscopic scale. Whilst good dispersion is essentially a matter of well-conrolled mixing in a melt this is not necessarily wholly sufficient. During the crystallization process that can accompany solidification, zone refining can occur leading to preferential aggregation of additive at spherulitic boundaries — Figure 18 shows this type of segregation of a UV absorber in polypropylene. The major effects are broadly summarized in Table 3.

5. FACTORS TO BE CONSIDERED IN THE ENVIRONMENT

In searching for a definition of weathering the researcher is faced with a spectrum ranging from the rigorous definition given by Stager.[16] 'The effect of all technoclimatic, mechanical, physical, chemical and electrical influences which under operating conditions, by their collective effect on the surface and in the interior, lead through irreversible processes to final deterioration', to specific aspects related to given applications. In the latter class would be included the definition employed by the British Plastics Federation[17] and which for the purpose of this paper we have taken as our standard: 'The effect on colour, appearance and physical properties of plastics brought about by their prolonged exposure to atmospheric and weathering conditions in this country (UK).'

The factors important to a material when it is exposed to natural weathering have been identified by Rugger,[18] Estevez,[19] and many others. Natural sunlight consists of radiation extending from and including UV, through the visible to the infrared. In order for the radiant energy to initiate reactions it must be absorbed and the frequency of the absorbed energy depends on the chemical structure of the group irradiated.[20,21] Furthermore, it is the wavelength of the radiation (expressed in energy per photon) rather than the total amount absorbed over a period of time (assuming that time to be longer than the decay time of the excited atom) which determines activation and reaction. Achhammer[20] has recorded that the energy to break bonds in molecules is of the order of 50−100 kcal/mole. This compares with the energy of 82 kcal/mole associated with radiation of wavelength 3500 Å which is the mid-point of the radiation comprising 5% of natural sunlight at the earth's surface.

Combinations of UV radiation, water, and infra-red radiation (heat) will often result in more advanced levels of degradation that UV radiation by itself. This is because heat, moisture (and oxygen for that matter) can modify the polymer or its additives so as to shift the UV absorption to those wavelengths at which photochemical degradation can occur. Temperatures increases can affect the degradation processes and Melchore[22] quotes that a $10°C$ increase in temperature will double the speed of the degradative reaction. The effect of alternating the thermal state of a plastic material can give rise to the development of surface crazing owing to excessive expansion and contraction, and is even more pronounced if the material contains 'frozen-in' stresses resulting from fabrication. Reactions between polymers and oxygen are very slow at normal ambient temperatures. The more serious problem would appear to be the synergistic effect with moisture and UV radiation at elevated temperatures. These aspects have also been discussed by Achhammer.[20]

Water plays an important role in the weatherability of plastics. It can in fact be beneficial by washing away soluble products of degradation which might otherwise catalyse decomposition. But generally its effect is detrimental in that it removes additives essential to the performance of the material such as soluble stabilizers and antioxidants. Also, polymers in which hydrolysable groups are formed as a result of oxidation are susceptible to degradation by water. The combined effects of sudden changes in humidity and temperature are particularly potent, and there is evidence[23] that water deposited in an environment producing condensation is the more active factor than straightforward wetting by rainfall, etc. in that both the onset and development of surface defects is more rapid under such conditions. The direct action of micro-organisms by mastication or the indirect action of the corrosion by their waste products can be severe. However, such attack is usually confined to materials buried in earth, submersed in water, or exposed to tropical climatic conditions.[24] Dust, industrial gases, salt, or fine-sand-laden sea air, etc. will also contribute to the degradation process, not only by chemical action but also by erosion.

Although, in general, exposure to UV radiation and water is considered to make the major contribution to degradation[18] it is the summation of all the effects which will determine whether or not a material is suitable for a given application. Extending this argument a stage further, it is well known that plastic materials react differently to the presence of chemicals in both the liquid and gaseous states. UPVC, for example, can be soaked in petroleum ether without any noticeable damage, but put it in an environment containing commercial petrol (in which there is a quantity of benzene) and the results are quite drastic. Again UPVC will not in the short term show deterioration in the presence of xylene, whereas many crystalline polymers will. So once more there is the importance of choosing the plastic to match the possible contamination of the atmosphere by chemicals in everyday use, but not normally considered to be components of the atmosphere.

6. THE INHIBITION OF SURFACE AND SUBSURFACE DEFECTS

The inhibition of the defects described previously are inseparable from the steps taken to improve weathering peformance. The literature covering the ageing and weathering of plastics was given an intensive review by RAPRA in 1970,[17] and much of what was said then is quite as relevant today, although advances have obviously been made in most areas.

Much of the basic work in the investigation of the effect of weathering was done many years ago; for example the potential benefit of using thicker materials was investigated by Gotfried and Dutzer,[25] Martinovitch,[26] and Gibson[27] amongst many others.

Whilst there are a wide range of factors to be considered in the inhibition of weathering defects these can be divided into three broad categories: of stabilization of the polymer against the effects of radiation (particularly UV), against the effects of heat, and by the reduction of stresses, both internal and externally applied. Some remedial steps are similar from polymer to polymer, others differ to a substantial extent and when an unfavourable reaction cannot be prevented from occurring then it is important to limit the size of the resulting defects to an extent where they are not noticeable. Orientation, or stress, is reduced by storage and thermal treatment in the thermoelastic range is often recommended prior to exposure, although it should be noted that the completely isotropic state is hardly ever attained.

(i) Polyolefins

Degradation of polyolefins by UV radiation is a photo-oxidative process involving chain scission. To prevent radiant energy, in particular UV radiation, being absorbed by the polymer, UV absorbers may be introduced to absorb preferentially this harmful radiation and re-emit the energy at some other non-destructive wavelength. The oldest and one of the most effective methods of UV stabilization is the addition of carbon black to the polymer. However, for this to be successful particular attention must be paid to the type of black used, its particle size, concentration, and dispersion. Its effectiveness as a light screen was demonstrated by Wallder[28] who showed that the useful life of polyethylene could be substantially increased if it contained 2% channel black; increasing the concentration provides no further improvement to light resistance.

Particle size is important.[28] For example, furnace blacks with a particle size 45–275 mμ show little effect, whilst channel blacks of particle size 10–30 mμ give considerable resistance to light degradation.[29] The effect of particle size on light attenuation is considered by Ambrose[30] and maximum attenuation is found to occur over the size range 20–70 mμ. Obviously, the optimum particle size of carbon black will increase as the wavelength of the incident radiation increases. Over the principal range of interest the optimum size is of the order of 100 mμ and Cocks and Metzger[31] emphasize that blacks of 70 mμ are the most effective for the attenuation of UV radiation. Nevertheless, in practice it is more usual to use particles some 5–10 times smaller, i.e. 10–20 mμ. This arises from the necessity to take into account the fact that such finely divided particles will not remain as discrete entities, but in general will agglomerate to form larger 'particles' of the size recommended by Ambrose, Cocks, and Metzger. Recently the use of zinc oxide has been suggested as a low-cost UV stabilizer, and its performance in synergistic systems has been compared with standard organic UV stabilizers.[32] No matter what material is used it should be remembered that any particulate additive put into a polymer to inhibit defects due to weathering will not protect the true surface itself; for a 2% addition of carbon black the surface zone will be affected to a depth of the order of 100 additive particle diameters.

The resistance to photodegradation of polyolefines can be considerably improved by the incorporation of soluble UV stabilizers. The mechanism of stabilization is complex and involves considerations arising from the preferential absorption of the UV radiation, deactivation of excited states in the polymer molecules, and the termination of free-radical reactions. Increasing the concentration of the stabilizer increases stability to photodegradation, but such increases are limited to the point where the polymer becomes saturated and the excess stabilizer diffuses to the surface and is washed or eroded away.

A very real and important phenomenon affecting not only this but other stabilization systems, particularly in polypropylene, has been illustrated by Curson[33] and Frank and Lehner.[34] This phenomenon is the rejection of additives by the growing crystal front, causing their redistribution and localized concentration during the crystallization of an initially homogeneous polymer melt. Computer simulation and prediction of the growth/rejection process has been carried out by Billingham et al.[35] Basically the overall result of this phenomenon is a greater or lesser degree of rejection of the additive to the spherulite boundaries. The degree of rejection is dependent upon the mobility of the additive molecules within the polymer, the rate of crystallization, the type of crystallinity, and degree of ordering of the crystallites being developed by the polymer. A typical illustration of this phenomenon is shown in Figure 18. It can be argued that such 'zone refining' is beneficial in that stabilizers are concentrated in regions expected to be rich in polymeric material, particularly susceptible to degradation. However, weathering, as we have seen, is essentially a surface defect and the effect on the surface of 'zone refining' in some instances, such as where there is surface nucleation, will be to deprive the surface of the necessary stabilization.

It is important to realize that not only the molecular structure of the stabilizer, but the nucleating effect of other additives and the thermal history of the polymer itself are all important factors in the final distribution of the stabilizer, the location of which has a fundamental bearing on the effectiveness of stabilization. In connection with this it should be noted that we have found with polypropylene in particular that variations in thermal history, cooling conditions, reheating, etc. can result in the polymer crystallizing in at least six different ways, including three different crystalline forms and variations of ordering of the crystallites within these forms.

The main requirements of antioxidants are that they must be capable of prohibiting oxidation, they must be permanent, compatible with the polymer, non-colouring, non-toxic, and they should not have any adverse effect on the mechanical or electrical properties of the polymer. Most of what has been said concerning the effect of crystallizing conditions of the polymer on the distribution of UV absorbers is applicable also to antioxidants. It is useful to know that there is often a synergistic reaction between UV absorbers and antioxidants, such that the former are often more effective in the presence of the latter. Carbon black functions as a mild thermal antioxidant in polyethylene, but it can also reduce the activity of some antioxidants and stabilizers. In some cases, particularly with polypropylene, the effect can be bad enough to warrant the exclusion of carbon black from the formulation.

The incorporation of certain pigments into polyolefins in concentrations from 0.5 to 2% can result in increased weathering resistance. Suitable pigments are those which absorb strongly in the UV region while exhibiting little or no absorption in the infrared. The effect of black and white pigments in polypropylene and polyethylene show clearly that finely divided, well-dispersed black is the most effective as a light absorber.[35] Surprisingly, titanium dioxide is shown to have a protective influence when used at concentrations of 1–2% in mouldings, but in thinner sections and films the well-known photo-catalytic activity of the pigment overcomes its opacifying effect and the final result can be poorer than unpigmented polymer.

Finally, a brief mention should be made of further implications of the variety of crystalline structures produced within bulk samples of these polymers, particularly polypropylene. Different types of spherulite will undoubtedly have different degrees of strain at their boundaries. Consequently, variations in thermal history giving rise to the development of a range of different structures will result in the incorporation of localized regions with varying degrees of strain, the highest of which will dominate the initial weathering properties of the plastic. Also, associated with the phenomenon of rejection, differing from spherulite type to spherulite type, is the rejection of inherently weak components of the polymer (e.g. low molecular weight species) to areas where they become the dominant component.

(ii) PVC

Although PVC is probably the most widely used plastic in external environments and is often considered as one of the most weather-resistant polymers available today, there is still a hint of the controversy illustrated by the statements made more than 12 years ago by Penn[36] who said that PVC is weather-resistant, and DeCoste et al.[37] who claimed that PVC is not inherently weather-resistant but that it derives its properties through formulation. It was Estevez[19] who really put the problem in its true perspective and said that the weatherability of PVC is the joint responsibility of the raw material suppliers and the fabricators.

Upon exposure to heat and light PVC is notoriously unstable, and this is evidenced by its rapid discoloration and serious stiffening. The situation is aggravated by the presence of plasticizers and other additives which are themselves prone to degradation.[38] For successful applications the formulation must contain certain additives capable of specific chemical reactions with the resin, plasticizer, decomposition products, etc. to prevent discoloration and loss of physical properties. According to Geddes[39] the apparent inferiority of the thermal stability of PVC is due to discoloration becoming unacceptable at very low levels of HCl loss (less than 0.1%) while in other polymers degradation must be quite considerable before the physical properties are appreciably affected. The initial colour of PVC depends on the initiator and other additives used during polymerization.[40] Colour development is closely related to the chemical changes occurring, particularly HCl evolution[41] and Kralova and Vesely[42] showed that the presence of UV absorbers and antioxidants have little effect, whereas the presence of an HCl acceptor improves colour stability during degradation.

When considering PVC for any application the most important question to be answered is, 'Which formulation will best fit the application, giving the best balance of properties at lowest cost?' In the choice of formulation the 'K' value (or molecular weight) of the polymer must be considered together with its type, suspension, emulsion, or bulk (mass). Generally speaking, the higher the 'K' value the higher are the processing temperatures required and the better are the mechanical properties. Although the 'K' value by itself may not affect weathering directly, its indirect effects through processing and properties must be considered.

Bulk polymerization gives, apart from its high initiator concentration, the purest polymer, but it is not normally recommended for weather-resistant applications. The reason for this could be in the claim made by Matthan et al.[43] that bulk polymer has a low thermal conductivity, thereby making its processing difficult to control and as a consequence its weather resistance unreliable. Suspension polymerization provides less pure products which nevertheless exhibit greater heat stability, less water absorption, better electrical properties, and greater clarity. Emulsion polymers are easiest to process, but are least pure.

Once upon a time it was particle shape and size of the granules which were thought to be the most important features of a polymer, but it is becoming more obvious that these are only two of many other features of the morphology of PVC which contribute to the performance of the plastic. In recent years in-depth investigations into the structure and gelation mechanisms of PVC[44] have shown that the problem is indeed extremely complex.

The requirements for effective stabilization of PVC are many and complex. As yet no single stabilizer has been found exhibiting all the desirable qualities, and it is therefore necessary to devise a stabilizing system paying particular attention to thermal stability. These are mainly HCl acceptors, but some may possess UV absorption characteristics. Their effectiveness is usually judged by how they prevent colour change during processing. Lead compounds are amongst those most commonly used in opaque compounds and their use in plasticized grades of PVC is described by Pearson and Morley[45] and in UPVC by NL Industries Inc.[46] They are least expensive, create opacity, and are toxic. Both the tribasic and dibasic salts often require the presence of lubricants to obtain optimum performance from them.[47] Some heat stabilizers exhibit antioxidant and light-absorbing properties. Dibasic lead phosphite is one of these, but generally it is necessary to incorporate a separate UV screen, usually light-shielding pigments. The best is small particle size carbon black.

Other stabilizers used are the Group II metal (e.g. barium and cadmium) salts of organic acids; these exhibit good compatibility and high efficiency at moderate cost. Combinations of barium and cadmium salts are particularly synergistic. Aromatic phosphites, epoxidized oils, and fatty esters also provide synergistic actions, at the same time serving as secondary plasticizers. For non-toxic uses the weaker calcium/zinc salts synergized with aromatic phosphites and epoxidized soya-bean oil are probably best. There are many other stabilizers on the market too numerous to mention and all having their individual advantages, but at present their inhibiting disadvantage for general use is cost.

Up to about 10 years ago information on the use of lubricants was scant, confusing, and contradictory. Advances made in recent years have firmly established a categorization on a graded scale, with the extremes being represented by classifying as internal and external lubricants, based on their effect on processability and also on compatibility ratings with the polymer. It is even envisaged that the multiplicity of lubricants common to early formulations may soon be reduced to a maximum of two, if not one. Two positive statements can be made in respect of lubricants and weathering. One is that the incidence of voiding can be varied by adjusting the lubricant level; the second is that since they play a major part in controlling the processing behaviour lubricants must also play a major part in controlling weathering behaviour.

Generally, fillers reduce cost, can improve insulation, improve resistance to yellowing, and improve abrasion resistance. They should not be used indiscriminately in vinyl plastics and judicious selection of materials and their concentration may lead to some properties being improved. Titanium dioxide gives an improvement in weatherability[48] owing to its absorption of nearly all the UV radiation, whilst in the same report silica fillers are recorded as having a pronounced deleterious effect on weathering. More recent work by But[49] has indicated that an increase in the ageing stability of chalk-filled compounds can be achieved by the addition of silica powder.

When plasticizers are included in the formulation a number of additional factors arise. They are known to have been leached out, volatilized, or oxidized independently of the polymer. As a plasticizer may itself impart improved weathering resistance[50] and synergize with stabilizers to improve their performance, its choice is important. An odd phenomenon noticed by Estevez[19] and Darby and Graham[51] is that the optimum life of plasticized PVC occurs at about 40 parts per hundred of resin plasticizer. The low mobility of the stabilizers at lower concentrations and the increased carbonyl content of the component (increasing probability of UV absorption) at higher concentrations is thought to account for this.

Processing aids are used to reduce the melt viscosity of a compound during processing, thereby facilitating processing without affecting the physical properties or heat-distortion temperature. Impact modifiers improve the impact strength without reducing the heat-distortion temperature. However, the use of impact modifiers does not allow the production of crystal-clear grades and impregnated formulations soon degrade and become more brittle than similar unmodified compounds aged for the same time. Good light stability of compounds incorporating acrylic impact modifiers are attainable in highly pigmented products, and Perry[52] stated that the use of acrylic impact modifiers in unpigmented applications requiring good light stability is not recommended. However, some acrylic processing aids do not have this adverse effect. ABS impact modifiers are found to degrade in a similar way to acrylics and therefore unpigmented formulations are not recommended for outdoor use. The best impact modifier to use for such applications is chlorinated polyethene.

Apart from providing colour, pigments must be permanent and provide some protection. Black formulations weather best and colourless or transparent weather worst. Nearly all pigments confer some degree of weather resistance,[53,54] their nature and concentration having a major effect.

(iii) PMMA

Another of the most weather-resistant polymers is PMMA. Its good light transmission is retained, probably because the reaction between the polymer and UV radiation results in chain scission which does not give rise to discoloration. If colour does develop it is probably because of degradation of additives. Certain coloured samples are found to darken more readily than to fade, and again this is due to the properties of the dyes rather than those of the polymer.

The prime cause of subsurface voiding and surface crazing is the presence of tensile strain in the surface of the sheet, coupled with scission degradation of the polymer chains. The strain is induced by the stripping process when cast acrylic sheet is removed from the glass plates forming the cell in which the material is polymerized. Such strain is reduced considerably by the use of specific additives in the process, but it is not eliminated. Other sources of strain are the shaping and forming techniques and polishing. In all instances a final heat treatment is recommended, but it is acknowledged that such treatment is not always carried out by the fabricators or users.

7. CONCLUSIONS

A great deal remains to be done before we have a truly comprehensive account of failure, superficial or otherwise, over the size range from ångströms to micrometres to millimetres, and especially over what governs the transitions from submicro cavitation to a growing craze and, from that, to a crack and ultimately to component failure — the sophisticated experimental techniques and imaginative theories likely to be developed will surely be found in academia rather than industry.

One feels, however, a sense of dissatisfaction when a technical problem is unsolved and there is no scientific consolation in the realization that as few end users seem prepared to pay a really significant premium for longevity as there are suppliers prepared to guarantee it. Modest premia, yes, sufficient to encourage stabilizer manufacturers to produce a trickle of new, limited volume and therefore expensive products; in fact chemical protection is a fairly expensive business — hence the popularity of black.

The essentials seem clear, however, and barring some completely unforeseen, and almost unforeseeable, development in the techniques of inhibition, the plastics industry will continue to tolerate polymers prone to superficial defects when naturally aged. If nothing can stop the eventual formation of defects in such materials their extent in depth can be markedly reduced by correct protective formulation; visual evidence of their presence can also be masked by sensible pigmentation. Indeed, there remains much truth in the old adage, 'What the eye does not see . . . '

REFERENCES

1. H. H. Kausch, J. A. Hassell, and R. I. Jaffee, (eds.) *Deformation and Fracture of High Polymers* Plenum Press, New York (1973).
2. P. L. Pratt (editor-in-chief), *Fracture 1969*, Chapman & Hall London (1969).
3. P. G. Faulkner and J. R. Atkinson, *Plast. and Polym.*, **40**, 109 (1972).
4. R. P. Kambour and R. W. Kopp, *J. Polym. Sci.*, 7, Part A2, 183 (1969).
5. P. Beaham, M. Bevis, and D. Hull, *Phil. Mag.*, **192**, 1267 (1971).
6. S. N. Zhurkov, V. S. Kuksenko, and A. I. Slutsker, *Fracture 1969*, ed. P. L. Pratt, p. 531, Chapman & Hall, London (1969).
7. S. N. Zhurkov and V. S. Kuksenko, *Int. J. Fract.*, **11**, 629 (1975).
8. D. E. Winkler, *J. Polym. Sci.*, **35**, 3 (1969).
9. J. B. Howard, *S.P.E. Trans.*, **4**, 217 (1964).
10. J. Wolf, *Gas*, **87**, 433 (1964).
11. M. Kitagawa, *J. Polym. Sci., Polym. Phys. Ed.*, **14**, 2095 (1976).
12. R. L. Aldleman, 'Use of elastic light scattering to determine solid-state polymer structure', Ph.D. Thesis, London University (1974).
13. H. H. Frey, *Kunststoffe*, **53**, 103 (1963).
14. D. C. F. Couzens, *Polymer Rheology and Plastics Processing*, in Conference Proceedings (Eds P. L. Clegg, F. N. Cogswell, D. E. Marshall, and S. G. Maskell), British Society of Rheology, published by Plastics and Rubber Institute (1975).
15. R. A. V. Roff and K. W. Doak (Eds) *Crystalline Olefine Polymers* 366, Wiley, New York (1964).
16. H. Stager, *Kunststoffe*, **49**, 589 (1959).
17. 'Ageing and weathering of plastics', *Literature Review*, Compiled by J. Mattham *et al.* RAPRA (1970).
18. G. R. Rugger, *Mat. in Des. Eng.*, **59**, 236 (1964).
19. J. M. J. Estevez, *Plast. Inst. Trans.*, **33** , 89 (1965).
20. B. G. Achhammer reported in *IUPAC International Symposium of "Ageing of Plastics"*, Dusseldorf (1959); see also B. G. Achhammer and G. M. Kline, *Materie plastiche*, **26**, 783 (1960).
21. R. C. Hirt, N. Z. Searle, and R. G. Schmitt, *SPE Trans.*, **1**, 21 (1961).
22. J. A. Melchore, *I & E Prod. Res. & Dev.*, **1**, 232 (1962).

23. R. J. Taylor, ICI Ltd. (Plastics Div.), private communication.
24. 'Microbiological deterioration of rubbers and plastics', RAPRA, Information Circular No. 476.
25. C. Gottfried and M. J. Dutzer, *J. Appl. Polym. Sci.*, 5, 612 (1961).
26. R. J. Martinovitch, *Plast. Technol.*, 9, 45 (1963).
27. R. E. Gibson, *Weston Plast. Tech. Rept.*, 8, No. 4, 37 (1971).
28. V. D. Wallder *et al.*, *Ind. Eng. Chem.*, 42, 2320 (1950).
29. W. L. Hawkins, R. H. Hansen, W. Matreyek, and F. H. Winslow, *J. Appl. Polym. Sci.*, 1, 37 (1959).
30. J. F. Ambrose, *Brit. Plast.*, 30, 446 (1957).
31. C. G. Cocks and A. P. Metzger, *7th Annual Symp. on Tech. Prog. in Comm.*, Wire and Cables, Asbury Park, New Jersey (1958).
32. Anon., *Plast. Technol.*, 21, No. 10, 11 (1975).
33. A. D. Curson, 'Distribution of additives in polymers by UV microscopy', unpublished presentation. 'Micro 72', Oxford (1972).
34. H. P. Frank and H. Lehner, *J. Polym. Sci.*, C, 31, 193 (1970).
35. F. H. TeTigue and M. Blumberg, *Appl. Poly. Symposia*, Vol. 4, in Ed. Mussa R. Kamal, *Weatherability of Materials*, p. 175, Wiley, New York (1967).
36. W. S. Penn, *Plastics in Building*, Maclaren, London (1964).
37. J. B. DeCoste, J. B. Howard, and V. T. Wallder, *Ind. Eng. Chem. & Eng. Data Series*, 3, 131 (1958).
38. J. B. DeCoste and V. T. Wallder, *Ind. Eng. Chem.*, 47, 314 (1955).
39. W. C. Geddes, 'The mechanism of PVC degradation', *Techn. Rev. No. 31*, RAPRA., May 1966.
40. J. Sterpak and C. Jirkal, *Chem. Listy*, 59, 1201 (1965).
41. G. Y. Ocskay, *et al.*, *Kinet. Mech. Polyreactions. Int. Symp. Macromol. Chem. Prepr.*, 5, 157 (1969).
42. A. Kralova, and R. Vesely, *Anal. Fys. Me Tody Vyzk Plastic Pryskyric Proc. Conf.*, 2, 568 (1971).
43. J. Matthan, M. Wiechus, and K. A. Scott, 'Ageing and weathering of plastics', *Literature Review*, RAPRA, Part 6 (1970).
44. M. W. Allsopp, A. D. Curson, and M. J. Hitch, unpublished.
45. D. Pearson and J. Morley, *Brit. Polym. J.*, 6, 13 (1974).
46. See, *Plast. Technol.*, 22, 15 (1976).
47. W. S. Penn, *PVC Technology*, Maclaren, London (1966).
48. J. G. Hendricks, *Plast. Techn.*, 1, 81 (1955).
49. T. S. But *et al.*, *S.B. T.R. Vses Nauch-Issled. Inst. Nov. Stroit. Mater.*, 25, 98 (1969).
50. Z. Korez, *et al.*, *P.R. Zakrem Towarozn Chem. Wyzsza Szk. Ekon, Poznaniu, Zesz, Nauk*, 1, 223 (1971).
51. J. R. Darby and P. R. Graham, *Mod. Plast.*, 39, 148 (1962).
52. N. L. Perry, *Mod. Plast.*, 40, 156 (May 1963).
53. M. Mielke, *Plastverarbeiter*, 26, 73 (1975).
54. G. Benzing, *Plast. Mod. Elast.*, 26, 219 (1974).

18

Photodegradation and Photo-oxidation of Polymer Surfaces

B. Rånby

The Royal Institute of Technology, Stockholm, Sweden

1. INTRODUCTION

Surface effects in the photodegradation and photo-oxidation of polymers are frequently observed and reported, but have only recently been studied and described in basic terms. When exposed to sunlight or artificial UV-irradiation, surface coatings of paints containing pigment particles are known to suffer reduced surface gloss and gradually 'chalk', i.e. lose pigment and other particles from the coated layer by surface erosion.[1] Changes in wetting and adhesion as a result of exposure are also studied and reported.[2] There are many papers in the literature describing such phenomena for different paints. There are also several studies published on photodegradation of plastic surfaces in relation to ageing effects of the materials.[3] The purpose of this paper is to describe methods available to study surface effects on polymeric materials caused by photochemical reactions on exposure to UV and visible light, and report some results obtained as examples.

2. EXPERIMENTAL METHODS

(i) Optical and electron microscopy

This is used to study changes in surface geometry (morphology) during degradation. Optical microscopy has limited resolution (about 1 μm). It gives information about formation of coarse cracks, crevices, bubbles, etc. which may be correlated with surface gloss.[4] Electron micrographs of surface replicas, e.g. using polymer gels or melts for replication, have high resolution ($\sim$5–10 nm).[5] They are difficult to prepare and may contain artefacts from the preparation. Scanning electron microscopy (SEM) is a most versatile method to use for surface studies.[6] The resolution is of the order 10–20 nm. SEM gives excellent stereoscopic pictures of the surface morphology. Even crystallization phenomena can be observed.

(ii) Infrared absorption spectroscopy

When applied in transmission studies of thin films ($\leqslant 10$ μm) infrared absorption spectroscopy is sensitive ($\sim 1\%$), but usually not selective for changes located to the surface layer ($\leqslant 10$ nm) of the film. New IR methods based on attenuated total reflection (ATR) at film surfaces are most useful.[7] With ATR crystals of different materials, the depth of penetration of the IR beam can be varied from 0.1 to about 4 μm. This method gives both the amount and the location of modified groups in the polymer film, e.g. in surface oxidation (cf. Chapter 15).

(iii) Chain degradation

In surface reactions chain degradation can be estimated in various ways. Shorter chains formed by degradation in the surface layer may crystallize to a larger extent than the bulk of the material, as measured from the intensity of certain IR absorption bands. This may be measured using an ATR attachment applied to give a certain depth of penetration.[8] In some cases the photodegraded surface layer is soluble enough for a selective extraction of its content of low molecular weight polymer.

(iv) ESCA measurements

These are very useful for analysis of modified groups in very thin surface layers of a polymer sample. The depth of penetration in ESCA studies of polymers are of the order 1–3 nm, i.e. 3–10 molecular layers.[9] This makes the ESCA method extremely sensitive for small chemical changes in the top layer of a polymer film (thickness 10–20 μm), i.e. in cases where the modifications are limited to a few molecular layers at the film surface. Experimental evidence for such oxidative ractions will be reported.

(v) Comparison of methods

Comparison of methods for surface analysis shows widely different depths of penetration. IR absorption spectroscopy with ATR attachment measures at a minimum depth of about 100 nm, SEM has a resolution of about 10 (or 10–20) nm, while ESCA analysis penetrates to a depth of only about 1 (or $\leqslant 3$) nm. In this way the three methods are complementary.

3. STUDIES OF SURFACE DEGRADATION

A few typical studies of degradation and oxidation will be described to illustrate the results obtained for various polymer surfaces using the methods described above: polypropylene film,[10] polymer blends of ABS type,[11] surface coatings containing pigment particles,[12] and polydiene films (polybutadiene and copolymers).[13]

(i) Photo-oxidation of polypropylene film

Polypropylene is known to be sensitive to photodegradation when exposed to sunlight or UV-irradiation in air, also at wavelengths $\geqslant300$ nm. In a study of Blais et al.[10] a commerically extruded polypropylene film of 22 μm thickness was used. The polymer was predominantly isotactic in structure. The samples were acetone-extracted in a Sohxlet before use. ATR infrared spectroscopy was applied with a series of reflection elements, giving various depths of penetration in the measurements from 0.19 to 0.59 μm. The formation of hydroxyl groups by UV-irradiation ($\geqslant320$ nm) is predominantly at the surface (Figure 1) as measured from the IR stretching frequency band ($\sim$3400 cm^{-1}). With transmission IR

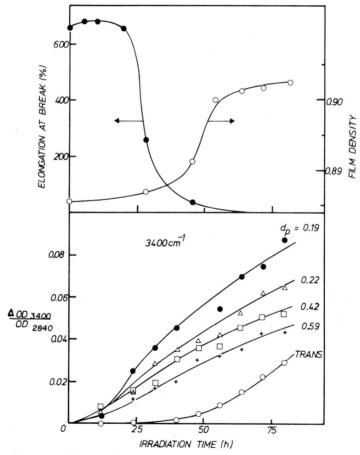

Figure 1. Effects of UV irradiation in air on film properties of isotactic polypropylene. The optical densities of IR spectra refer to OH groups (3400 cm^{-1}) and CH$_2$ groups (2840 cm^{-1}), and d_p = depth of penetration of the IR beam at 3400 cm^{-1} (data from Blaus et al.[10]).

384

analysis the OH groups can be measured only after long times of irradiation. With increasing oxidation, the 'elongation-at-break' of the film decreases sharply, i.e. increased brittleness. At the same time the density of the film increases, which indicates an increased degree of crystallinity. This is interpreted as due to degradation of the main chains. Also, the increased crystallinity is most extensive near the film surface (Figure 2). The intensity ratio (I) of IR absorption at 997 and 974 cm^{-1} measures the helical content of the cyrstalline polymer. The I-values

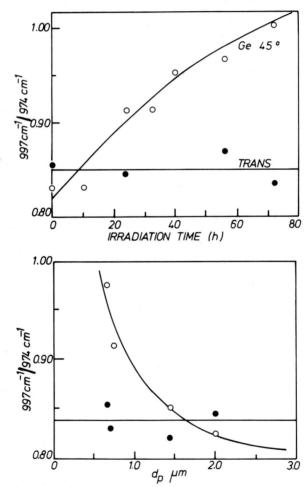

Figures 2 and 3. Effects of UV irradiation in air on polymer ordering in films of isotactic polypropylene. The IR absorbance ratio 997/974 cm^{-1} measures the content of helical structure in the polymer chains. The 'Ge 45°' values are measured by ATR technique (surface reflection) and the *trans*-values by IR transmission, d_p is the depth of penetration in μm.[10]

increase with increasing irradiation time in the ATR measurements (depth of penetration = 0.66 μm). The increase in I-values is a surface effect (Figure 3), going down to about 2 μm from the film surface.

(ii) Photochemical ageing of polymer blends (ABS type)

Polymer blends containing copolymers of styrene (S), acrylonitrile (A), and butadiene (B) are known to be sensitive to degradation by weathering, mainly due to photochemical degradation. The main types of ABS plastics are: (type A) blends of two copolymers (styrene–acrylonitrile and butadiene–acrylonitrile, respectively) and (type B) styrene–acrylonitrile blends copolymerized with poly-butadiene dissolved in the monomer blend. In a recent study[11] flat test pieces of two ABS materials (types A and B) were prepared by compression moulding and exposed outdoors for testing. After exposure, the test pieces were cut into thin sections using a microtome. IR transmission spectra of thin sections (Figure 4) from the surface and from the middle part of ABS test pieces exposed outdoors for six months (type A) show the effects. The surface section (20 μm) is extensively oxidized while the middle section is largely unchanged. The difference beween the two spectra indicates formation in the surface layer of hydroxyl (–OH), aldehyde (–CHO), ketone (>C=O), peroxide (–C–O–O–), and carboxyl (–COOH) groups and a simultaneous decrease in 1,4-*trans* double bonds and 1,2-vinyl bonds. The double bonds in the polybutadiene component have completely disappeared in the surface layer after six months' outdoor exposure (Figure 5). The OH groups are formed mainly at the film surface (Figure 6). Exposure of an ABS copolymer of

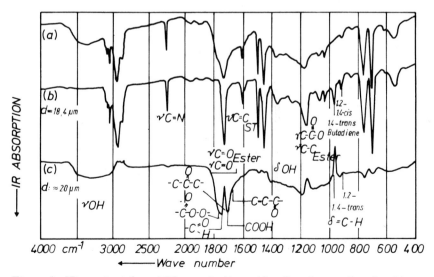

Figure 4. IR spectra of an ABS sample (type A), after six months of outdoor exposure: (a) surface layer, (b) middle of sample (1.5 mm thickness), and (c) difference spectrum (a) − (b) (from Priebe *et al.*[11]).

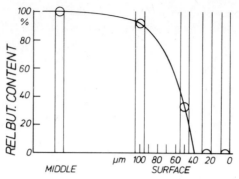

Figure 5. Butadiene content versus depth from surface in ABS (type A) sample after six months of outdoor exposure.[11]

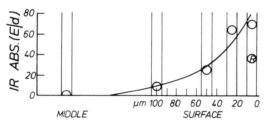

Figure 6. Increase in IR absorption of OH groups at decreasing depths from the ABS surface (type A) (exposed as in Figure 5).

type B gives similar effects, analysed from IR spectra (Figure 7). It is quite clear that the acrylonitrile and styrene units in ABS (type B) are very little affected by outdoor exposure in the surface layers where butadiene units disappear completely (Figure 8). The hydroxyl groups formed are strongly concentrated near the exposed surface (Figure 9). The measured concentrations of styrene, acrylonitrile, butadiene, and hydroxyl units at different depths from the surface are given in Table 1. It is evident that the photochemical sensitivity of ABS plastics is related to the oxidative reactions of the diene units in the polymer structures. The initiation of these processes may be by formation of free radicals, for example allylic radicals, or by singlet oxygen ($^1O_2$) addition to double bonds, giving hydroperoxy groups are shown in reactions (1) and (2), respectively.

$$+CH_2-CH=CH-CH_2+ \longrightarrow +\dot{C}H-CH=CH-CH_2+ \quad + H\cdot \quad (1)$$

$$+CH_2-CH=CH-CH_2++{}^1O_2 \longrightarrow +CH_2-\underset{\underset{OOH}{|}}{CH}-CH=CH+ \quad (2)$$

allylic hydroperoxide

These reactions are well established in the photodegradation of polydienes.[14]

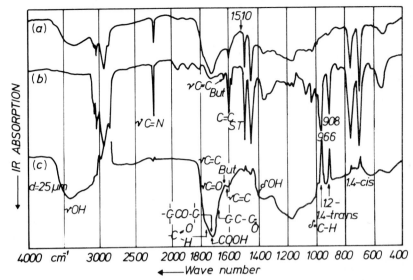

Figure 7. IR spectra of an ABS sample (type B) after six months of outdoor exposure: (a) surface layer, (b) middle of sample, and (c) difference spectrum (a) – (b).[11]

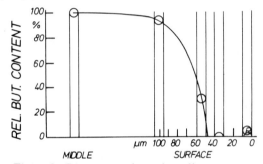

Figure 8. Decrease in butadiene concentration close to BAS (type B) sample surface after six months of outdoor exposure.[11]

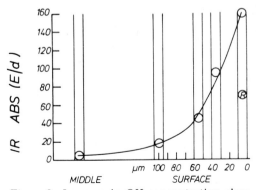

Figure 9. Increase in OH concentration close to ABS (type B) sample surface after six months of outdoor exposure.

Table 1 Relative content of acrylonitrile styrene and butadiene units and hydroxyl groups at different depths from the surface of an ABS plate (type B, grafted copolymer) exposed outdoors for six months (West Germany). Data based on transmission IR measurements.[11]

Depths from exposed surface (μm)	Acrylonitrile (%, rel.)	Styrene (%, rel.)	Butadiene (%, rel.)	IR abs. E/d (3400 cm^{-1})
Surface layer [a]	—	—	0	160
30	96	100	0	95
50	92	96	31	44
100	100	100	94	18
1500 [b]	100	100	100	~10
Back side (surface)	~100	~99	5	~70

[a] No even film could be prepared.
[b] Middle of plate.

(iii) Degradation of paint surfaces

From a comprehensive study of the degradation of pigmented paints and commercial plastic surfaces,[12] a series of results for an alkyd paint are shown as an example. The paint binder was a commercial air-drying alkyd, white pigmented with a 50 : 50 blend of anatase (chalking TiO_2 particles) and rutile (chalk-resistant TiO_2 particles) to pigment volume concentration of 15%. The pigment particle size

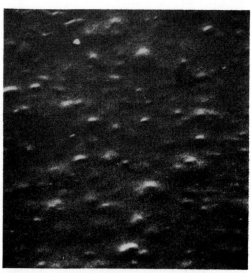

Figure 10. SEM picture of an alkyd paint surface, not exposed to oxidative degradation.[12] Gloss value 86 units. (Magnification 7700 diam.)

was ⩽0.1 μm, obtained by grinding on a laboratory calender mill. As a drying agent, a mixture of naphthenates was added (0.5% lead, 0.05% cobalt, and 0.02% calcium salts). The paint layers were applied on alod-oxidized Duralumin plates, dried in dust-free air for 30 days, and exposed in an Atlas weatherometer (Model DMC-RC) with a carbon arc illumination in 102 min cycles, followed by an 18 min salt-free water spray.

The original and the exposed paint surfaces were analysed in a Gardner gloss meter at 60° angle, and their morphology studied by SEM at a magnification of 7700 diameters.

The paint surface structure is shown in a series of SEM pictures. The unexposed surface (Figure 10) is slightly uneven, owing to pigment particles below the paint surface (gloss value 86). After 44 and 85 h exposure in a weatherometer, the gloss value decreases to 80 and 78, respectively, and surface pitting gradually appears (Figures 11 and 12). After long exposures (197, 306, and 350 h), the paint surface cracks up (Figures 13, 14, and 15) and pigment particles are gradually released (chalking). The gloss value decreases to 55, 1, and 1, respectively. The surface gloss retention as a function of exposure time in a weathermeter (h) and outdoors in Stockholm, Sweden (weeks) is shown in Figure 16. In this diagram 'p' means appearance of pits or pores and 'k' indicates chalking.

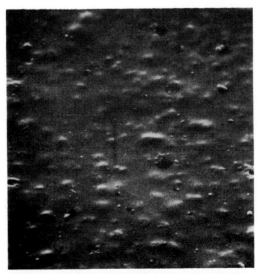

Figure 11. SEM picture of alkyd paint surface exposed to oxidative degradation in a weatherometer for 44 h, gloss 80 units. (Magnification, 7700 diam.[12])

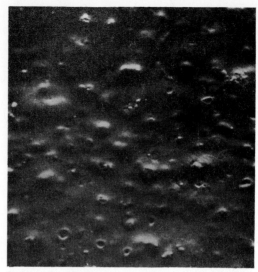

Figure 12. SEM picture of alkyd paint surface exposed to oxidative degradation in a weatherometer for 85 h, gloss 78 units.[12] (Magnification, 7700 diam.)

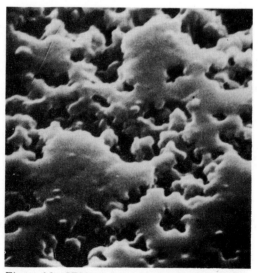

Figure 13. SEM picture of alkyd paint surface exposed to oxidative degradation in a weatherometer for 197 h, gloss 35 units.[12] (Magnification, 7700 diam.)

Figure 14. SEM picture of alkyd paint surface exposed to oxidative degradation in a weather-ometer for 306 h, gloss 1 unit.[12] (Magnification, 7700 diam.)

Figure 15. SEM picture of alkyd paint surface exposed to oxidative degradation in a weather-ometer for 350 h, gloss 1 unit.[12] (Magnification, 7700 diam.)

392

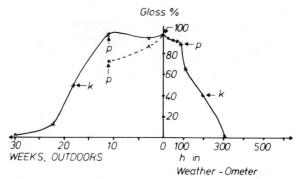

Figure 16. Surface gloss retention versus exposure in weatherometer (h) and outdoor exposure in Stockholm (weeks). Beginning pore formation at 'p' and beginning chalking at 'k'.[12]

A careful study of the SEM pictures shows that the 'pitting' of the paint surface starts at the pigment particles, especially where they form clusters. Titanium oxide (TiO_2) is known to have photochemical activity (a sensitizing effect) on the degradation, for example of 6,6-nylon fibers.[15]

A model of the photochemical degradation process of the pigmented paint layer is given in Figure 17. The effects of UV light and water spray causes pitting (Figure 17(b)). Water gradually washes away the degraded binder from pigment clusters. Free pigment particles on the degraded surface cause 'chalking' from the paint surface (Figure 17(c) and (d)). Eventually the whole surface is covered with pigment particles free from binder (Figure 17(e)). The SEM pictures have a resolution of 10–20 nm, which makes individual TiO_2 pigment particles visible ($\leqslant 100$ nm = 0.1 μm).

(iv) ESCA studies of singlet oxygen oxidation of polydiene surfaces

Oxygen in the excited singlet state ($^1O_2$) reacts with double bonds in alkane chains.[14] The excitation is made with a stream of pure oxygen ($^3O_2$) at low pressure, flowing through the cavity of a microwave generator. A film of polydiene rubber is rapidly oxidized when exposed in the stream.[16,17] The oxidation is limited to a very thin surface layer (1–3 nm) which is most conveniently studied by application of ESCA (electron spectroscopy for chemical analysis). The method was originally developed by Kai Siegbahn[18] and successfully applied to polymer surface studies by Clark and Feast.[9]

In this case we are primarily interested in analysis of the ratio oxygen/carbon in the surface layers. The ESCA signals arising from photoemission of the electrons in the C_{1s} and O_{1s} levels have distinctive binding energies, depending on local electronic environment (cf. Chapter 10), and this together with the investigation of relative intensities provides a means of studying the oxidized surface regions.

In our reported measurements, which are preliminary, charge effects caused

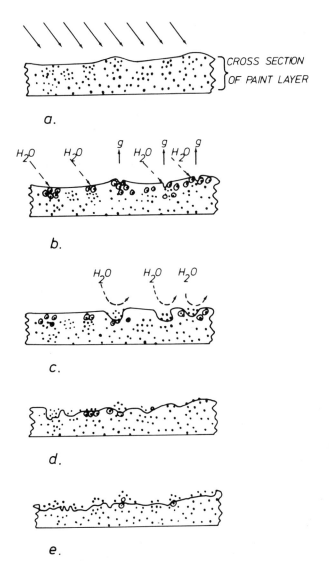

Figure 17. Schematic model of photochemical degradation of pigmented paint layer exposed in weatherometer or outdoors.[12]

irregular shifts of the ESCA lines which made further structure analysis impossible. (For a discussion of charging effects, cf. Chapter 16.)

Two polydiene samples were studied.

(1) *Cis*-1,4-polybutadiene, a commercial sample of high structural regularity, prepared by organo-Li catalysts.

(2) Styrene–butadiene copolymer (25 : 75 wt. %), a commerical sample prepared by anionic polymerization. The butadiene units are largely in *cis*-1,4 form.

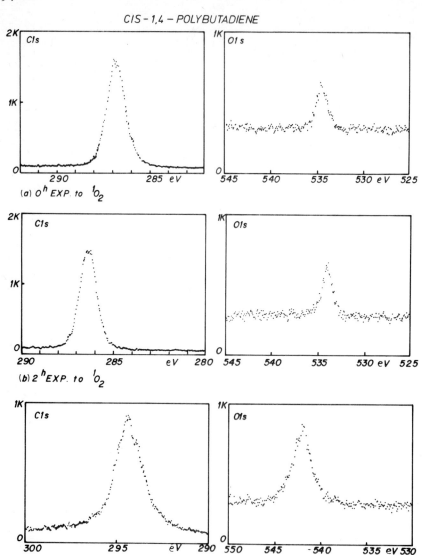

Figure 18. ESCA spectra of thin films of *cis*-1,4-polybutadiene as prepared (a), after exposure to an oxygen stream containing singlet oxygen ($^1O_2$) for 2 h (b), and 24 h (c).[13] (No correction has been made for sample charging.)

The singlet oxygen microwave generator used in this work operates at 1 Torr oxygen pressure and at a frequency of 2450 MHz (a wavelength of about 10 cm). This gives about 10% singlet oxygen ($^1O_2$) in the oxygen stream.[16] The film samples were prepared by evaporation in nitrogen atmosphere of a cyclohexane solution on thin aluminium foil.

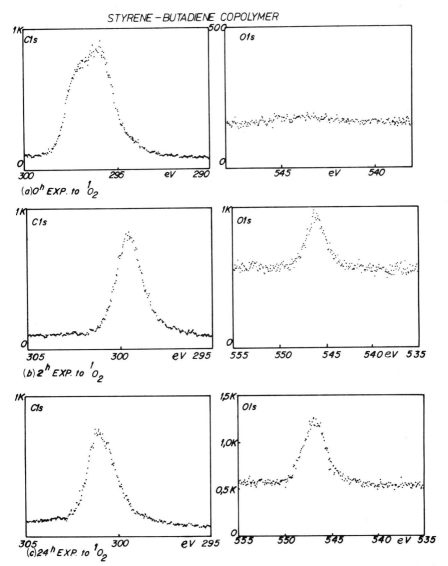

Figure 19. ESCA spectra of thin films of styrene—butadiene copolymer as prepared (a), after exposure to an oxygen stream containing singlet oxygen ($^1O_2$) for 2 h (b), and 24 h (c).[13] (No correction has been made for sample charging.)

The ESCA recordings were made with a Hewlett—Packard instrument (Model HD 5960 A) in the Institute of Physics, University of Uppsala, Sweden by Dr. N. Martensson. The spectra for *cis*-1,4-polybutadiene in Figure 18(a) indicate that the original sample also contained oxidized groups to some extent. The [O]/[C] values are given on a molar basis. Exposure to $^1O_2$ for 2 and 24 h increased the oxygen

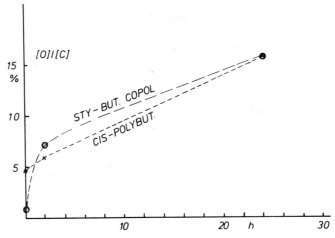

Figure 20. Relative oxygen content ([O] / [C] mole %) in surface layer of thin films of *cis*-1,4-polybutadiene and styrene—butadiene copolymer, as prepared (0 h) and after exposure to an oxygen stream containing singlet oxygen for increasing times (h).[13]

content gradually (Figure 18(b) and (c)). The shifts on the electronvolt scale are intepreted as charging effects.

The thin films of styrene—butadiene copolymer had originally a very low oxygen content (Figure 19(a)). Exposure to $^1O_2$ for 2 and 24 h increased the oxygen content gradually. The [O]/[C] data as a function of $^1O_2$ exposure time are given in Figure 20. The ESCA spectra do not permit an analysis of the chemical structure of the oxidized groups, but they could all be of the expected type, i.e. hydroperoxide groups. These investigations are in progress in our laboratories.

ACKNOWLEDGEMENTS

My thanks are due to Professor Kai Siegbahn and Dr. Nils Martensson, Institute of Physics, University of Uppsala, Sweden, for the ESCA measurements and helpful discussions in relation to this study.

REFERENCES

1. S. B. Twiss *et al.*, *Official Digest (Philadelphia)*, **28**, 93 (1956); **30**, 7, 788 (1958).
2. E. Fischer and K. Hamann, *Farbe und Lack* (Hannover), **63**, 209 (1957).
3. D. J. Carlsson and D. M. Wiles, *Macromolecules*, **4**, 174 (1971).
4. M. Hess, *Paint Film Defects,* Chapman & Hall, London, 1965.
5. E. G. Bobalek *et al.*, *Official Digest (Philadelphia)*, **27**, 984 (1955).
6. L. E. Brooks, P. Sennett, and H. H. Morris, *J. Paint Technology (Philadelphia)*, **39**, No. 511, 472 (1967).
7. J. P. Luongo and H. Schonhorn, *J. Polymer Sci.*, A-2, **6**, 1649 (1968).

8. D. J. Carlsson and D. M. Wiles, *Macromolecules*, **4**, 179 (1971).
9. D. T. Clark and W. J. Feast, *J. Macromol. Chem. – Revs. Macromol. Chem.* **C12** (2), 191 (1975), cf. Chapter 16.
10. P. Blais, D. J. Carlsson, and D. M. Wiles, *J. Polymer Sci.*, A-1, **10**, 1077 (1972).
11. E. Priebe, P. Simak, and K. Stange, *Kunststoffe*, **62**,:2, 105 (1972).
12. E. Bendel, *Degradation of Plastics and Paint Surfaces*, Natl. Swed. Building Research (Stockholm). Report R14 (1974).
13. B. Rånby and B. Terselius, unpublished-data.
14. B. Rånby and J. F. Rabek, *Photo-oxidation and Photodegradation of Polymers*, pp. 199 and 274, Wiley, London, 1975.
15. G. S. Egerton and K. M. Schah, *Textile Res. J.*, **38**, 130 (1968).
16. B. Rånby and J. F Rabek, *ACS Symposium Series*, No. 25, 391 (1976).
17. J. F. Rabek and B. Rånby, *Polymer Eng. Sci.*, **15**:1, 40 (1975).
18. K. Siegbahn *et al.*, *Nova Acta Regiae Soc. Sci. Ups. (Sweden)*, Ser. IV, 20 (1967).

19

Tribological Characteristics of Polymers with Particular Reference to Polyethylene

D. Dowson

The University of Leeds

1. INTRODUCTION

Polymeric materials have found wide application as bearing materials in recent years. Such bearings are generally self-contained, self-lubricating, and generally accommodating to the detailed geometrical form of the harder, sliding components which they support. There are probably more polymeric bearings produced than any other type and their characteristics attract much attention.

Thermosetting resins were introduced as bearing materials in the 1930s, with the thermoplastics following some 20 years later. Phenolic, and to a lesser extent epoxy, resins have been used extensively, generally with cloth or fibre reinforcement. In many cases solid lubricant fillers such as graphite and molybdenum disulphide have been used to improve the tribological characteristics of thermosetting materials.

Polyamide resins were in the forefront of the introduction of thermoplastic materials for tribological applications, nylon gears and bearings having been used successfully in many applications. At a later stage Fibreglass filler was introduced to improve the mechanical properties of the material, and graphite or molybdenum disulphide was used to improve the resistance to wear. Polytetrafluoroethylene (PTFE) has proved to be remarkable and by far the most significant fluorocarbon in the tribological field. PTFE has a coefficient of friction lower than graphite and indeed lower than any other solid. The range of 0.05–0.1 which is frequently quoted for the coefficient of friction of PTFE is attributed to the smooth molecular structure of the material. The material is widely used in low-friction applications, but its relatively poor mechanical strength frequently necessitates the use of fillers such as glass fibre. In many cases metallic powders are used to improve the thermal and electrical properties of the material and solid lubricants like graphite and molybdenum disulphide have been used to increase the resistance to wear.

The nylons and fluorocarbons are the most widely used thermoplastics in tribological situations, but other polymers have been recognized for specific and important roles in recent years. These materials include the polyimides, polyacetals, polycarbonates, polypropylene, and polyethylene.

The wear rate and hence the life of polymeric materials has often been related to the product of the mean pressure on the bearing and the sliding velocity; the so-called pV factor. The justification of this approach is found in the fact that the effective operating temperature of a given bearing configuration is directly related to pV. The utility of this approach is clear, but the oversimplification of a complex problem embodied in the procedure and the need to introduce a number of restrictions has introduced substantial criticisms in recent years. In general, it is more satisfactory to relate bearing life to well-established wear equations in which the volume of material removed by wear is expressed as a function of the applied load, sliding distance, and mechanical properties of the material.

In this chapter attention will be focused upon the tribological characteristics of polyethylene; one of the less widely used polymeric bearing materials. The reason for this choice is that medium- to high-density ultra high molecular weight polyethylene (UHMWPE) is now extensively used in total joint replacements for the human body. Most of the discussion of the tribological characteristics of UHMWPE will thus be related to this important and interesting application.

It is convenient at this juncture to introduce the notation to be used in subsequent sections:

Notation

e = strain at fracture (change in length/original length),
k_1 = dimensionless wear coefficient $(3HV/PL)$,
k_2 = dimensional wear coefficient (V/PL),
p = mean pressure (P per unit projected area),
s = limiting stress,
H = hardness,
L = sliding distance,
P = applied load,
R_a = arithmetic mean deviation (surface roughness),
V = volume of material removed by wear,
μ = coefficient of friction

2. FRICTION AND WEAR OF POLYMERS

When a hard material, like steel, slides over a relatively soft material, like a polymer, two important physical actions govern the tribological behaviour of the system. The first is *adhesion* and the second *deformation*. Furthermore, it is well established that in many cases a thin film of polymer is transferred to the hard counterface, where it is retained by physical bonding. The real contact and adhesion between the surfaces thus takes place largely between the bulk polymer and the polymeric transfer film. The transfer film as a thickness which is typically in the range

10^{-8} – 10^{-5} m and may be uniform or lumpy in distribution. Under more severe conditions, and particularly at higher sliding speeds, thicker and generally more uniform transfer films are established. The influence of the transfer film upon friction and wear has been discussed in general terms by Bowden and Tabor[1] and in relation to polythene by Briscoe and Tabor.[2]

The presence of a transfer film usually means that the coefficient of friction of metals sliding on polymers is similar to that of the polymer sliding on itself at low sliding speeds. The same is not true for wear under engineering conditions, since the poor thermal conductivity of polymers results in high surface temperatures and reduced strength when two blocks of polymer slide over each other.

(i) Friction

Under relatively mild contact conditions, and particularly at low speeds, adhesion appears to play a dominant role in the friction process when metals slide over polymeric materials. A dissipative action associated with deformation of the asperities can also be important and Bowden and Tabor have drawn attention to the fact that this is associated more with internal hystereses than with the well-known ploughing action observed when hard metals abrade softer materials.[1] Shooter and Tabor first demonstrated that the simple concept of the role of adhesion in the friction process provided a reasonable explanation of the observed behaviour,[3] but more recent work has drawn attention to the importance of molecular structure and viscoelasticity.

The very low friction exhibited by PTFE and high-density polyethylene has been attributed by Pooley and Tabor to the smooth molecular profiles of these materials.[4] Other polymers, like polypropylene, low-density polyethylene, and polychlorotrifluorethylene have relatively rough or irregular profiles. Polymer surfaces are subjected to high strains in the sliding process and orientation of the long molecules can influence friction. The friction tends to fall as the molecules align themselves with the direction of sliding and to revert to a higher value if the sliding direction changes.

Most polymers exhibit viscoelastic behaviour and the friction therefore exhibits time-dependent characteristics. It is important to keep this point in mind when comparing the results of friction tests on identical polymers carried out on different machines and possibly at different speeds.

Polymers have provided valuable additions to the range of low-friction materials available to engineers, but it is their wear characteristics that are of particular interest in relation to total replacement joints.

(ii) Wear

The progressive loss of material from the surfaces of sliding components can seriously impair the functioning of machine elements and might even lead to failure of machinery. In the case of polymers, abrasive, adhesive and fatigue wear mechanisms can be identified and a useful review has been presented by Lancaster.[5]

Abrasive wear is normally associated with relatively rough counterfaces. The hard asperities penetrate the polymer and remove material by a micro-machining or shearing process as illustrated in Figure 1(a). Ratner et al.[6] have argued that the resistance to abrasive wear should be related to the product of the limiting stress s and elongation e for a material, and they derived the following wear equation to take account of this proposal.

$$V \propto \frac{\mu PL}{H(se)} \tag{1}$$

Adhesive wear results from fracture of the material on planes beneath the initial interface between the polymer and the counterface in the regions of real contact between surfaces prominences. A small piece of the polymer is plucked from the bulk material to form a loose wear particle or to be attached to the counterface to contribute to the transfer film, as illustrated in Figure 1(b). The process can take place between the metal and the polymer or between the transfer film and bulk polymer. As the transfer film builds up, the surface topography of the counterface changes, but in due course an equilibrium situation may be reached in which the amount of material removed from the bulk polymer by adhesive wear is equal to the rate of subsequent detachment of wear particles from the transfer film. Archard[7] has derived the following relationship for the volume of material removed by adhesive wear V:

$$V = k_1 \frac{PL}{3H} \tag{2}$$

The use of this equation is restricted by the uncertainty surrounding the value of the mean flow stress or hardness H, particularly for a viscoelastic material under conditions representative of the sliding contact. For this reason some investigators prefer to combine the constant 3, the hardness H, and the dimensionless wear coefficient k_1 in a dimensional wear coefficient k_2 defined by

$$V = k_2 PL \tag{3}$$

Fatigue wear has long been recognized as an important process in some rolling and sliding conjunctions involving metals, but evidence of a similar effect in polymers has recently been presented by Dowson et al.[8] Surface fatigue wear in polymers probably results from repeated stress cycles applied to material associated with asperity interactions as illustrated in Figure 1(c). It is observed after a substantial period of rubbing and hence might not be detected in short-term tests. The onset of this form of surface degradation is related to load, and the wear particles appear in the form of thin flakes detached from sites of small cracks in the polymer surface whose orientation is roughly perpendicular to the direction of sliding.

The wear of polymers is a complex process involving one or more of the three mechanisms described above. The relative importance of each process will depend upon the load, speed, and environment and might even change with time in a given conjunction. When a polymer slides over a hard counterface some abrasive action

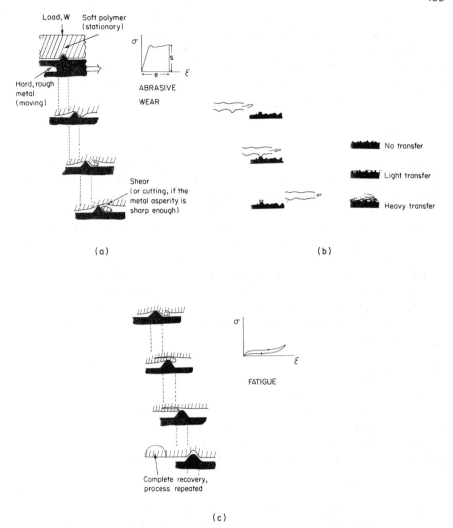

Figure 1. Diagrammatic representation of wear Mechanisms for polymers. (a) Abrasive wear, (b) adhesive wear, (c) surface fatigue.

can be expected in the initial stages. Adhesive processes generally assume increasing importance and in due course dominate the wear behaviour of the polymer. If the sliding points persist for a long period, fatigue wear might supplement the adhesive action and increase the total rate of wear. However, if the counterface is relatively rough, the rate of removal of polymer by abrasive and adhesive actions might be so great that the surface does not have time to develop surface distress associated with fatigue.

3. BACKGROUND TO THE USE OF POLYMERS IN TOTAL JOINT
REPLACEMENTS FOR THE BODY

Since Thomas Gluck introduced ivory ball and socket replacements for the hip in 1890 a wide range of natural and man-made materials has been used for replacement components in load-bearing synovial joints. Philip Wiles introduced the first metal-on-metal joint replacement for the hip in 1938. Both the replacement femoral head and acetabular cup were made of stainless steel and within two years a cobalt–chromium alloy femoral head was introduced in the United States. Acrylic acetabular cups and femoral heads were used in the 1940s, but the wear rate was excessive and the mechanical strength inadequate. A valuable review of the history of hip-joint replacements has been presented by Scales.[9]

In 1950 a polyethylene acetabular cup was used at the Royal National Orthopaedic Hospital, Stanmore, Middlesex, but it was worn through after 18 weeks. In the 1950s nylon acetabular cups were found to have an unacceptably

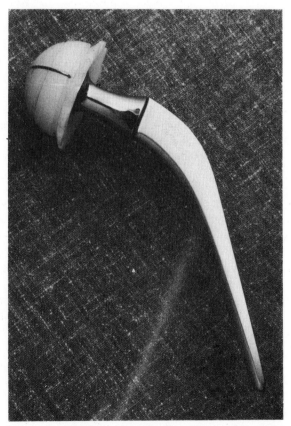

Figure 2. Charnley total hip joint replacement (stainless steel femoral component with UHMWP acetabular cup).

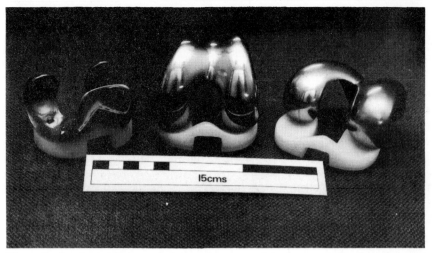

Figure 3. Leeds total knee joint replacement (chrome–cobalt alloy femoral component with UHMWPE tibial component).

high wear rate. Metal-on-metal total joint replacements, using either stainless steel or vitallium, developed rapidly at this time with methyl-methacrylate being used to fix the components in the bones. It was recognized that if metal-on-metal combinations were employed, identical materials had to be used in order to avoid severe corrosion problems. This represented unsound tribological practice and the high friction which ensued created severe fixation problems.

In 1959 John Charnley introduced his metal-on-plastic total hip joint replacement in an attempt to reduce friction and to alleviate the fixation problem. The femoral component was made of stainless steel and the acetabular cup of PTFE. The friction was greatly reduced by this combination of materials, but the poor wear characteristics of PTFE caused the acetabular cups to wear out in about three years. In 1964 Charnley turned from PTFE to UHMWPE for the acetabular cups, and the latter is now well established as the polymer which is used almost universally in total joint replacements in the body. Some metal-on-metal prostheses are still used, but the metal-on-polymer combination now occupies about three-quarters of the market. Illustrations of current forms of metal-on-plastic total replacement hip and knee joints are shown in Figures 2 and 3. It is therefore important to understand the tribological characteristics of UHMWPE and to consider factors which influence its rate of wear.

4. WEAR TESTING

There are two main reasons for carrying out wear tests on materials. The objective might be to ascertain the basic mechanisms of wear for a particular combination of materials, or the more restrictive yet equally elusive determination of the rate of wear to facilitate the estimation of useful life of engineering components.

The three main procedures adopted for these studies are:

(a) Monitoring the performance of full-scale units during normal functioning.
(b) Component testing in bearing or joint simulators.
(c) Laboratory testing of specimens under well-controlled conditions.

In the case of replacement joints for the body *in vivo* monitoring is generally achieved by periodic X-ray examinations which indicate the rate of penetration of the metal component into the polymer. Joint simulators designed to reproduce the loads, motions, and environments experienced by total joint replacements are generally complex and expensive devices. They are, nevertheless, most valuable in the evaluation of prototype joints, and it is a matter of some concern that they are not more widely used before prostheses are inserted in the body. It is, however, the third procedure which forms the basis for the work described in this chapter.

In more fundamental studies of the wear process, it is desirable to use test arrangements which provide known, reproducible, and often steady conditions of load, speed, environment, and contact conditions. Different investigators have employed various conjunction arrangements, but one form of equipment that has grown in popularity is the pin-on-disc machine. Sometimes a ring-on-disc configuration is used, but the results presented here were obtained from a number of pin-on-disc machines designed to provide steady or reciprocating rotary motion or linear reciprocating motion.

Figure 4. Tri-pin-on-disc wear-testing machine.

(i) Pin-on-disc machine

The general form of a tri-pin-on-disc friction and wear testing machine is shown in Figure 4. One of the test materials, usually the metal, is machined in the form of a disc and located on pegs on a turntable driven at constant speed in the

Figure 5. Pin holder for tri-pin-on-disc wear-testing machine.

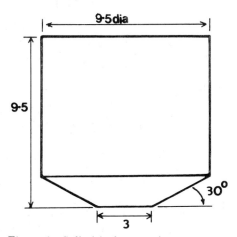

Figure 6. Cylindrical wear pin geometry.

environmental chamber shown in Figure 4. Three test pins, usually the polymer, are mounted in a frame as shown in Figure 5 and loaded against the turntable by means of static weights or a hydraulic system. The frame is prevented from rotating by means of three cantilevers whose deflection indicates the friction torque on the assembly.

The form of a typical wear pin is shown in Figure 6. The wide-angle (120°) truncated cone arrangement provides substantial mechanical support for the wear face which has a nominal circular form of initial area $8-10$ mm^2.

(ii) Measurement of wear

It is possible to measure wear by monitoring either dimensional changes, such as the diameter of the wear face, or weight loss. The former has much appeal and is simple, but with polymers the results generally show considerable scatter. The reasons for this are that many polymers are dimensionally unstable, such that changes in temperature or humidity can influence the dimension being monitored, irrespective of wear. For this reason the measurement of weight loss is the preferred procedure, but even then considerable care has to be exercised.

Many polymers absorb moisture from the atmosphere and this implies that changes in humidity will lead to changes in weight. Furthermore, if the pins are lubricated, the uptake of lubricant might exceed the weight of the wear particles in the initial stages of the test. For this reason a control pin is placed close to the wear specimens, but is not subjected to wear. Weight changes of the control pin are then deducted from the weight changes recorded for the test pins in order to estimate more correctly the total loss attributable to wear.

Before each test, the polymer pins and metal counterfaces were thoroughly cleaned, the latter ultrasonically. The wear pins were removed periodically, typically after about 20 km of sliding, carefully cleaned with a tissue, left for three hours at 20°C and a relative humidity of 40%, and then weighed a number of times on an accuate balance. A knowledge of the mean loss of weight and the density of the polymer enabled the volume of material removed by wear to be ascertained. The density of the polymer was determined by means of a Techne density column made up with a distilled water/isopropyl alcohol mixture; calibration being effected by glass floats.

5. THE WEAR OF SURGICAL GRADE ULTRA-HIGH MOLECULAR WEIGHT POLYETHYLENE AGAINST STAINLESSS STEEL

(i) The polymer

The UHMWPE used in the experiments was RCH 1000, surgical grade polyethylene, manufactured by Hoechst A.G. and supplied by High Density Plastic Ltd. Todmorden, in the form of a large slab (1 m x 0.5 m x 0.055 m). All the wear pins were machined from this block into cylindrical form with a truncated conical end as shown in Figure 6.

Table 1 Physical and mechanical properties of surgical grade UHMWPE (RCH 1000)

Molecular weight	$3.5 \times 10^6 - 4.0 \times 10^6$	
Density (kg/m^3)	940	
Vickers hardness number (20°C)	5.3 ± 0.3	
(1 VHN = 9.81 MPa)		
Bulk shear stress (MPa)	24	
Tensile properties		
Gauge length (mm)	25.4	25.4
Strain rate (s^{-1})	3.28×10^{-3}	3.28×10^{-2}
Yield stress (MPa)	20.3	24.9
Fracture stress (s) (MPa)	28.4	30.6
Fracture strain (e) ($= sl/l$)	5.67	4.74
Product (se) (MPa)	161	145

The average molecular weight determined by light scattering was quoted to be in the range $3.5 \times 10^6 - 4 \times 10^6$. X-ray, tensile, and hardness tests carried out in the four corners and the centre of the slab failed to detect any variation in properties. The hardness of the polymer was measured by means of the Vickers pyramid hardness test, and shear and tensile properties were measured on an Instron testing machine. A summary of the mechanical properties of the polymer is presented in Table 1. X-ray diffraction photographs indicated that the UHMWPE was isotropic with the accepted unit-cell dimensions. The rate of absorption of water was monitored over a period of 500 h and the results are shown in Figure 7.

The wear face on the pin was machined in a lathe and the spiral machining marks are shown in Figure 8(a). Initial values of centre line average R_a obtained on a Taylor Hobson Talysurf III ranged from 0.7 to 1.5 μm. One of the reasons for

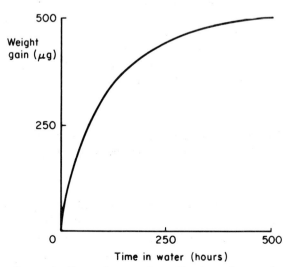

Figure 7. Rate of water absorption by polymer pin.

410

Figure 8. Wear face of polymer pin (magnification 100x). (a) Initial turned surface, (b) worn face after 300 km sliding, load 155 N, speed 0.24 m/s.

selecting a truncated conical from of wear pin was that the wear face would be relatively small, typically 8–10 mm^2, so that any influence of initial machining upon wear rate would be restricted to the early stages of the wear experiments. The polymer wear face after 300 km of sliding at a speed of 0.24 m/s and a normal load of 155 N per pin is shown in Figure 8(b).

(ii) The stainless steel counterfaces

All the counterfaces were machined from surgical grade austenitic stainless steel, EN58J, supplied in bar form. The discs were circular, of diameter 90 mm, and the pin configuration presented a wear track diameter of 76.2 mm. The counterfaces were surface ground to a finish of about 0.13 μm R_a and then lapped on a Logitech lapping machine using a leather cloth and 6 μm diamond paste until the required surface finish was achieved.

(iii) Microscopy

Optical and electron microscopes were used to examine the surfaces of both the polymer pins and the counterfaces throughout a wear test. A Vickers M55E optical microscope was used in both incident and transmission modes, the sections used in transmission work having a thickness in the range 10–20 μm.

A Cambridge Stereoscan 600 was used to study the polymer and counterface surfaces. When scanning electron microscopy failed to provide sufficient resolution of the fine detail of the wear faces, transmission electron microscopy of replicas of the worn surfaces was adopted. A two-stage replication technique using Bexfilm and gold/palladium shadow coating was used for transmission electron microscopy.

(iv) Test conditions

Wear tests have been carried out on tri-pin-disc machines of the form shown in

Figure 4, together with a six-station linear reciprocating wear-testing machine. In all cases the polymer pins had the form shown in Figure 6.

(a) *Load.* The loads applied to each pin ranged from 25 to 156 N and all were applied by dead weights.

(b) *Speed.* The steady sliding speed used for most of the tests on the tri-pin-on-disc machine was 0.24 m/s, which is representative of the conditions in large, heavily loaded joints in the human body, although in some cases lower speeds ranging down to 0.1 m/s were used. The maximum sliding speeds in the reciprocating wear-testing machine were in the range 0.26–0.31 m/s.

(c) *Environment.* All the tests were carried out in a dust-free environment, the test specimens being contained in a bell-jar on the tri-pin-on-disc machine and in a Perspex box on the linear reciprocating machine. In the latter case dry, filtered, compressed air was fed continuously into the environmental chamber. Most of the tests were carried out under unlubricated conditions, but in some cases the specimens were immersed in water.

(d) *Temperature.* The disc surface temperatures and on some occasions the bulk polymer pin temperatures were recorded by means of thermocouples.

(e) *Friction.* The friction torque on the pin holder in the tri-pin-on-disc machine was monitored by measuring the deflection of restraining cantilevers by means of induction tranducers. Strain-gauged proof rings inserted into the driving rods enabled the force of friction between the polymer pins and steel counterfaces to be estimated in the linear reciprocating machine.

(v) The transfer film

A transfer film of polymer built up on the stainless steel counterfaces in all tests. The appearance of this film after 250 and 1000 km of sliding at a speed of 0.24 m/s and a load on each pin of 155 N is shown in Figure 9. The banded nature of the

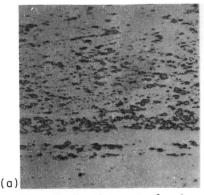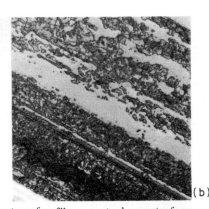

Figure 9. Development of polymer transfer film on steel counterface.
(a) 250 km sliding distance, load 155 N per pin (magnification 550x);
(b) 1000 km sliding distance, speed 0.24 m/s (magnification 700x).

412

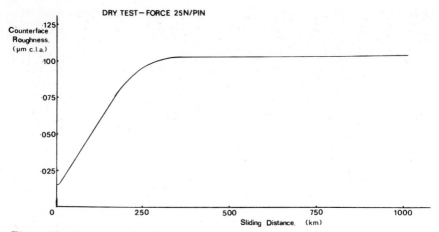

Figure 10. Variation of counterface roughness with time. Polymer: UHMWPE; counterface: stainless steel(EN58J).

transfer film was always evident, but as the tests progressed the bands broadened to present a more uniform appearance.

The stainless steel counterfaces presented a very good initial surface finish as a result of careful lapping. The initial centre line average R_a was nearly always less than 0.02 μm, generally about 0.015 μm, and in some cases as low as 0.013 μm. Sometimes the R_a value for the counterface would reduce in the initial stages of running, presumably due to infilling of the depressions by polymer, but the surface roughness would then increase as the banded transfer film depicted in Figure 9 was formed. In a test in which the applied load was 25 N per pin the counterface roughness reached an equilibrium value after about 300 km of sliding, as shown in Figure 10.

At high loads and relatively high interface temperatures, the central part of the transfer film showed signs of a brown discoloration thought to be indicative of polymer degradation. No discoloration of the wear pin surface was noted. It is interesting to note that Vinogradov et al.[10] have reported that oxidative degradation of polypropylene on steel enhanced the resistance to wear.

(vi) The morphology of wear

As wear progresses the surface of the polymer presents a number of characteristic features revealed by microscopy which have been described by Brown[11] as wear grooves, pulls, and smears and crescents.

Wear grooves roughly parallel to the direction of sliding were found on all the pins, although some minor grooves inclined at an angle to the sliding direction were also evident in a number of cases. A typical example of such a groove is labelled (c) in Figure 11.

A common feature evident on micrographs of the worn polymer surface was interpreted as a localized region of polymer pulled and often smeared in the

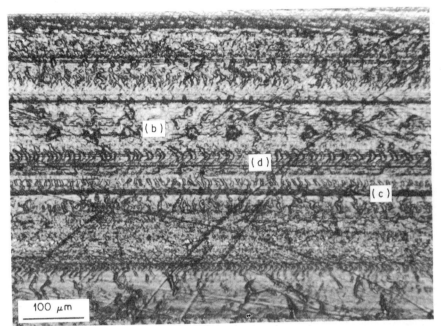

Figure 11. Morphology of wear. Note features (b) pulls and smears, (c) wear groove, (d) crescents.

direction of sliding. Such features are labelled (b) in Figure 11. Bands of raised crescent-shaped features were also observed, as illustrated by region (d) in Figure 11. These markings were roughly perpendicular to the direction of sliding and of irregular spacing. A more uniform rippled formation was also seen in bands, the wavelength being less than 10 μm.

(vii) Wear rates

It is well known that the volume of material removed by wear shows erratic behaviour in the initial period of sliding. The polymer UHMWPE proved to be no exception, and it is customary to ignore the results obtained during this initial phase which corresponds roughly with the removal of the surface layers associated with the machining and surface preparation procedures.

For a given load, speed, and environment, the volume of material removed by wear after the erratic initial stages was found to be directly proportional to sliding distance. The general form of the relationship between the volume of material removed by wear V and sliding distance L is shown in Figure 12. The important features are two linear relationships with the wear taking place at a faster rate in the later stages (region B) than in the initial period (region A) of sliding.

Microscopy reveals that adhesive wear takes place in both regions A and B, but that surface degradation associated with a fatigue process supplements the process in region B. The surface degradation evident in region B, but not in region A,

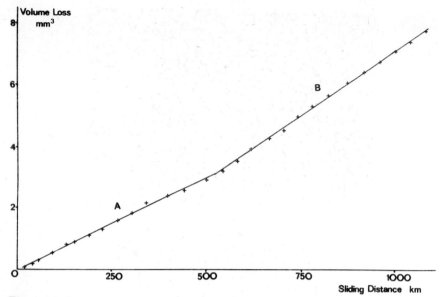

Figure 12. Volume of material removed by wear under dry conditions as a function of sliding distance. Ultra-high-density polyethylene sliding on stainless steel (EN58J) (load 25 N per pin, sliding speed 0.24 m/s).

consists of small cracks, roughly perpendicular to the direction of sliding. A representative picture of these cracks is shown in Figure 13(a) and it can be seen that they are a few micrometres long but very narrow. Sometimes flakes of material leave the surface of the polymer during this fatigue process, one of the shallow hollows left behind on the surface being shown in Figure 13(b).

The linear relationships between V and L are entirely consistent with the basic equations of wear (equations (2) and (3)). The linear characteristics also confirm the important point that the volume of material removed by wear V depends upon load P and not pressure. Since the pin has a conical form, the wear face increases in size throughout the test and hence the mean pressure reduces whilst the load remains constant. If wear depended upon pressure and not load, the relationship between V and L shown in Figure 12 would not be linear.

(a) *Dry wear.* Results from 14 unidirectional wear tests on isotropic UHMWPE carried out on tri-pin-on-disc machines under dry conditions are shown in Table 2. It can be seen that although the loads, initial surface finishes, and, over a narrow range, sliding speeds, varied from one test to another, the calculated coefficient k_2 defined by equation (3) was reasonably constant. The order of magnitude of k_2 was 10^{-7}, the range being from 0.6×10^{-7} to 3.5×10^{-7} mm^3/Nm in region A, with an average value of 2.5×10^{-7} mm^3/Nm. In region B the range of values recorded for nine of the tests was 2.4×10^{-7} to 4.2×10^{-7} mm^3/Nm with an average of 3.4×10^{-7} mm^3/Nm. The corresponding average values for k_1 in regions A and B were 3.3×10^{-8} and 5.3×10^{-8}, respectively.

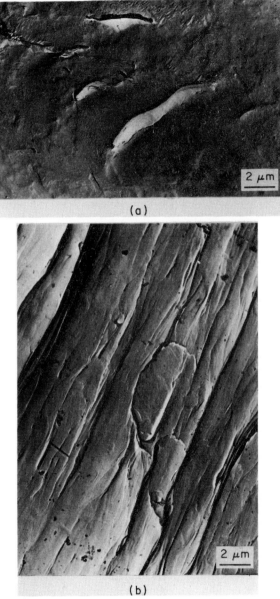

(a)

(b)

Figure 13. Surface features indicative of fatigue wear
(region B). (a) Cracks, (b) spalls.

The coefficient of friction, which showed little correlation with load or initial counterface roughness, was found to have an average value for these 14 tests of 0.44.

(b) *Wear in distilled water.* Since synovial fluid is regenerated in many joints fitted with total joint replacements, it is of interest to see how the presence of liquids

Table 2 Dry, unidirectional, wear tests on isotropic UHMWPE sliding against stainless steel EN58J (tri-pin-on-disc machine)

Test	Load/pin (N)	Sliding speed (m/s)	Initial counterface roughness $(\mu m)R_a$	Total sliding distance (km)	Sliding distance at onset of region B (km)	Wear coefficients (k_2) (mm^3/Nm)	
						Region A	Region B
1	25.0	0.24	0.018	1084	440	2.5×10^{-7}	3.4×10^{-7}
3	44.6	0.24	0.036	214	—	2.6×10^{-7}	—
4	44.9	0.24	0.015	214	—	1.2×10^{-7}	—
6	65.2	0.24	0.015	492	130	1.3×10^{-7}	2.9×10^{-7}
7	65.1	0.24	0.015	685	480	2.2×10^{-7}	4.1×10^{-7}
8	85.0	0.24	0.015	364	180	2.7×10^{-7}	3.4×10^{-7}
9	105.2	0.24	0.020	217	110	2.6×10^{-7}	3.3×10^{-7}
11	145.2	0.24	0.013	156	97	2.3×10^{-7}	3.6×10^{-7}
12	155.5	0.24	0.018	34	—	1.3×10^{-7}	—
2	33.7	0.24	0.094	675	400	3.0×10^{-7}	3.5×10^{-7}
5	45.0	0.24	0.120	520	145	0.6×10^{-7}	2.4×10^{-7}
10	126.4	0.24	0.091	189	100	2.3×10^{-7}	4.2×10^{-7}
3	44.6	0.24	0.036	214	—	2.6×10^{-7}	—
4	44.9	0.24	0.015	214	—	1.2×10^{-7}	—
13	44.7	0.16	0.020	114	—	3.5×10^{-7}	—
14	44.7	0.10	0.018	147	—	3.0×10^{-7}	—

influences the rate of wear of UHMWPE when rubbing against steel. Synovial fluid, which is essentially a dialyzate of blood plasma, degenerates rapidly upon exposure to the atmosphere. It is essentially a watery substance with a viscosity determined by the concentration of hyaluronic acid and it is therefore reasonable to explore the wear characteristics under the limiting, low-viscosity conditions of distilled water.

It was found that the wear rates were reduced drastically by the presence of distilled water, but the general features of water designated by regions A and B in Figure 12 were still evident. The average values of k_1 and k_2, determined for region A when the specimens were bathed in distilled water, were found to be 1.36×10^{-9} and 0.87×10^{-8} mm^3/Nm, respectively.

(c) *The Wear of polyethylene sliding on polyethylene.* A few preliminary experiments have demonstrated that polyethylene suffers excessive wear when sliding over identical material in bulk, although the friction changes by less than a factor of two.

The wear rates recorded for UHMWPE pins sliding over a disc of identical material under dry conditions ranged from 0.9×10^{-4} to 3×10^{-4} mm^3/Nm. These wear coefficients are about three orders of magnitude greater than those recorded for UHMWPE sliding on steel under otherwise similar conditions.

6. THE INFLUENCE OF MOLECULAR ORIENTATION, COUNTERFACE ROUGHNESS, AND SILICONE IMPREGNATION UPON THE WEAR OF POLYETHYLENE

In section 5 the basic features of the wear of the polymer which is invariably used in total joint replacements for the body were considered in some detail. The useful life of a prosthesis is governed by the rate of wear of the polymeric component and there is, therefore, an incentive to seek even better materials or procedures which can be used to improve the wear resistance of polyethylene. In the present section we consider three potentially significant developments which might lead to improvements in the wear rates of polyethylene.

(i) Molecular orientation

Drawn UHMWPE has been used to investigate the effects of orientation upon wear. Low and intermediate draw ratios of 1.72 : 1, 2.26 : 1, and 2.56 : 1 were achieved by tensile drawing, whilst intermediate and high draw ratios of 2.2 : 1 and 3.4 : 1 were produced by hydrostatic extrusion. In the latter case a 750 MPa hydrostatic extrusion press manufactured by Fielding and Platt was used. The polymer billet was heated to 100°C and hydrostatically extruded through a 15° semi-angle, 15.8 mm diameter die. Some physical and mechanical properties of the material produced by hydrostatic extrusion are compared with those of the isotropic material in Table 3. It can be seen that the anisotropic material exhibits higher yield and fracture stresses, but lower ductility than the isotropic material in the direction parallel to drawing, but that the trend is reversed in the perpendicular direction.

Table 3 Physical and mechanical properties of isotropic UHMWPE and UHMWPE formed by hydrostatic extrusion

	Isotropic UHMWPE		Extruded UHMWPE			
Material						
Draw ratio	—		2.2:1		3.4:1	
Density (kg/m^3)	940		937		934	
Orientation	—		Parallel	Perpendicular	Parallel	Perpendicular
Gauge length (mm)	12.7	4.5	12.7	4.5	12.7	4.5
Strain rate (s^{-1})	6.56×10^{-2}	7.4×10^{-2}	6.56×10^{-2}	7.4×10^{-2}	6.56×10^{-2}	7.4×10^{-2}
Yield stress (MPa)	28.2	27.7	—	18.9	—	—
Fracture stress (s) (MPa)	32.2	32.9	65.4	13.9	111.2	—
Fracture strain (e) (sl/l)	5.31	3.84	1.89	4.90	0.83	—
Product (se) (MPa)	171	126	124	68	92	—

The experimentally determined wear coefficients k_2 on pins in which the direction of drawing, and hence orientation, was parallel to the direction of sliding, were slightly less than the values obtained for isotropic UHMWPE, the values being 0.7×10^{-7} and 0.9×10^{-7} mm^3/Nm, respectively. For pins produced with orientation perpendicular to the sliding direction the wear rates were much worse than for isotropic material, the value of k_2 being as high as 10.7×10^{-7} mm^3/Nm.

The present limited experiments suggest that molecular orientation does little to enhance the wear resistance of UHMWPE and might make it much worse if the direction of orientation cannot be aligned with the direction of sliding at all times.

(ii) Counterface roughness

There are indications of an interesting and potentially important influence of counterface roughness upon the wear of UHMWPE. In general, the wear rate or relatively soft materials sliding against hard counterfaces decreases as the surface quality of the counterface is improved. Buckley[12] reported an unusual effect in which UHMWPE exhibited a minimum rate of wear as the surface finish of the counterface on a disc-and-shoe wear-testing machine was progressively improved. The optimum surface topography was observed on a ground steel surface having an initial roughness of 0.37 μm (r.m.s.).

Swikert and Johnson,[13] who also worked at the NASA Lewis Laboratory in Cleveland, Ohio, also found that a cobalt chromium replacement femoral head produced by liquid honing to a surface finish of 0.11 μm (r.m.s.) yielded a minimum rate of wear.

Dowson et al.[14] have examined the influence of initial surface roughness upon the wear rate or UHMWPE in a linear reciprocating machine. A typical set of results is plotted in Figure 14, and it can be seen that the wear coefficient exhibits a minimum at a surface finish of about 0.4 μm (R_a).

The evidence available to support the view that there is an optimum surface finish which yields minimum wear is somewhat limited, although it has come from two different establishments and three different forms of wear-testing machines. A more detailed characterization of the counterface surface topography is called for before firm conclusions can be drawn, since different machining processes were used to generate the surfaces used in each of the three experiments cited. It is possible that the minima correspond to changes in surface preparation procedures.

Metallic components for total joint replacements are normally expensive and since they are produced with a high-quality finish, often corresponding to roughnesses to the left of the minimum shown in Figure 14, there are good economic as well as scientific reasons for exploring these suggestions in more detail.

(iii) The wear of silicone-impregnated polyethylene

The marked reduction in the wear of UHMWPE when dry contact conditions give way to lubrication by distilled water has already been noted. Silicone fluid has been used extensively as a lubricant and in a number of surgical procedures, and

420

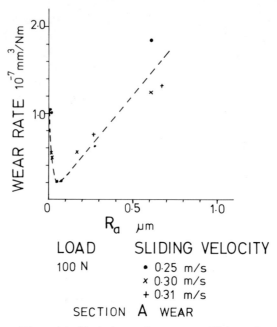

Figure 14. Variation of wear coefficient k_2 with counterface roughness R_a. Ultra-high-density polyethylene sliding on stainless steel (EN58J).

appears to be acceptable within the body. A number of experiments have therefore been carried out to determine the mechanical and tribological characteristics of polyethylene impregnated with various proportions of silicone fluid.

Abouelwafa *et al.*[15] have reported a study of low-density polyethylene containing up to 10% by weight of silicone fluid. The fluid was polydimethylsiloxane, a stable, low-volatility fluid of density 975 kg/m³ and kinematic viscosity 0.03 m²/s (30 000 cS), supplied by Dow Corning.

Blocks of LDPE were formed by compression moulding of LDPE granules which had been mixed mechanically with silicone fluid. The silicone fluid and polyethylene were mutually insoluble, and in the moulded product the silicone formed a dispersed second phase of droplets up to 5 μm in diameter. Samples were produced which contained 1%, 2%, 5%, and 10% of silicone by weight, and the mechanical properties were determined in the normal way. Wear tests were carried out on a tri-pin-on-disc machine at a steady sliding speed of 0.24 m/s and a load on each pin of 52 N. The variation of volume of material lost by wear V with sliding distance L is shown in Figure 15. The initial wear rates were high in all cases, but after about 30 km of sliding a near-linear relationship was established. The slopes of the curves after 20 km of sliding were used to estimate the wear coefficients k_2 recorded with other mechanical properties in Table 4.

The results show that the wear rate is reduced by about one order of magnitude

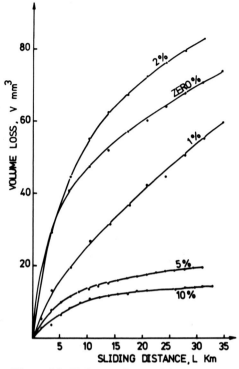

Figure 15. Volume of material removed by wear as a function of sliding distance for silicone-impregnated low-density polyethylene.

by the addition of 5% or 10% of silicone fluid. This dramatic effect is accompanied, however, by a reduction in the mechanical strength of the material and the balance of advantage has to be assessed carefully for each application. It remains to be seen whether silicone-impregnated UHMWPE shows similar characteristics and whether any increased resistance to wear is offset by reduced creep resistance. Both wear and creep will influence the rate of penetration of a metallic femoral head into a polymeric acetabular cup.

7. THE WEAR OF PROSTHESIS

During the development of a total replacement knee joint Seedhom et al.[16] demonstrated that wear data derived from laboratory tests could be used to predict the life of a prosthesis in the body.

Brown et al.[17] examined six polyethylene acetabular cups which had been in service for periods ranging from one to five years and found direct evidence of wear processes similar to those observed in pin-on-disc machine tests. Calculations showed that fatigue wear corresponding to region B in Figure 12 should occur after

Table 4 Physical, mechanical and tribological properties of low density polyethylene (LDPE) and LDPE containing silicone fluid of viscosity 0.03 m^2/s (30 000 cS)

	LDPE	LDPE + 1% silicone	LDPE + 2% silicone	LDPE + 5% silicone	LDPE + 10% silicone
Density (kg/m^3)	926	926	925	924	922
Vickers hardness number	2.4	2.2	2.2	2.1	1.7
Bulk shear stress (MPa)	10.2	9.5	9.1	8.9	6.8
Elastic limit (MPa)	5.7	5.7	4.9	4.6	3.9
Fracture stress (s) (MPa)	13.7	12.9	12.2	9.3	6.9
Fracture strain (e) ($= sl/l$)	1.78	1.73	1.59	1.29	1.81
Product (se) (MPa)	19.4	18.0	15.5	11.1	11.7
Wear coefficient (k_2) (mm^3/Nm) (after 20 km sliding)	2.11 × 10^{-5}	2.61 × 10^{-5}	2.53 × 10^{-5}	0.29 × 10^{-5}	0.18 × 10^{-5}

about five to ten years in the life of a total hip joint replacement. Surface degradation similar to that shown in Figure 13 has since been detected in a hip joint after eight years of service.

A careful study of worn acetabular cups by Dowling et al.[18] has revealed the complex nature of wear in total joint replacements. Many of the features observed in the laboratory tests are also seen on the acetabular cups, but there is evidence to suggest that wear is but one factor to be considered in calculating the total rate of penetration of the femoral head into the acetabular cup.

8. SUMMARY

Polymeric materials are now widely used in tribological situations and useful introductions to the subject have been presented by Lancaster,[19] Lee,[20] and Dowson et al.[21] In this chapter we have examined in some detail the tribological features of UHMWPE: the material widely used in total joint replacements for the body.

The general features of polymeric wear are outlined, wear equations are presented, and current approaches to wear testing have been described. Attention is focused upon the utility of pin-on-disc wear-testing machines.

Factors affecting the nature and rate of wear of UHMWPE when rubbing against stainless steel (EN58J) under dry conditions are then discussed in some detail and recent experimental findings are reported. The wear mechanism is associated with the formation of a thin transfer film of polymer on the steel, and significant features of the morphology of wear revealed by optical and electron microscopy are described.

Wear rates are quoted for the dry rubbing of UHMWPE on steel, rubbing under distilled water, and UHMWPE rubbing on itself under dry conditions. Attention is drawn to an increase in wear rate after a period of rubbing which appears to be caused by a form of surface fatigue.

Factors which might modify the rate of wear of polyethylene on steel and which are the subject of further investigation such as molecular orientation, counterface roughness, and silicone impregnation are discussed.

Finally, evidence that the features of UHMWPE wear noted in laboratory tests are also present in total replacement joints after a period of service in the body, is noted.

This detailed account of recent studies of the tribological characteristics of UHMWPE against stainless steel demonstrates the great merit of polymeric materials for particular bearing applications. About 75% of total joint replacements for the body now utilize a metal-on-polymer combination of materials.

REFERENCES

1. F. P. Bowden and D. Tabor, *The Friction and Lubrication of Solids*, Oxford University Press, London (1964).
2. B. J. Briscoe and D. Tabor, 'Self-lubricating polythenes', *Colloques Internationaux du C.N.R.S.*, No. 233, *Polymères et Lubrification* (1974).

3. K. V. Shooter and D. Tabor, 'The frictional properties of plastics', *Proceedings of the Physical Society*, **865**, 661–671 (1952).

4. C. M. Pooley and D. Tabor, 'Friction and molecular structure: the behaviour of some thermoplastics', *Proceedings of the Royal Society*, **A329**, 251–274 (1972).

5. J. K. Lancaster, 'Abrasive wear of polymers', *Wear*, **14**, 223–239 (1969).

6. S. B. Ratner, I. I. Farberova, O. V. Radyukevich, and E. G. Lur'e, 'Connection between wear resistance of plastics and other mechanical properties', in *Abrasion of Rubber* ed. D. I. James, MacLaren, London, p. 135 (1967).

7. J. F. Archard, 'Contact and rubbing of flat surfaces', *Journal of Applied Physics*, **24**, No. 8, 981–988 (1953).

8. D. Dowson, J. R. Atkinson, and K. Brown, 'The wear of high molecular weight polyethylene with particular reference to its use in artificial human joints', in *Advances in Polymer Friction and Wear*, Vol. 5B, ed. Lieng-Huang Lee, Plenum Press, pp. 533 –548 (1975).

9. J. T. Scales, 'Arthroplasty of the hip using foreign materials: a history', The Institution of Mechanical Engineers, *Proceedings 1966–67*, Volume 181, Part 3J, pp. 63–84 (1967).

10. G. Vinogradov, V. Mustafaev, and Y. Y. Podolsky, 'A study of heavy metal-to-plastic friction duties and of the wear of hardened steel' *Wear*, **8**, 358–373 (1965).

11. K. J. Brown, 'A study of the wear of polyethylene', Ph.D. Thesis, The University of Leeds (1975).

12. D. H. Buckley, 'Introductory remarks – friction and wear of polymeric composites', in *Advances in Polymer Friction and Wear*, ed. Lieng-Huang Lee, pp. 601–603, Plenum Press (1974).

13. M. A. Swikert and R. L. Johnson, 'Simulated studied of wear and friction in total hip prosthesis components with various ball sizes and surface finishes', *NASA TN D-8174* (1976).

14. D. Dowson, J. M. Challen, K. Holmes, and J. R. Atkinson, 'The influence of counterface roughness on the wear rate of polyethylene', *Proceedings of the 3rd Leeds–Lyon Symposium on Tribology on 'The Wear of Non-Metallic Materials'*, Mechanical Engineering Publications, to be published.

15. M. N. Abouelwafa, D. Dowson, and J. R. Atkinson, 'The wear and mechanical properties of silicone impregnated polyethylene', *Proceedings of the 3rd Leeds–Lyon Symposium on Tribology on 'The Wear of Non-Metallic Materials'*, Mechanical Engineering Publications, to be published.

16. B. B. Seedhom, D. Dowson, and V. Wright, 'Wear of solid phase formed high density polyethylene in relation to the life of artificial hips and knees', *Wear*, **24**, No. 1, 33–51 (1973).

17. K. J. Brown, J. R. Atkinson, D. Dowson, and V. Wright, 'The wear of ultra-high molecular weight polyethylene and a preliminary study of its relation to the *in-vivo* behaviour of replacement hip joints', *Wear* **40**, 255–264 (1976).

18. J. Dowling, J. R. Atkinson, and D. Dowson, 'The characteristics of UHMWPE acetabular cups worn in the human body', to be published.

19. J. K. Lancaster, 'Basic mechanisms of friction and wear of polymers', *Plastics and Polymers*, December, 297–306 (1973).

20. L. H. Lee (Editor), *Advances in Polymer Friction and Wear, Polymer Science and Technology*, Vol. 5A, 5B, Plenum Press, New York and London (1974).

21. D. Dowson, M Godet, and C. M. Taylor (eds.), 'The wear of non-metallic materials', *Proceedings of the 3rd Leeds–Lyon Symposium on Tribology*, Mechanical Engineering Publications, to be published.

Index